Exploring Educational Administration

Titles of Related Interest

Evers and Lakomski/Knowing Educational Administration: Contemporary Methodological Controversies in Educational Administration Research

Hodgkinson/Administrative Philosophy: Values and Motivations in Administrative Life

MacPherson/Educative Accountability Policies for Educational Institutions and Systems

Exploring Educational Administration

Coherentist Applications and Critical Debates

by

COLIN W. EVERS
Monash University, Australia

and

GABRIELE LAKOMSKI
University of Melbourne, Australia

PERGAMON

An Imprint of Elsevier Science

U.K.	Elsevier Science Ltd, The Boulevard, Langford Lane, Kidlington, Oxford OX5 1GB, U.K.
U.S.A.	Elsevier Scienc Inc., 660 White Plains Road, Tarrytown, New York 10591-5153, U.S.A.
JAPAN	Elsevier Science Japan, Tsunashima Building Annex, 3-20-12 Yushima, Bunkyo-ku, Tokoya 113, Japan

Copyright © 1996 Elsevier Science Ltd

All rights Reserved. No part of this publication may be reproduced, stored in a retrieval system or transmitted in any form or by any means: electronic, electrostatic, magnetic tape, mechanical, photocopying, recording or otherwise, without permission in writing from the publishers.

First edition

Library of Congress Cataloging in Publication Data

Exploring educational administration: coherentist applications and critical debates/[edited] by Colin Evers and Gabriele Lakomski. —1st ed.
    p. cm.
Includes bibliographical references and indexes.
ISBN 0-08-042766-9 (hardcover)
1. School management and organization. 2. School management and organization—Research. I. Evers, C. W. II. Lakomski, Gabriele.
LB2805.E94 1996   96-20577
                                                                         CIP

British Library Cataloguing in Publication Data

A catalogue record for this book is available from the British Library

---

Exploring educational administration: coherentist applications and critical debates
1. School management and organization. 2. Educational planning. I. Evers, Colin W. II. Lakomski, Gabriele
371.2

ISBN 0080427669

ISBN 0 08 0427669

Printed and bound in Great Britain by Redwood Books Ltd

Contents

PREFACE	viii
ACKNOWLEDGEMENTS	x
INTRODUCTION	xiii

Part I: Naturalistic Coherentism in Educational Administration

Chapter One

Recent Developments in Educational Administration — 1

Chapter Two

Educational Administration as Science: A Postpositivist Proposal — 14

Chapter Three

Towards Coherence in Administrative Theory — 29

Part II: Naturalistic Extensions

Chapter Four

Educational Organizations as Systems — 41

Chapter Five

Schooling, Organizational Learning and Efficiency in the Growth of Knowledge — 57

Chapter Six

Leadership and Learning: From Transformational Leadership to Organizational Learning — 71

Chapter Seven

Policy Analysis: Practical Reason or Empirical Science? — 87

Chapter Eight

Administrative Decision-Making as Pattern Processing   104

Chapter Nine

Educating the Brain   115

Part III: Beyond Subjectivism and Humanism

Chapter Ten

On Theory in Educational Administration: Beyond Greenfield's Subjectivism   129

Chapter Eleven

Cognition, Values and Organizational Structure: Hodgkinson in Perspective   142

Chapter Twelve

What Price Democracy? An Examination of Arrow's Impossibility Theorem in Educational Decision-Making   154

Part IV: Critical Debates

Chapter Thirteen

Explaining and Improving Educational Administration — ***Donald Willower***   165

Chapter Fourteen

The Salvation of Educational Administration: Better Science or Alternatives to Science — ***Peter Gronn and Peter Ribbins***   176

Chapter Fifteen

On Knowing: Cultural and Critical Approaches to Educational Administration — ***Richard Bates***   189

Chapter Sixteen

The Epistemological Axiology of Evers and Lakomski: Some Un-Quinean Quibblings — ***Christopher Hodgkinson***   199

Chapter Seventeen

Response to Commentaries   211

Chapter Eighteen

Three Dogmas of Materialist Pragmatism: A Critique of a Recent Attempt to Provide a Science of Educational Administration — ***Trevor Maddock***   215

Chapter Nineteen

Three Dogmas: A Rejoinder   238

Chapter Twenty

Knowledge, Certainty and Openness in Educational Administration —
Martin Barlosky   247

Chapter Twenty One

Response to Barlosky: Methodological Reflections on Postmodernism   262

Chapter Twenty Two

A Rejoinder to Evers and Lakomski — ***Martin Barlosky***   271

Conclusion:

Doing Educational Adminstration — Preliminary Reflections on a Theory of
Practice   279

AUTHOR INDEX   286

SUBJECT INDEX   289

Preface

When we completed our first book *Knowing Educational Administration* in 1991 we knew that there had to be at least one sequel, if not several. Having laid out the epistemological argument for developing a new science of educational administration — the main theme of the book — and demonstrated some of its scope and applicability, we realized that we needed to provide more evidence, argument, and examples of what such a new science would look like. *Exploring Educational Administration* represents further developments in our thinking especially in regard to how we could strengthen the case for a natural science of educational administration, and provide examples from the field that would speak to practitioners as well as researchers.

Although much in this collection has been published before, and despite different audiences and purposes, all chapters support the common theme of extending and strengthening the case for a natural science of educational administration. Such a case is normally thought difficult to make for two reasons arising out of a mistaken identification of science with positivism. First, science is thought to be a very restricted form of knowledge, omitting consideration of values and also human subjectivity. And second, science is thought to be so highly structured and concerned with generality that it is unable to connect usefully with the complexity and particularity of practice.

The explorations in the pages that follow attempt both to strengthen that framework, especially in debates against critics, and to apply it to issues in administrative practice. The main developments involve the use of natural science to understand human subjectivity. In this endeavour we begin to employ research from the neurosciences, particularly neural network models of human cognition. Theories of cognition have played a major role in administrative studies over the last fifty years, with Herbert Simon's view of bounded rationality perhaps being the most influential. One response to this tradition in our work has been to see human fallibility as functioning as an argument for promoting learning, in individuals and in organizations. Hence our interest in learning models of leadership and organizational design. Another response has been to explore ways of representing knowledge of practice that are less dependent on the so-called symbolic paradigm. Our efforts to take seriously neural network accounts of how the non-symbolic representation of knowledge of practice might fruitfully be conceived and applied to an understanding of administrative phenomena is reflected in several of the studies contained here.

In more traditional approaches to science of administration, it is customary to regard science as providing a model, or example, of how theory is to be constructed. For us, it is not the supposed form of science that is important, but its content. Our naturalistic epistemology draws on what science can tell us about the nature of knowledge, and even about the nature of theory itself. On this latter point, we draw on ideas in information theory to propose ways in which social science and natural science might be unified within the one theoretical framework. We attempt to meet one of the main consequences of a radical subjectivist version of educational administration by responding that the natural and the social lie on the same continuum. Another result is that if the continuum is defined by the extent to which the various claims in a theory are context-dependent, we are able to acknowledge, within administrative theory, the specificities and particularities of practice.

Many of the central features of our methodology and approach to educational administration have been debated in the literature. These debates have helped shape our work, from encouraging greater clarity through to forcing substantial extensions of our program. The last part of our book is given over to the critical reactions of others and our responses. In including these exchanges, we are choosing to emphasize the somewhat provisional, experimental and open ended nature of our research. The essays in this volume are very much explorations. We think they have created the possibility for a more complete and systematic account of administrative practice to be given from our perspective — but that is another story, yet to be told.

Colin W. Evers and Gabriele Lakomski
Melbourne, 1996

Acknowledgments

In the preparation of the essays for this book we have benefited from much critical advice. Five were originally presented as conference papers: Chapter 2 at the International Intervisitation Program meeting in Manchester in 1990; Chapters 5 and 12 at a conference organized by the Monash University Centre for School Decision-Making and Management, held at Woodend, Victoria, in 1988; Chapter 9 at the Annual Conference of the Philosophy of Education Society of Great Britain in London, in 1990; and Chapter 10 at the International Intervisitation Program meeting in Toronto, in 1994. We owe thanks to all who participated in these audiences for their thoughtful comments. We are also grateful for the financial support for conference travel given by Monash University and the University of Melbourne.

Many of these chapters had their origins in requests from editors, publishers and learned societies. We would therefore like to thank the following for the opportunities they provided: Barbara Barrett, Judith Chapman, Michael Connelly, Peter Gronn, Ruth Jonathan, Yvonne Martin, Peter Ribbins, Jim Ryan, Ross Thomas, and Elizabeth Weiss.

We owe a special debt to Martin Barlosky, Richard Bates, Peter Gronn, Christopher Hodgkinson, Trevor Maddock, Peter Ribbins, and Don Willower for their willingness to engage our ideas in print, and for allowing us to reprint their contributions.

A number of publishers have generously given permission to reproduce previously published material, as follows: Chapters 1 and 6, The Australian Council for Educational Administration (Victoria); Chapter 2, Longman; Chapters 3 and 4, Allen & Unwin (Australia); Chapters 5 and 12, Taylor & Francis; Chapter 7, Elsevier Science; Chapter 8, The Australian Council for Educational Administration; Chapter 9, The Philosophy of Education Society of Australasia; Chapters 10, 13, 14, 15, 16, and 17, The British Educational Management and Administration Society; Chapter 11, *The Journal of Educational Administration and Foundations*; Chapters 18 and 19, MCB University Press Limited; Chapters 20, 21, and 22, The Ontario Institute for Studies in Education.

Finally, we include bibliographic information on all our own source materials, especially since we have made numerous revisions to the originals for incorporation into this book, as well as full publication details of the work of other contributors:

EVERS C.W. (1995). Recent developments in educational administration, *Leading and Managing* 1(1), pp. 1–12. (Chapter 1)

EVERS C.W. AND LAKOMSKI G. (1991). Educational administration as science: A postpositivist proposal, in P. Ribbins *et al.* (eds.) *Developing Educational Leaders*, pp. 97–114. (London: Longman). (Chapter 2)

EVERS C.W. AND LAKOMSKI G. (1995). Towards coherence in administrative theory, in C.W. Evers and J.D. Chapman (eds.) *Educational Administration: An Australian Perspective*, pp. 97–109. (Sydney: Allen & Unwin). (Chapter 3)

LAKOMSKI G. AND HAYNES F. (1995). Educational organizations as systems, in C.W. Evers and J.D. Chapman (eds.) *Educational Administration: An Australian Perspective*, pp. 97–109. (Sydney: Allen & Unwin). (Chapter 4)

EVERS C.W. (1990). Schooling, organizational learning, and efficiency in the growth of knowledge, in J.D. Chapman (ed.) *School-based Decision-making and Management* pp. 55–70. (London: Falmer Press). (Chapter 5)

LAKOMSKI G. (1995). Leadership and learning: from transformational leadership to organizational learning, *Leading and Managing* 1(3), pp. 211–225. (Chapter 6)

LAKOMSKI, G. (1991). Policy analysis: practical reason or empirical science?, *International* Journal of Educational Research, 15(6), pp. 537–551. (Chapter 7)

EVERS C.W. (1994). Administrative decision-making as pattern processing, in F. Crowther *et al.* (eds), *The Workplace in Education, ACEA Yearbook 1994*, pp. 266–275. (Sydney: Edward Arnold). (Chapter 8)

EVERS C.W. (1990). Educating the brain, *Educational Philosophy and Theory*, 22(2), pp. 65–80. (Chapter 9)

LAKOMSKI G. AND EVERS C.W. (1994). Greenfield's humane science, *Educational Management and Administration*, 22(4), pp. 260–269. (Portions of Chapter 10)

EVERS C. W. AND LAKOMSKI G. (1993). Cognition, values and organizational structure: Hodgkinson in perspective, *Journal of Educational Administration and Foundations*, 8(1), pp. 45–57. (Chapter 11)

LAKOMSKI G. (1990). What price democracy? An examination of Arrow's impossibility theorem in educational decision-making, in J.D. Chapman (ed.), *School-based Decision-making and Management* pp. 71–83. (London: Falmer Press). (Chapter 12)

EVERS C.W. AND LAKOMSKI G. (1993). Response to commentaries, *Educational Management and Administration*, 21(3), pp. 185–187. (Chapter 17)

EVERS C.W. AND LAKOMSKI G. (1994). Three dogmas: a rejoinder, *Journal of Educational Administration*, 32(4), pp. 28–37. (Chapter 19)

EVERS C.W. AND LAKOMSKI G. (1995). Response to Barlosky: methodological reflections on postmodernism, *Curriculum Inquiry*, 25(4), pp. 457–465. (Chapter 21)

WILLOWER D.J. (1993). Explaining and improving educational administration, *Educational Management and Administration*, 21(3), pp. 153–160. (Chapter 13)

GRONN P.C. AND RIBBINS P. (1993). The salvation of educational administration: better science or alternatives to science, *Educational Management and Administration*, 21(3), pp. 161–169. (Chapter 14)

BATES R. (1993). On knowing: cultural and critical approaches to educational administration, *Educational Management and Administration*, 21(3), pp. 171–176. (Chapter 15)

HODGKINSON C. (1993). The epistemological axiology of Evers and Lakomski: some un-Quinean quibblings, *Educational Management and Administration*, 21(3), pp. 177–184. (Chapter 16)

MADDOCK T. (1994). Three dogmas of materialist pragmatism: a critique of a recent attempt to provide a science of educational administration, *Journal of Educational Administration*, 32(4), pp. 5–27. (Chapter 18)

BARLOSKY M. (1995). Knowledge, certainty and openness in educational administration, *Curriculum Inquiry*, 25(4), pp. 441–455. (Chapter 20)

BARLOSKY M. (1995). A rejoinder to Evers and Lakomski, *Curriculum Inquiry*, 25(4), pp. 467–474. (Chapter 22)

Introduction

Since publication in 1991 of our book *Knowing Educational Administration: Contemporary Methodological Controversies in Educational Administration Research*, the debate in educational administration and policy studies has lost none of its passion concerning issues of ethics and values, of appropriate leadership, and of whether any of these could be addressed from within a scientific framework. Theoretical diversity continues to characterize the field, and logical empiricist models of behavioural science, dominant for so long, are more and more being challenged by a range of alternatives. Amongst these are varieties of subjectivism, critical theory, cultural perspectives, and most recently, post-modernism. Further examples appear in a Special Issue of the *Educational Administration Quarterly*, entitled 'Nontraditional Theory and Research', guest edited by Daniel Griffiths (1991, pp. 262–451).

The field thus continues to ferment which is in our view a healthy development because it attests to seriousness of purpose in addressing the fundamental problems of theory and practice. But there is also a sense in which we can no longer speak of 'intellectual turmoil' (Griffiths) in the field, since some theoretical boundaries appear to have been drawn reasonably clearly that have resulted in a kind of stand-off between the parties. What we mean by this is that there is widespread agreement on the weaknesses of traditional science approaches, commonly simply labelled as 'positivist', which assumed a quite narrow account of knowledge and how it can be justified. The unfortunate outcome of such narrowness was that much was ruled out of administrative theory and practice that is important to it, such as how to decide between competing values, and how to decide and defend what one ought to do when faced with competing alternatives in daily practice. And it was not only the case that administration as a discipline and practice was thus afflicted — science itself suffered since logical positivist methodology ruled out much that was of value in it as well.

As a consequence of the criticisms raised by philosophers of science such as Feyerabend, Hanson, and Kuhn, conceptions of science have developed that are post-positivist and that, in turn, provided the justification for changes in educational research methodology, commonly identified as 'research paradigms' (following Kuhn). The current state of play, then, is a kind of stand-off between alternative and widely accepted modes of research that run alongside traditional methods of acquiring and defending knowledge. Insofar as the alternative modes or research paradigms in educational administration resulted from and incorporate some of the substantive criticisms made of logical positivist and empiricist forms of science from within the parent disciplines, they represent a positive development. Yet in doing so, they also import unwarranted conclusions and theoretical difficulties inherent in Kuhn's paradigms view in particular, with serious consequences for the theory and practice of educational administration.

While administrative science, positivistically conceived, eliminates values from its purview, modes of research developed as alternatives either seriously limit or eliminate science altogether from its conceptions. Administrators, then, are left with accepting the traditional positivist science of administration, which is able to deliver knowledge of a restricted, quantitative, kind, but which is unhelpful when it comes to making a decision between, say, closing down a school, or having it merge with a neighbouring one because of a drop in enrolments. On the other hand, an advocate of one of the alternatives to administrative science, say, a cultural theorist, may well be able to focus on the various and competing values involved in the closure versus merger issue, without being able to decide between them rationally because values can neither be correct nor incorrect since they are a matter of personal belief or preference. Putting the resultant quandry in pointed form: choosing between the two options of conducting research in educational administration means that we are left with either (a) accepting a traditional notion of science that is value-free, or (b) accepting an alternative to science that is devoid of facts. In either case we deprive ourselves of valuable knowledge, scientific and non-scientific, needed to think about and conduct the business of administration effectively. Fortunately, there is a way to resolve this difficulty that does not force us to give up on any knowledge whatsoever. On the contrary, it allows us to collect it from wherever it can be got.

Both kinds of attempts at excluding knowledge, in our view, stem from the pervasive failure to draw a sharp line between positivism and science. Positivism in all its forms is a cluster of philosophical theories about the nature and methodology of science. Whether these theories are indeed sound is an issue that needs to be raised separately. We believe that the version of positivism that underwrites traditional science of educational administration, i.e. logical empiricism, has fundamental and systematic flaws, especially in respect of its epistemology. We have thus advocated and defended a particular post-positivist theory of science that is able to accomodate ethics, values, and human subjectivity, and have subsequently placed ourselves squarely on the side of science. The solution we offer to the current stand-off is hence neither a return to traditional science, nor the acceptance of multiple paradigms with their many world views which fragment the research enterprise, but to develop a new science of administration. Our new science is justified by a coherentist epistemology that is the best available alternative to foundational theories of knowledge, shared by (all varieties of) positivism, as well as some of its critics.

Knowing Educational Administration (Evers and Lakomski 1991) defends and applies this coherentist epistemology and offers a fresh perspective on a number of important methodological debates in educational administration. Since epistemology shapes both the structure and content of substantive theories, our perspective has quite systematic and far-reaching consequences for theory and research in the field. One important consequence is a shift to naturalism because we maintain that administrative theories cohere with natural science. Another is a more generous view of what counts as evidence and a resulting holism that shuns partitions on knowledge. Since both these features are central elements of our perspective we call our position *naturalistic coherentism* to reflect their importance.

While *Knowing Educational Administration* needed to be explicit and detailed in laying out the epistemological argument for naturalistic coherentism, and although we also put it to use in terms of applying it both to the Theory Movement and its modern successors, to a range of alternatives, as well as research methodology, it remained relatively unspecific on its implied naturalism. The most important element of our naturalism is the acknowledgement that human beings have to acquire all their knowledge from infancy onwards. While we are born with certain physical predispositions, we have to *learn* about the world around us. In the building up of our vocabulary to describe the world and thus to represent it, we construct our theory of the world and also construct meaning.

This modest point of human beings as learners exerts powerful constraints on theory construction in educational administration insofar as any claims made have to be compatible with, or *cohere,* in our terminology, with the human capacity to acquire it. Traditional theory of knowledge, for example, fails on this score because it does not satisfy the learning requirement. In technical terms, it fails to be self-referential. What is stipulated to be knowledge, on the logical positivist or empiricist account, just could not have been learnt because of its implied theory of mind/brain and its conception of how humans acquire knowledge (this was basically what we argued in Chapter 2 of *Knowing Educational Administration*). This difficulty also shows up in various branches of applied social science such as the literature dealing with leadership, to take just one example.

Leaders are learners like the rest of us and so must somehow have learnt what goes into becoming leaders. Yet, discussions about leaders tend to emphasize personality characteristics such as 'charisma' or 'vision', as well as the possession of supposedly superior knowledge about whatever problem they are facing. What is not explained at all, leaving the whole process mysterious, is by means of which human capacities these features could have been acquired. The old adage that 'leaders are born not made' expresses this conundrum well. Our view on the matter is quite different. Left in the realm of the mysterious, we will not be able to explain 'leadership', and subsequently plan for leader development. This is one very important practical outcome of a causal explanation, one that coheres with our actual human capacities to acquire and process knowledge. What is of fundamental concern in our naturalistic coherentism, following Dewey's maxim, is that there is nothing as practical as a good theory, and that the proof of the pudding is in the eating. Causal explanations are our best warrant of changing practice effectively. A second important issue needs to be raised here: the presumed primacy of language as the medium for acquiring and processing knowledge.

The standard model of cognition and information processing is one that assumes that our brains, being highly complex information processors, operate in the fashion of a 'sentence-cruncher', to use Patricia Churchland's colourful phrase. But according to contemporary neuroscience the story of how we acquire and process knowledge is quite different. What happens when we read a sentence, for example, is not a word-by-word linear 'recognition', and a corresponding matching of words with their templates in our brains, but rather the instantaneous recognition of a *neuronal pattern*. The architecture of the human brain is such that it is something of a confederation of processing systems, or *neural nets*, whose modus operandi is that they function in a manner called 'parallel distributed processing' (Rumelhart and McClelland 1986a, 1986b). This means that stimuli received from the environment can be processed at once all over the brain, depending on what the stimuli are. It is this particular feature that allows us to recognize a face 'in a flash', to operate a car while thinking about a million other things, and to be a good practitioner. 'Good practice', for example, is often the ability to execute something well without our being able to explain verbally how it is done (Evers 1996; Lakomski 1996). In everyday life we talk about having a hunch or following our intuition to make sense of this ability. Recent models of brain functioning go a long way towards unravelling this mystery. Language, whether in everyday terms or the specialist languages of philosophers and scientists with their emphases on formal, logical relations, is not the primary medium of knowledge acquisition and processing, but is a second-order human ability, which developed late as an evolutionarily useful mode of 'outward' representation of 'inner' and as yet insufficiently understood processes. While there seems to be little disquiet over neuro-scientific explanations regarding more mundane human activities, the level of skepticism rises sharply where such issues as human subjectivity and culture are included as contenders for neuro-scientific explanation. Human subjectivity, our values, feelings and beliefs have long been seen as the prerogative of the interpretive social sciences, with their attendant methodological procedure of understanding or *'verstehen'*, which is uniquely suited to get at our inner mental furniture. In our view, everything around us needs to be interpreted, or is theory-laden, including our understanding of ourselves as interpreting beings who have feelings and beliefs. There is no special privilege attached to understanding, and beliefs, reasons, and values, belonging to folk psychology (Churchland 1988), are short-hand expressions of complex neuronal processes we do not yet fully understand. It is puzzling to us that at a time where scientists are beginning to unravel many of the traditional mysteries of what goes on inside a person's head, that is, begin to find causal accounts for human action, our naturalistic program is considered to be 'reductionist' in the sense of '*de*-humanizing'. Suffice it to note here, then, that our program is considered to be controversial

In our second book, *Exploring Educational Administration*, we offer an initial attempt at spelling out some of the implications, applications and consequences of *naturalistic coherentism* for the practice of educational administration. The book demonstrates how a post-positivist science of administration can deal with traditional and vexed issues in the field. We also offer here, for the first time in consolidated form, the various critical debates, spread across a number of international journals, that have surrounded *Knowing Educational Administration* since it was published. Similarly, the discerning reader would have noted that our own work, written since 1991, and reproduced here has been previously published in a range of journals. It seemed to us optimal to collect this work in one volume in order that the impact and reach of naturalism in all manner of traditional problems and issues in educational administration could be properly demonstrated and discussed. And since we also subscribe to the cardinal virtue of conservativeness as one of our coherence criteria, we thought it far more economical to offer in one place what otherwise has to be searched for in many. Insofar as *Exploring Educational Administration* follows on from *Knowing Educational Administration*, it should be read as one in a series. At the same time, however, we also think that our second book should be accessible to the reader unfamiliar with our first book. It is for this reason that we have structured the text in such a way that enough of the epistemological machinery is presented to enable the reader to make the naturalistic applications and extensions intelligible. The text is laid out in four distinct parts.

Part I, *Naturalistic Coherentism in Educational Administration*, consists of three chapters and presents the theoretical rationale and foundation for the book. It retells some of the coherentist story and also offers further philosophical developments and expansions of our *naturalistic coherentism*. In particular, we spell out some more of the detail of what is meant by coherence in administrative theory. This, *inter alia*, takes the form of examining the strengths and weaknesses of traditional and alternative conceptions of educational administration, especially in regard to their relationship to practice. We provide some discussion and extensions regarding the role of theory both for understanding the nature of practice and as a guide to decision-making.

Part II, *Naturalistic Extensions*, offers the reader some preliminary applications of the naturalism side of coherentism in terms of explaining some of the exciting work done in the neurosciences, information theory, and control theory in the context of educational administration. The work presented in this section is original and of a pioneering kind insofar as we demonstrate the scope and explanatory power of naturalistic coherentist criteria across a diverse range of phenomena in the social sciences. We examine the utility of the *systems metaphor* in organizational theory; present a novel *organizational learning* model which, based on a realistic account of fallible human learning, and incorporating feedback loops, maximizes error correction and furthers the growth of knowledge in organizational structure; critically analyse a currently prominent conception of leadership, that of *transformational leadership*, and demonstrate how the quantitative methodological tools used outrun the claims made on behalf of the model; offer an original account of *policy analysis* that demonstrates the incoherence of the sentential paradigm at the heart of the traditional conception of policy analysis and policy studies; introduce some of the technical literature on realistic models of brain processes, 'connectionism'; and present a novel account of administrative decision-making as pattern processing. We conclude this section with an explicit and more technical discussion of the currently best models of human brain processing, the neural network account, which has been more or less implied in the preceding chapters, and in our work in general.

Part III, Beyond Subjectivism and Humanism, offers new commentary on, and further developments of, theoretical–philosophical issues raised by the work of Th. Greenfield and Christopher Hodgkinson in particular. Both writers have been concerned with what are fundamental problems in contemporary social science, such as the notion of subjectivity and values. From our naturalistic point of view, we can offer some solutions to problems they have raised and left unanswered. With regard to Greenfield's work, for example, we argue that his subjectivism, to become fully developed, needs a naturalistic account of values. Reconsidering Hodgkinson's values taxonomy which underpins his theory of leadership, we argue that the isomorphy between his stratified leadership/followership structure is unsatisfactory because it prohibits the growth of knowledge. The latter is paramount in our naturalistic theory of reason, which issues in an appropriately open organizational design that furthers human/organizational learning. We conclude this section by examining democratic decision-making in the context of difficulties raised by the economist Kenneth Arrow, which cast doubts on the decisiveness of decisions reached via democratic decision procedures.

Part IV is entitled *Critical Debates*. This section is a collection of critical commentary on our work as it was published in the following three journals: *Educational Management and Administration*, *The Journal of Educational Administration*, and *Curriculum Inquiry*. We reproduce the three sets of debates here in full. We believe that this section is of particular value because the debates, conducted with some of the most eminent scholars in the field, single out various contentious issues in our work and discuss these in detail. Most important of these is our conception of 'coherence', and the definition and utility of our coherence criteria, singly and in concert; our relationship to empirical work; our interpretation of Greenfield's subjectivism; the structure of our ethical theory, and worries over our purported neglect of discussing cultural, economic and political issues as the relevant context for administrative work. These debates help clarify, and allow us to correct where necessary, various aspects of our new science that, we think, will be most helpful to students, practitioners, and professors of educational administration alike.

Our Conclusion, entitled *'Doing Educational Administration — Preliminary Reflections on a Theory of Practice'*, sums up the major contribution of this book. Since all good things come in threes, as confidently predicted by the well-known German proverb, we also foreshadow our third book in the series which will be a fully developed account of a naturalistic–coherentist theory of administrative practice.

References

Churchland P.M. (1988). *Matter and Consciousness*. Cambridge, Mass.: MIT Press, revised edition).
Evers C.W. (1996). Philosophy of education: A naturalistic perspective, in D.N. Aspin (ed.) *Logical Empiricism and Post-Empiricism in Philosophy of Education*. (London and Cape Town: Butterworths).
Evers C.W. and Lakomski G. (1991). *Knowing Educational Administration: Contemporary Methodological Controversies in Educational Administration Research*. (Oxford: Pergamon Press).
Griffiths D. (1991). (ed.) Nontraditional theory and research, Special Issue of *Educational Administration Quarterly*, pp. 262–451).
Lakomski, G. (1996). Tacit knowledge in teacher education, in D. N. Aspin (ed.) *Logical Empiricism in Educational Discourse*, Vol. 2 (*Advanced Volume*). (London and Cape Town: Butterworths).
Rumelhart D.E. and McClelland J.L. (eds.) (1986a). *Parallel Distributed Processing*. Vol. 1. (Cambridge, Mass.: MIT Press).
Rumelhart D.E. and McClelland J.L. (eds.) (1986b). *Parallel Distributed Processing*. Vol. 2. (Cambridge, Mass.: MIT Press).

1
Recent Developments in Educational Administration

Over the past twenty years there has been a substantial growth in the range of theories about educational administration. Although much theorizing was accomplished in the period from the early 1950s to the early 1970s — the principal period of educational administration's establishment as a discipline — it was dominated by a vision of educational administration as a science (Griffiths 1964, Hoy and Miskel 1991, pp. 1–34, Evers and Lakomski 1991, pp. 46–75, Lakomski and Evers 1995, pp. 1–17). Consequently, many of the newer approaches have been formulated specifically in opposition to science as a model for administration. Indeed, the debate for and against the possibility of a fruitful science of educational administration has, we think, provided the main framework for theoretical discussion, and development, in the field since the late 1940s.

In what follows, we shall briefly review some of this debate, touching on both theoretical issues, and also different associated conceptions of practice. For different views about the nature of practice are also embedded in theoretical perspectives, whether these are acknowledged or not. We conclude by defending an organization learning model of practice.

Science in Educational Administration

The rise of science in educational administration in the United States during the 1950s was the result of an organized intellectual movement to replace 'naive empiricism' with rigorous theorizing. Known as the Theory Movement, and including scholars of the calibre of Andrew Halpin and Daniel Griffiths, and sponsored with very large grants from the Kellogg Foundation, it sought to develop knowledge for the improvement of educational practice (Moore 1964). However, by 'knowledge' was meant claims structured into theories, and by 'theory' was meant a hypothetico-deductive structure of the kind championed by the self-proclaimed logical empiricist philosopher, Herbert Feigl (1974).

According to logical empiricist canons, scientific theories have three main elements:

1. They consist of a set of topic specific empirical claims arranged in a hierarchy. At the bottom, as it were, are very particular claims; singular observations about this event or that phenomenon. The further up the hierarchy one goes, the more general the claims become, perhaps then applying to all schools, then all organizations, then all living systems, and finally all physical objects. (Logical empiricists were much influenced by the example of physics, often using it as a model for all legitimate knowledge.)
2. They are testable, and are justified by a process of empirical testing. Essentially, the claims (or hypotheses) in the hierarchy are used to deduce claims lower down, including singular observation reports. Since observation reports form the foundation of empirical evidence then hypotheses are confirmed if the observations they imply are made, and disconfirmed if contrary observations are made. Roughly speaking, a theory is more justified that another if it has more confirmations and fewer (or no) disconfirmations. This process of deduction and testing is known as the hypothetico-deductive method.
3. All key terms in the theory admit of empirical definition; that is, definition with reference to some set of observations. This requirement appears difficult to achieve with so-called theoretical terms: for example, terms in physics such as 'electron', or 'quark', which refer to unobservables. However, the requirement is interpreted as a demand for such terms to be given operational definitions, which usually amounts to the specification of some empirical measurement procedure (Hoy and Miskel 1991, pp. 3–4).

By modelling administration on a philosophy, in turn heavily indebted to an idealised version of physics, the writers of the Theory Movement hoped to provide two main advantages for practitioners. The first would be a rigorous procedure for improving the knowledge base of the profession. When pitted against anecdotal evidence and the 'folk wisdom' of practitioners, the hypothetico-deductive method could be relied on systematically, over time, gradually to grind out more and more justified claims about educational administration. It was expected ultimately to deliver true generalisations about schools, bureaucracies, teaching, leadership, and organizational design, and to locate these truths within justified claims on more general matters such as the behaviour of social systems (Griffiths 1959, pp. 21–46).

The second main advantage was that the knowledge thus provided would be in a form that could readily be applied. Thus consider the problem of making good educational decisions within the constraints of a fairly typical decision structure.

1. Some goal, G, is thought to be desirable; say reducing levels of absenteeism at school, or gaining better school results in VCE studies.
2. Some means, M, is reckoned to be the most efficient way of achieving G.
3. Therefore, since M brings about G, M should be implemented.

Although in no way a rigorous structure, it is suggestive of where administrative theory enters the practitioner's scene. There is first the empirical question of what causes G to occur. And second, because the usual measures of efficiency are empirical — e.g. time, money, use of equipment — there is the calculation of alternatives to single out M. With the value of a hierarchical theory design, ascending from the particular to the general, residing precisely in its capacity to link events of type G to events of type M, a good administrative theory should be able to provide advice on options for the effective realisation of educational aims. Much traditional social science research in educational studies meshes with this rough model of decision. Most often through the use of surveys and questionnaires, variables are postulated, or discovered, and statistically significant correlations established that can then be winnowed down into useful causal generalisations (Lakomski 1989). The literature on school effectiveness, for example, is replete with this kind of reasoning. The selection of the most efficient option draws even more heavily on the detail of theory, because it is the causal consequences of options that form the basis of any other comparisons (such as cost).

The most serious challenge for any traditional science of educational administration that has claims to be relevant is to produce useful true generalizations that will enable the explanation, or even prediction, of particular social and organizational phenomena; in short, to come up with true, non-trivial laws, akin to those found in the natural sciences. This challenge has been met by appealing to systems theory, which tries to classify collections of interrelated parts into those that possess very general empirical properties. Consider, for example, some of the properties of open systems. They are said to be to *self-regulating*, maintaining some sort of *equilibrium* through *feedback*, which also contributes to an ability to 'maintain themselves in *steady states*' (Griffiths 1964, p. 116).

As a guide for practice, the systems metaphor suffers from two, ironically opposed, sorts of difficulty (see Chapter 4). First, the high level of generality to which systems theory aspires encourages the use of mutually interdependent sets of definitions; that is, the key terms of the theory are defined with reference to one another, with the result that systems have the properties they do more by definition than by empirical discovery. The most general claims of the theory end up looking true but trivial because all the relevant phenomena are described in terms of the theory's interdependent terminology (Evers and Lakomski 1991, pp. 67–72). Thus any systems flowcharts designed to display the operation of a school can end up grouping and classifying features into precisely those that are to do with feedback, self-regulation, stability, and the like. Secondly, because this terminology derives meaning from other contexts, certain empirical possibilities can be neglected, or even ruled out. The obvious consequence is an emerging gap between, on the one hand, administrative practice where experience (theorized from these other contexts) can include structural conflict and division, instability and the politics of destabilization, and power struggles among groups with opposing interests and, on the other, systems theory with its more benign characterization of organizational life.

Combining the methodological quest for administrative theory to develop its own law-like generalizations with the very context specific demands of successful practice has been a task that has so far proved to be beyond the resources of traditional science of educational administration.

Is there a more modest view of 'theory', less restrictive than the one logical empiricists developed from more austere examples, that can contribute to the construction of useful theories in educational administration? Until recently, answers to this question have been predicated on the assumption that all science is fundamentally positivist (or logical empiricist). Researchers have therefore either persevered with trying to construct a science of administration that meets these demands, or they have seen the task as totally misconceived and have pursued models of theory quite different to science.

Values in Administration

One obvious limitation of traditional approaches to theory is that they exclude values. They maintain a sharp distinction between fact and value. The reason values are excluded is a quite straightforward consequence of the background empiricist theory of knowledge. A theory is known to be true, or is more likely to be true, if it is testable, has been tested and strongly confirmed, and contains empirically well-defined terms. But, it is alleged, values are not empirically testable or definable operationally. For example, one can test and perhaps confirm hypotheses about school vandalism, noting patterns of correlation among defined variables. It is, however, something else again to justify the claim that such vandalism is wrong. The idea here is that one can observe destruction, or damage to property, or even the suffering behaviour of others without observing something called wrongness. No matter how carefully we observe, it is only facts that are ever seen and never values. For the empiricists of the Theory Movement, values were inner subjective reactions to the facts, reflecting preferences, not facts themselves (For an overview of ethical issues in administrative theory see Evers 1990, Evers and Lakomski 1994.)

Within the terms of our simple decision structure, it means that the role of theory is restricted largely to dealing with different means for achieving goals. For practitioners looking for advice on what goals are desirable or worth achieving, the most traditional decision theory offers is an analysis of preferences, of what people might desire under certain circumstances. However, there is a substantial difference between aggregated facts about group preferences and whether these preferences are to be valued, or are morally appropriate.

Even the business of choosing means from among alternatives can be a moral matter. First, the most efficient solution to a problem can be unethical; being fair, following due processes, and acting equitably can simply eliminate alternative means before they get to be considered on grounds of efficiency. Second, choosing to be efficient is itself a moral issue. For example, the Canadian practice of maintaining an expensive bilingual policy places certain social outcomes ahead of efficiency on moral grounds. Similarly, in 1993, when the Victorian state government removed about 8500 teachers from the teaching service and closed about 200 schools, an important policy consideration was cost efficiency: small schools run at a much higher per student cost than large schools. However, it is a normative decision, not an economic decision, whether the disadvantages of higher cost per student outweigh the advantages of, let us suppose, closer community involvement, greater student welfare, and less student travelling time. (And for a government there are other alternatives to be weighed up normatively outside the education budget). Many decision schemata hide these moral issues when it comes to theory, but they cannot be avoided in practice where they are faced day by day.

Two theories of administration have been quite influential in offering a moral perspective on administration. The first, developed by the Canadian scholar Christopher Hodgkinson (1978, 1983, 1991) declares administration not to be a science at all, but rather, a humanism. This is because, for Hodgkinson, science deals only with factual matters whereas administration is value-laden (Chapter 13; Evers 1993, Hodgkinson 1991). Hodgkinson also maintains that decision-making is central to administration. Because knowledge of logic and value constitute the essentials of decision, administrator training will involve some training in philosophy where these matters are dealt with most systematically. This training need not be in academic philosophy, although he does identify the Oxford Philosophy, Politics and Economics (PPE) degree as an example. Training can also include a focus on Eastern philosophies and contemplative practices.

Given the kind of very detailed and specific demands made upon school leaders, how can a more abstract, philosophic perspective be of use? Hodgkinson addresses this issue by defending a stratification of organizational life. Administrators are at the top of the hierarchy, those with managerial functions are next, and the rank and file are at the bottom. All the familiar tasks of calculating means, of juggling budgets, of working out costs and benefits remain, but at the managerial level. But administration is above management and looks to setting goals and purposes, determining broad direction, and establishing the basic organizational values and over-riding priorities. Administrators are responsible for an organization's vision. If principals leading Victoria's Schools of the Future feel that the massive increase in managerial tasks resulting from decentralisation of functions is crowding out deliberation on educational values then, from Hodgkinson's perspective, the school has been restructured in a way which diminishes its moral autonomy (Hodgkinson 1991, pp. 156–158).

Critical theory provides a second influential approach to values in administration. Richard Bates (1995) is a prominent Australian theorist whose work has helped define a critical approach to administrative issues, while the most systematic presentation of critical theory in educational administration can be found in the work of William Foster (1986). Although there are a number of versions of critical theory (Maddock 1995), the most important derives from the earlier work of Habermas, with its critique of science. For writers such as Bates and Foster, science represents only one form of knowledge. For a more complete account of the kinds of knowledge that arise out of human experience, we need also to take into account the knowledge required for interpreting and understanding human communication. And in addition to having an interest in being able to manipulate and control the world, which science provides, and in being able to communicate and interpret communications, we also have an interest in being free from domination; we have an emancipatory interest.

Patterns of oppression and domination have many causes and are subject to many different analyses. However, one consequence of oppression is to rob certain people of a voice in the process of communication. Thus oppression on the basis of race, gender, class, or religion can be expected to be reflected as unequal participation in the speech community. This is not simply a matter of being denied a right to speak. It includes diminished access to the knowledge necessary to make an informed contribution. For a contribution can be unequal by being uninformed.

Critical theory is critical of science of administration because it applies the framework of manipulation and control to the social sphere — to the manipulation and control of people as means to unscrutinized ends. Science is claimed to be not so much value-free as value-ignorant, especially of its own unacknowledged values. The reason this managerialism is unacceptable from a moral point of view is because those being manipulated down the hierarchy do not have an equal voice in the decisions being made which affect them. Instead, critical theory recommends participatory and democratic organizational designs; that is, designs that will instantiate at the organizational level, the values required to approximate as closely as possible the ideal speech situation.

The justification of these values of freedom of speech, social and political equality, tolerance, fairness, and truth-telling, to name a few, does not proceed by defending certain consequences. For example, they are not justified by appeals to greater efficiency, or by considerations of human happiness, or maximizing utility. Rather, they are defended by claiming that they are presupposed by any act of communication, however imperfectly realised. Just as the person who demands a defence of reason is presupposing some standard of rationality with which to assess a proffered answer as an adequate defence, so the person who demands a defence of these values is already presupposing them in engaging in the communicative practice of asking for a defence. This form of argument from presupposition is known as a transcendental argument and it is employed by Habermas in various guises in his work early and late (Lakomski 1988). It is derived ultimately from the writings of Kant, who made it famous in his *Critique of Pure Reason*. Although we would want to support many of the values championed by critical theorists, transcendental arguments are extremely difficult to mount and are rarely successful. The main reason for failure is that what ends up being the source of presupposition is not some phenomenon like speech, but rather a *theory* of speech. But theories are fallible, often wrong, and often can be improved upon, especially by recent advances in science. It would therefore be more useful to be able to appeal to our best current theories, particularly scientific theories, to defend our values. Unfortunately, partitioning off values from science blocks this approach.

Subjectivity in Administration

While some alternatives to science have focussed on providing a much needed ethical perspective on administrative problems and issues, another cluster of alternatives has focussed on the importance of human subjectivity. Thomas Greenfield is the best known writer in educational administration to champion this approach (Greenfield and Ribbins 1993). For the practitioner, the effect of Greenfield's arguments can be seen most clearly if we consider again our familiar decision structure on going from goals, via means, to implementation. The core of the structure for traditional science of administration is the business of producing reliable knowledge concerning which means produce which ends; that is, producing a sound causal model complete with appropriate generalizations and relevant empirical facts for each situation.

Greenfield has two main criticisms of this enterprise. first, he attacks the alleged objectivity of any model, or theory, that is justified by the hypothetico-deductive machinery of empirical testing. And second, he argues that social science is fundamentally distinct from natural science, and that as a result, social science is irreducibly subjective. His reasoning is worth exploring because of its very great influence on the field. (For a very detailed discussion see Evers and Lakomski 1991, pp. 76–97.)

Against objectivity in natural science, his strategy is to blur the distinction between observation and theory. Drawing on work in the philosophy of science, he claims, quite rightly, that all observation involves some degree of interpretation. But interpretation is a matter of viewing phenomena through the lens of some theory, whether it be the ordinary commonsense categories embedded in natural language and everyday practices, or more revisionary categories that are the product of the specialised sciences. Where the data themselves are recognized as theoretical, the notions of confirmation and disconfirmation become compromised. The theory informing the nature of data may be less reliable than the theory the data is being used to refute. Another problem Greenfield draws attention to is that theories are always underdetermined by data. Any number of different theories will fit a finite data set. Any number of different curves can be drawn through the same set of points. This implies that the same data (points) are confirming different theories (curves). A final difficulty arises out of the complexity of empirical test situations, especially in social science. One is rarely sure exactly which part of the theory is being tested when it takes the whole of the theory, or a large network of claims, to imply some testable observation. Greenfield concludes that appeals to data are never sufficient for sound theory choice; in fact, that there is no such thing as objective theory choice. We make our choices by augmenting these resources with our own inner contribution of taste, value, subjective preference, and will.

For the practitioner, this means there are simply limits to what can be known about alternative means. Even when all the data are in we will have no closure of decision. Greenfield's advice is to cultivate the subjective, to broaden one's experience in range and diversity, to be guided by one's values and wider vision of life in general. Literature, art, music, and theatre are as useful in shaping the fulfilment of organizational purposes and goals as any of the traditionally conducted empirical studies.

Greenfield's defence of the distinctness of social science from natural science further reinforces this point. Unlike many other objects of natural science, human behaviour can be redescribed as partly a function of. thoughts, intentions, meanings and understandings. Thus a person, in acting, may be responding to an interpretation of the meaning of what another person may have intended by their own action. So instead of trying to use statistical regularity among behaviours as a basis for predicting social behaviour, as traditional conceptions of science would recommend, Greenfield (following Weber) suggests trying to understand the network of reasons, motivations, and intentions that underwrite behaviour. The assumption that people are engaging in meaningful action rather than caused behaviours is a much more powerful methodological tool for accounting for organisational life. Of course, meaning, interpretation, and the like , have long since parted company with an objectivity grounded in empirical testability and the apparatus of operational definition. Rather, we are dealing with the irreducibly subjective. (Greenfield's most incisive formulation of these issues can be found in Greenfield and Ribbins 1993, pp. 1–25.)

Greenfield's work articulates nicely with approaches which focus on the importance of organisational cultures, providing some clear examples of how school reform can fail if it attends to no more than budgets and organizational structures. An understanding of history, tradition, culture, and ritual can therefore be of great importance in leading to long-term reform.

Chaos and Postmodernism

Recently, a new argument, based on chaos theory, has arisen against the value of scientific approaches to administration. The example of the butterfly effect in meteorology, where the flapping of butterfly wings in one part of the world may ramify down the causal chain to produce a substantial weather impact on another part of the world, suggests limits to the predictive use of theories. And in the absence of predictive theories, there can be no reliable analysis of competing means to achieve goals. In short, it is suggested that administration has its butterfly equivalents (Green and Bigum 1993).

Without detailed mathematical modelling of administrative contexts, the claim is more an unproven conjecture than an argument, but some background may help make this point clearer. The behaviour of large systems, especially in the social realm, has always been difficult to predict, or model accurately. However, most researchers assumed that this was because of two problems: a difficulty over 'noise' or uncontrolled influences, and a difficulty over complexity, or too many variables. So it was believed that large systems are in fact deterministic, but appear to behave randomly because of these problems. Unfortunately, we now know that apparently random behaviour can attend quite simple systems with no noisy influences. It turns out, though, that this behaviour is the result of the system being non-linear, or able to be modelled by a non-linear set of equations. Linearity is a technical but fairly straightforward notion. It just means that any change in one variable will produce a constant change in another variable; that is, there is no change in the rate of change. Now viewed as a dynamical system, with parts in motion under the influence of forces, the world as physics describes it is highly non-linear. For example, gravity produces non-linear attractions among objects with mass. Chaos theory might better be viewed therefore as an attempt to extract the regularities that underlie the apparently random behaviour of complex systems. (This is essentially the position reached by Griffiths, Hart, and Blair 1991.)

Nevertheless, there has been some attempt to theorise the current social condition in terms of non-technical, or popular notions, chaos. In opposition to the basic assumptions of systems theory, the world appears to be characterized by uncertainty, instability, complexity, and uniqueness. The management of disorder and change seems to constitute the more appropriate demand on both theorists and practitioners. Decentralization of previously large bureaucratic structures into autonomous self-governing parts is typical the postmodern trend from highly ordered systems towards more disordered structures. Hargreaves (1994, pp. 83–85) describes such changes for teachers as involving 'occupational flexibility and technological complexity ... doubt and insecurity... moral and scientific uncertainty ... and organizational fluidity'.

Postmodern analysis has also been linked to the powerful subjectivist critiques of traditional science to produce varieties of relativism. The result is the so-called postmodern condition of fluid unstructured, even disordered, social change, together with no possibility of a true or accurate account of these phenomena. Indeed, we have a prevailing scepticism of the very categories of truth and accuracy.

While the resulting focus on theories of change is a welcome corollary of postmodern studies in education, the intellectual value of the theories is compromised by relativism. (Hargreaves avoids the problem by distinguishing between postmodernism and postmodernity. He believes in the latter, which is a feature of current societies, but not the former, which is an attack on knowledge.) In our view, a more fruitful way to theorize administrative knowledge is required; a way that moves beyond the old debates about science of administration and its alternatives.

Focussing on Coherence

Over the last ten years or so, we have been working on such an alternative. We are in favour of a new science of administration; not a view where science is conceived positivistically, or along logical empiricist lines, but rather one where science is justified holistically, by appeals to systematic coherence. Greenfield's criticisms of traditional science are correct inasmuch as appeals to empirical evidence — the so-called empirical foundations of knowledge — are never sufficient to justify theory choice. However, the move to invoke subjectivism, as Greenfield does, or to give up on the problem of justification as postmodernists do, or to partition science off from the rest of knowledge as critical theorists sometimes do, is premature. In our view, there is arguably more to justification than empirical evidence. There is also consistency, and simplicity, and comprehensiveness, and fecundity, and learnability. These are super-empirical virtues of any good theory, and together with empirical adequacy they form part of a coherence theory of justification. (For a systematic presentation of these ideas see Evers and Lakomski 1991.)

A coherence model of justification permits the development of a postpositivist science of educational administration that is broad enough to include values and considerations of human subjectivity. (Not everyone believes this claim. For some vigorous debates see Part IV; Maddock 1995, Evers and Lakomski 1994.)

Consider the problem of defending a particular set of values. We shall begin obliquely by endorsing postmodernist denials of certainty: all knowledge is fallible. Fallibility does not imply, however, that all knowledge is equally worthless. In a complex and uncertain world it may be sufficient to do better than chance. In reaching the goal of getting to the other side of a busy road, for example, a decision when to cross informed by the belief 'avoid colliding with moving vehicles' is likely to produce more satisfactory results than a decision based on coin tossing. Success in getting around in the world is a statistical matter, a matter of behaving non-randomly in appropriate circumstances. Whatever else it may be, a good theory will be a representation of knowledge that is able to capture the statistical structure of experience, both past and hopefully, future experience. Since we are unlikely to just hit on a good theory by chance, it is useful to have procedures for changing our theories that will do better than coin tossing. Given that all knowledge is fallible, the next best thing is to be an efficient learner. The design that appears ubiquitous in natural learning systems is some version or other of trial and error, of working to minimize the gap between expectancy and experience.

Learning flourishes more under some social arrangements than under others.

By developing social relations of inquiry that permit a systematic approach to reducing the gap between theoretically motivated expectation and theory-laden observation, and by adjusting our theories in a coherent way when gaps emerge, we have a broad procedure for promoting learning that will do better than coin tossing. Even the early learning of language, with its subjectivist intentional idiom and talk of meaning and interpretation, is an exercise in extracting patterns from the daily flux of experience. For if human behaviour were random, interpretation would not be possible.

Now effective social relations of inquiry also require some ethical infrastructure. In promoting the growth of knowledge, alternative viewpoints need to be encouraged and tested, feedback should be promoted, tolerance needs to be exercised, there is a premium on freedom of speech and association, and also equality and respect for persons in order to promote wide participation in the epistemic community. These Deweyan values are not absolute, but they arguably cohere best with any non-random approach to getting around in the world, solving problems, or making decisions.

To return to our schema for decision-making, when it comes to organizational practice, ensuring the capacity for improving knowledge of both means and goals comes before organizational designs for making or implementing decisions. That is, organizational learning should be a basic feature of organizational design. Arguments for the fallibility of all knowledge apply as much to leaders' visions and directives as to any other source of knowledge. The very strong focus on the principal's leadership role in currently popular managerialist models of schooling seems to depend on a view of leadership that is not attentive enough to the frailties of human knowledge.

Just as coherence considerations can be invoked for justifying both scientific and value claims, so coherence considerations apply to the realm of the subjective. However, let us begin by observing that doing epistemology is not an a priori exercise. Rather theories of knowledge can be improved by being informed about the details of human knowledge acquisition. Our best theories of learning are therefore useful in structuring epistemologies, and as we have seen, epistemologies exert a powerful influence on the structure and content of substantive theories, such as those dealing with educational administration. (The key features of logical empiricism are, after all, epistemological claims.)

Our approach to administrative theory is scientific because our coherence epistemology needs to cohere with natural science, with scientific theories of human cognition. Until recently, most models of cognition were not scientific in this sense. Decision-making, for example, was characterized as a quasi-deductive relation among symbolic representations of knowledge, much along the lines of the decison structure we have been using throughout. However, in the last eight years there has been a major development in cognitive science leading to a shift to computer models of information processing in the brain. These 'neural networks' are now being used to model all sorts of cognitive phenomena in a way that promises to be much more realistic that the old, but familiar, symbol crunching models. (For an introduction to these ideas see Chapters 8 and 9.) The distinction between pattern processing and symbolic manipulation is perhaps the most important difference between the new and the old approach. Since neural networks represent knowledge of practice in much the same way as they represent knowledge of symbols, the new cognitive science also promises some insights on education for practice.

Our major interest in this approach, though, is that it may yield deep insights into human subjectivity in a way that will permit our theories of the natural world to cohere with theories of the social world. (Greenfield's approach leaves the acquisition of meaning and understanding something of a mystery.) On a coherence view of theory, it is now more feasible to view culture, ritual, organization, and language as patterns of association and communication, and to see their learning as meshing smoothly with a general theory of cognition as pattern processing. Needless to say, as things now stand, this is the beginning of a research program rather than the end of one.

Conclusion

The question of the relationship between administrative practice and administrative theory is a vexed one. An important school of thought sees theory as being shaped primarily by the demand to make sense of research findings gleaned from practice and the study of practitioners. However, we now know that the activity of practitioners is itself shaped by theories, whether acknowledged or unacknowledged. That is, administrators augment their experience in all sorts of ways in order to understand novel situations, or to make assumptions on the outcomes associated with altering what has been familiar practice therefore undermining the basis of accumulated experiential wisdom. All of this makes the current ferment in administrative theory crucial. In the absence of an emerging consensus, a valuable touchstone for good theory is still the fulfilment of theoretically motivated expectation. For as Dewey pointed out long ago, there is nothing quite so useful as a good theory. We would suggest, therefore, that many of the most interesting theoretical disputes in educational administration could be resolved, and our knowledge of matters advanced, if administrative practice became a more self-conscious arena for their trialing and testing. Such a possibility is open before us.

References

Bates R. (1995). Critical theory of educational administration, in C.W. Evers and J.D. Chapman (eds.) *Educational Administration: An Australian Perspective*. (Sydney: Allen and Unwin).

Evers C.W. (1990). Ethics and educational administration, in T. Husen and N. Postlethwaite (eds.) *International Encyclopedia of Education*, Supplementary Vol. 2. (Oxford: Pergamon Press).

Evers C.W. (1993). Hodgkinson on moral leadership, *Educational Management and Administration*, 21(2), pp. 259–262.

Evers C.W. and Lakomski G. (1991). *Knowing Educational Administration: Contemporary Methodological Controversies in Educational Administration Research* (Oxford: Pergamon Press).

Evers C.W. and Lakomski G. (1994). Educational administration: ethical and philosophical issues, in T. Husen and N. Postlethwaite (eds) *International Encyclopedia of Education*. (Oxford: Pergamon Press, second edition).

Feigl H. (1974). *Inquiries and Provocations: Selected Writings 1929–1974*. (Boston: Reidel).

Foster W.P. (1986). *Paradigms and Promises*. (Buffalo: Prometheus Press).

Green B. and Bigum C. (1993). Governing chaos and postmodern science, information technology and educational administration, *Educational Philosophy and Theory*, 25(2), pp. 79–103.

Greenfield T.B. and Ribbins P. (1993). *Greenfield on Educational Administration*. (London: Routledge).

Griffiths D.E. (1959). *Administrative Theory*. (New York: Appleton–Century–Crofts).

Griffiths D.E. (1964). The nature and meaning of theory, in D.E. Griffiths (ed.) *Behavioural Science and Educational Administration*. (Chicago: University of Chicago Press).

Griffiths D.E., Hart A.W., and Blair B.G. (1991). Still another approach to administration: Chaos theory, *Educational Administration*, 27(3), pp. 430–451.

Hargreaves A. (1994). *Changing Teachers, Changing Times*. (London: Cassell).

Hodgkinson C. (1978). *Towards a Philosophy of Administration*. (Oxford: Blackwell).

Hodgkinson C. (1983). *The Philosophy of Leadership*. (Oxford: Blackwell).

Hodgkinson C. (1991). *Educational Leadership: The Moral Art*. (Albany: SUNY Press).

Hoy W.K. and Miskel C. (1991). *Educational Administration: Theory, Research and Practice*. (New York: Random House, third edition).

Lakomski G. (1988). Critical theory, in J.P. Keeves (ed.), *Educational Research, Methodology, and Measurement: An International Handbook*. (Oxford: Pergamon Press).

Lakomski G. (1989). The journal of educational administration: mainstream, tributary or billabong?, in G. Harman (ed.) *Review of Australian Research in Education*. No.1. (Armidale: ACER).

Lakomski G. and Evers C.W. (1995). Theory in educational administration, in C.W. Evers and J.D. Chapman (eds.) *Educational Administration: An Australian Perspective*. (Sydney: Allen and Unwin).

Maddock T.H. (1995). Science, critique and administration: The debate between the critical theorists and the materialist pragmatists, *Educational Management and Administration*, 23(1), pp. 58–67.

Moore H.A. (1964). The ferment in school administration, in D.E. Griffiths (ed.) *Behavioural Science and Educational Administration*. (Chicago: University of Chicago Press).

2

Educational Administration as Science: A Postpositivist Proposal

Recent debates in educational administration have often focussed on purported limits to scientific views of administration and on the provision of alternatives. For example, scientific approaches are said to be incapable of dealing with ethical issues; yet the practice of administrators, managers and policy analysts is irreducibly value-laden, being routinely concerned with questions of what *ought* to be done or what is the *right* course of action to advise or follow. Or, since organizational behaviour involves vast networks of intentional human activity, no adequate understanding of organizations seems possible without some appeal to human subjectivity, to the *interpretations* people place on their own actions and those of others. Yet scientific models of administrative behaviour in the name of objectivity, seek to eschew interpretations, intentions and the inner life of agents in general. In the realm of research too, case studies, cultural studies and ethnographic methods seem able to deliver important detailed knowledge about administrative processes. But again, these are methodologies that are hard to place within a scientific tradition of controlled experiment or statistically significant reproducible results.

Such major differences between the orientation of traditional scientific approaches to educational administration on the one hand and the focus and direction of often quite systematic rivals or alternatives on the other, clearly lie behind much of the perceived intellectual turmoil in the field. Educational administration, in common with most of the applied social sciences, has experienced a growing appreciation of the methodological weaknesses inherent in positivist construals of science and its methods. Ironically, however, the natural sciences go from strength to strength and hardly any part of modern life remains untouched by the application or use of some aspect of natural science: for example, medicine, transport and communications, to name just a few. So at a time when natural science has never been more successful in explaining and predicting phenomena and in enhancing our understanding of the world, paradoxically its methods and content are increasingly being questioned or even denied in the social sciences.

We suggest a resolution of this paradox as it arises in educational administration. In our view, the paradox is generated principally by the still widespread but mistaken belief that *positivism*, in its many varieties, can be equated with *science*. But in philosophy positivism, in all its main forms, is now regarded as false, its key tenets clearly refutable. The most plausible current developments in philosophy of science and theory of knowledge reflect post-positivist views. Our point, therefore, is that while many of the criticisms of administrative science in educational administration are sound, they are directed at a narrow target. They discredit only positivist versions of administrative science. An alternative post-positivist science of educational administration, we believe, is not only possible but theoretically and practically desirable. The argument we employ for this conclusion has the following broad structure. First, we acknowledge the importance theoretical writers have attached to epistemology or theory of knowledge, by arguing that the weaknesses critics have identified in traditional administrative science flow from *foundationalist* epistemological assumptions embedded in traditional views. Instead, we claim that the proper justification of knowledge is structured by *coherentist* considerations such as theoretical simplicity, consistency, comprehensiveness, conservativeness and fecundity. If the justification of scientific claims, including administrative claims, proceeds according to the coherentist canons of our holistic epistemology, then the scope of science is very much broader than is usually conceived and will fail to sustain significant distinctions between fact and value, the subjective and the objective, and the alleged 'paradigms' of educational research.

Second, we note that major recent criticisms of traditional administrative theory assume theories that in turn share foundationalist epistemological structures. Critics tend to argue for more or different foundations for knowledge to supplement the deficiencies of positivist science. We advance our case here by offering some coherentist epistemological criticism of the Theory Movement, critical theory, and administrative decision-making.

Educational Administration and the Theory of Knowledge

Since the mid 1970s, educational administration, as an area of study has undergone a fundamental transformation. Although traditional views of science still dominate understandings of theory, research and administrative practice, there are now systematic alternatives to this approach. As a result, educational administration is now theoretically much richer, more diverse and complex than at any other time in its short history.

These developments have not occurred without controversy. For example, following a relatively brief period of intense, indeed unprecedented, academic debate in journals, books and conferences, Daniel Griffiths (1979, p. 43) remarked that:

if educational administration is not in a state of intellectual turmoil, it should be, because its parent, the field of organizational theory, certainly is.

Griffiths could well have added a number of related areas of applied social science to make his point such as policy analysis (Garson 1986), educational studies and educational research methodology (Phillips 1987), and social theory (Giddens 1982). These areas, too, were in turmoil, and for much the same reason. The traditional scientific view of knowledge was increasingly perceived to be inadequate as a basis for social science because it ignored values, human subjectivity, and the social and political context in which organizations exist and in which administrative practices occur.

It is unlikely that objections such as these would have been so effective in reshaping the agenda of educational administration were it not for the existence of alternative philosophical perspectives on the nature of knowledge that could function as frameworks for rival systematic conceptions of administration. And in our view, what has made the alternatives seem credible is the work done in the 1960s by Thomas Kuhn (especially his 1962), Paul Feyerabend (1981), and other philosophers of science, which showed, successfully, that traditional views of scientific knowledge are inadequate even for the physical sciences.

The importance for administrative theory of philosophy in general and theories of knowledge, or epistemologies, in particular has been widely acknowledged by writers such as Greenfield (1975), Griffiths (1979), and Willower (1981). In a recent analysis of the state of educational administration conducted as part of a review of contributions to the *Handbook of Research on Educational Administration*, Willower (1988, pp. 730–731) identifies six trends representing '... directions in which educational administration as field of inquiry appears to be moving', and he completes the list by remarking: 'The sixth trend is a turn towards philosophy, and especially towards epistemological questions'.

In our view, Willower is perfectly correct. The only caveat we would want to enter — and one with which we expect he would entirely agree — is that philosophy, especially epistemology, has *always* been significant, though perhaps not widely recognised as such until of late.

Epistemology and Administrative Knowledge

We argue that *all* major developments in educational administration, from the rise of the Theory Movement (see Culbertson 1981) in the late 1940s onward, have been driven by philosophical considerations. Although the reasons for this are complex and vary with the particular developments in question, the general pattern is clear enough. For *any* set of organised interrelated claims that purports to be knowledge, such as a theory of administration, is subject to constraints that apply to all knowledge claims. However, within philosophy it is epistemology that deals with questions concerning the nature of knowledge, what makes claims knowable, and how they may be justified. Our central argument is that what epistemology counts as a satisfactory justification imposes powerful constraints on the content and structure of administrative theory. Or, in other words, the structure of justification, as specified by epistemology, determines much of the overall framework in which theorising in administration is conducted.

Three major developments in epistemology may be cited to illustrate this point. *Logical empiricism*, which developed out of, and partly in opposition to, the logical positivism of the Vienna Circle, provided the first systematic philosophical influence. (For an overview see Achinstein and Barker 1969.) In particular, it is Herbert Feigl's version of logical empiricism that has been critical in the development of the Theory Movement (Feigl 1974), which in turn has shaped much of mainstream educational administration throughout the 1960s, 1970s, and the 1980s. In it may be found the bases for separating fact from value and observation from theory, for employing the methodological constraint of operational definitions, and for seeing administration theory as a classical hypothetico-deductive structure with laws at the top and facts at the bottom (see Feigl 1953).

The second development is the *paradigms approach* associated with the work of Kuhn and Feyerabend. Originally formulated as a systematic critique of logical empiricist views of scientific knowledge, their work — and especially Kuhn's, since it is written in non-technical language — has functioned increasingly to underwrite attacks on objectivity in the social sciences, and to promote varieties of relativism and subjectivism. It has been able to do this because, crucially, a paradigm is supposed to contain within itself the standards for its own assessment (see Kuhn 1962, pp. 109–110). In this intellectual climate, if alternative views of administration are construed as either different paradigms, or as developing within different paradigms, then they are presumed to enjoy some methodological immunity from objections arising from one particular paradigm, say a systems scientific view. At the extreme, different paradigms are said to be incommensurable, or unable to be compared or adjudicated (Kuhn 1962, p. 150).

Two important results of the Kuhn–Feyerabend critique of logical empiricism are alleged to support this extreme view as well as a number of other familiar subjectivist conclusions. The first result is that empirical adequacy is not a sufficient criterion for deciding the merits of competing theories: the same empirical foundation may adequately confirm any number of different theories. The second is that what counts as empirical evidence is partly determined by theory. Observations are said to be theory-laden, mainly because the vocabulary used to describe observations is also part of a wider theoretical vocabulary.

Consistent with the paradigms approach, a further conclusion drawn from these results is that science is significantly non-empirical, that considerations of empirical adequacy place no essential constraints on the construction of scientific theories. Thomas Greenfield (1978, p. 8), whose writings have been largely responsible for the subjectivist turn in educational administration, needs to presume something as strong as this in order to say:

> The process of truth making in the academic world...does not differ

materially from what goes into truth making in the world at large. Truth is what scientists agree on or what the right scientists agree on. It is also what they can get others to believe in.

And this, in turn, yields talk of reality being mind-dependent, of us inhabiting different worlds or their being multiple realities, all of which outrun any empirical evidence for distinguishing them. It also suggests a certain methodological infirmity when it comes to the question of evidence for adjudicating the merits of different interpretations of human behaviour, the stuff of hermeneutics, ethnographies and cultural studies of organizational life (see Evers 1988).

The third, most recent approach, and the one we prefer, acknowledges the soundness of certain key results arising out of the 1960s critiques of logical empiricism; notably the underdetermination of theory by observation and the theory ladenness of observation. However, the correct conclusion to be drawn from these results is not a flight from objectivity and realism. Rather, it is the admission that there is more to evidence than observation or the establishing of mere empirical adequacy (Churchland 1985). After all, any theory can be made to square with empirical findings if we are prepared just to go on adding statements to it. What is more, a contradictory theory will square with any finding whatsoever. On this third approach, which we endorse, theory choice needs to be guided by a consideration of the extra-empirical virtues possessed by theories. These virtues of system include simplicity, consistency, coherence, comprehensiveness, conservativeness, and fecundity, though they are often referred to collectively as coherence considerations or as the elements in a coherentist account of epistemic justification (Quine and Ullian 1978, Williams 1977).

In the following we want to demonstrate briefly how the constraints on justification imposed by our preferred coherentist or holistic epistemology can be used to reshape and redefine the substantive content of educational administration in the direction of a new science of administration.

Foundational Epistemologies and Coherence Justification

Since coherentist epistemologies are best seen as responses to the problems of foundational theories of justification it will be useful to begin by sketching a classical solution to the problem of knowledge. How do we know anything at all? What is knowledge and what makes it possible?

Within the classical empiricist tradition the candidates for immediate knowledge — the foundations — have been, in decreasing order of strictness, sense data, first person sensory reports, and observation statements (Hooker 1975). However, as the history of philosophy readily attests, the hope of justifying all knowledge in this way is fraught with difficulty, especially in view of some of its consequences for science. For example, many of the objects posited by physics are unobservable, at least directly — such things as time, curved space, electrons, and quanta — and are known only through economical theorizing about more gross observable consequences. Worse still, the lawlike universal generalisations characteristic of our best theories, seem to require as evidence an infinite number of observations. Yet only a finite range of observations is ever available for justifying claims of the form 'all X are Y'. It would be nice if we had some sound principle of induction that would enable us reliably to infer from a finite set of observations to an infinite set of past, present, and future events of the sort that scientific laws can delimit; but no such principle has ever been forthcoming (Popper 1963, especially ch. 1). Because so little of what ordinarily passes for reliable knowledge can be deduced from empirical foundations, classical empiricism functioned more as an attack on knowledge, with scepticism the end result, rather than a rational reconstruction.

The crucial methodological worry here is that the knowledge claims ruled out appear more reliable than the epistemology that rules them out. This is because an epistemology, in specifying conditions for claims to count as knowledge, also embodies a theory of the powers of the mind (Churchland 1987). For what we can know will depend, to some extent, on our cognitive capacities, our skills for learning and, in general, what sort of creature we are. One weakness of classical empiricism is that it embodies a singularly implausible empirical psychology of learning. For example, the process of learning from perception is not one in which a passive mind more or less faithfully records copies of sensory images, permuting them (or their decomposable components) according only to the laws of logic. Our current most sophisticated neurological theories of sensory information processing tell a vastly more complex story of human knowledge acquisition (Churchland 1986).

A further methodological worry with the classical view is that it appears to be unknowable on its own account of itself. To see this, recall that the epistemology makes general or universal claims concerning all (human) knowing. One its own terms, either these are known directly or indirectly. But classical empiricist epistemology in all its generality cannot itself be a sensory experience, that is, part of the foundations, if for no other reasons that only a limited number of relevant observations are ever possible. Nor can it count as derived knowledge because of the problem of induction, the problem of using finite observational evidence to infer a general claim. On our view, these methodological problems suggest that the epistemology is *incoherent*.

In response, logical empiricism, in common with twentieth century varieties of positivism, reversed the earlier classical relationship between theory and foundational evidence. For logical empiricists like Hempel (1965) or Feigl (1974) observation statements are deduced from theories rather than the other way around. That is, logically, theories imply observations. The relationship between theory and observation is therefore one of *testability*.

Roughly speaking, theories, as networks of general and particular empirical statements, are supported by evidence to the degree that the observation statements they imply are *confirmed* or the tests are successful. As a method for testing hypotheses by matching deduced observation statements against actual observations, logical empiricism is sometimes identified with the hypothetico-deductive method.

For present purposes, two clear difficulties, which lead to revisions along coherentist lines, may be briefly noted. The first, mentioned earlier, is that the same finite observational base may equally confirm different theories, as pointed out by Karl Popper (1959, p. 266). Piling up more and more confirming instances, he argued, is of little value when it takes only one disconfirming observation to falsify a theory. It is falsification that is crucial for promoting the growth of knowledge, for improving our theories, not confirmation. Theory change is driven by counter-examples, unexpected observations and predictions that are shown to be false. And our best theories are those that have been subject to the most severe testing but have not been falsified.

However, more is required for excellence of theory than just passing severe tests, for it is never individual hypotheses that are tested but, rather, whole networks of statements (Hesse 1970). And if networks or conjunctions of statements are needed to deduce observation statements for testing, a counter-example, or unexpected observation, shows at most only that one or more statements in the network are faulty. It does not, by itself, show which *particular* statements are in need of revision, as Quine (1951, p. 43) has argued. We can even adjust the troublesome observation statement if we feel that those parts of our theory under threat are more reliable than the theory implicit in making the observation.

How then are we to choose the best theory from among an infinite number of empirically adequate alternatives, all equally supported by whatever they deem to be a foundation for knowledge? Our suggestion is to choose the most *coherent* theory, that is, the one that enjoys more than any other the extra-empirical virtues of system (see Lycan 1988, BonJour 1985). To see how a coherence approach can work over the question of choosing theories of knowledge, consider again the two methodological weaknesses we noted in classical empiricism. One was that the epistemology could not explain how it could ever be known. It failed to be comprehensive over the matter of self-reference. In seeking to adjudicate on the status of all knowledge claims, it assumed an external vantage point that it could never know to be true. Other things being equal, therefore, we would prefer an epistemology that was truly comprehensive; one whose embodied psychology of knowledge acquisition renders it knowable.

Once we see an epistemology as itself a set of knowledge claims, we can ask whether the claims it rules out as knowledge are more reasonable or plausible than the epistemology that rules them out. For classical empiricism this is an acute methodological issue since it attempts to disqualify all of the most characteristic features of good science on the strength of a very modest empirical psychology used to select foundations for knowledge. It renders this particular programme of foundational justification incoherent by robbing it of its point. The way to avoid the problem is to require an epistemology to embody our most powerful and sophisticated theories of knowledge acquisition. But if we are appealing to our best natural science of human learning to justify knowledge there is no need to bother with foundations. We just appeal to science outright to justify and explain how scientific knowledge is possible. In our view, there is no knowable epistemically secure and privileged vantage point from which the whole of knowledge can be adjudicated. There is just our most coherent scientific practice. Epistemology becomes naturalized, as Quine (1969) suggests, and falls into place as a part of psychology.

How then can we apply coherence criteria to theories of educational administration? In general we require consistency; we would aim for more comprehensive theories — those able to explain more phenomena rather than fewer, and with fewer anomalies, counter-examples, and falsifying instances rather than more. We would prefer simplicity to complexity in the sense of using the least amount of explanatory apparatus to account for the largest range of phenomena. We prefer theories that do not outrun their own explanatory resources, that do not posit distinctions for which there is, on their own terms, no evidence. Finally, we require that administrative theories be *learnable* in the sense that they meet the following two demands that were applied specifically to epistemologies: first, that they cohere with the broad demands of our best naturalistic accounts of human learning, and second, that they are not inconsistent with more reliable bodies of knowledge elsewhere in our total or global world view. The net effect of these demands is to require administrative theory to be a part of the most coherent global theory we can construct. We end up with a science of administration to the extent that this global theory also includes our most reliable scientific knowledge (Evers 1988).

Applying these general considerations is always a matter of detailed critique of particular issues. Coherence justification, because of its global character, is just a more intricate and difficult business than foundational justification. However, since foundationalism is mistaken, there is really no serious alternative. The following applications, because of their brevity, are only methodological guides. Nevertheless, they do go some way towards illustrating the use of a coherentist methodology.

The Theory Movement

In the late 1940s an increasing number of scholars doing research in educational administration sought to develop a more systematic and rigorous basis for their work and findings. (For a historical overview see Moore 1964.) As an antidote to the so-called 'naked empiricism' (Halpin 1958, p. 1) of fact finding and anecdote collection assumed typical of the field, a number of attempts were made to establish a theoretical structure for administrative theory as it was then being applied in educational studies. The notion of theory that found favour, as we noted earlier, was Herbert Feigl's logical empiricist account of scientific theory, and so what became characteristic of the Theory Movement was the attempt to structure administrative theory and research according to the strictures of Feigl's vision of science and its methods. The results, always energetically pursued, met with varying degrees of success. We here consider one.

Of the many ways in which the epistemological doctrines of logical empiricism shaped the early development of the movement, none is perhaps so counter-intuitive and unrepresentative of ordinary administrative thought and practice as the removal of values from the scope of administrative theory. Administrative theories that disqualify themselves from addressing the value question have, however, a theory/practice problem: theory fails to be relevant for a large part of administrative practice.

This methodological infirmity arises if it is believed that every empirically significant term in a theory is meaningful because it corresponds to some specific range of sensory experience. Terms like 'chair' and 'table' readily satisfy this condition; terms like 'good' or 'just' appear problematic. Similarly, we have a fair idea of what counts as favourable or unfavourable evidence for testing the claim 'there is a chair in my office'; for a claim like 'that person was treated unjustly' there is a difficulty. Essentially the difficulty is this. All the sensory evidence we may ever gather for the claim will merely describe how the person was in fact treated, the facts of the matter, as it were. But the injustice is not a fact there to be observed. It is not some kind of object that produces sensations of injustice. Rather, so the story goes, the basis for our judgment of injustice resides in our subjective response to the observational evidence. So if cognitive significance resides in term-by-term correspondence with specific sensory experiences, or even in testability, then, as Hooker (1975, p. 191) remarks in his critique of empiricism's theory of language, '... empiricists, like positivists, offer no cognitive content to ethics, aesthetics, religion, metaphysics, or indeed to philosophy'. Stripped of its cognitive content, moral deliberation and judgment collapse down into mere affective preference.

We will outline here just one line of response to this argument. We can begin by noting that if the argument is sound our knowledge of scientific concepts like electron or quantum is as problematic as our knowledge of moral concepts since neither many of the theoretical terms of science nor the sentences in which they figure correspond to any definite range of observations. The demand that all concepts be operationally defined — that is, defined as the operations to be performed in some test — is the traditional way of meeting this difficulty (Hempel 1966, pp. 88-97). Hence the common practice of attempting to give operational definitions in traditional science of administration (for examples see Griffiths 1959, pp. 75-91). However, strictly speaking, every purported definition admits of an infinite number of alternative possible operations, which would make a scientific vocabulary potentially limitless, and hence unlearnable. To be sure, many of the differences would seem trivial; for example distinguishing 'length$_1$' as measurement with a wood rule from 'length$_2$' as measurement with a plastic rule. But the distinction between trivial and non-trivial differences is a *theoretical* distinction, drawn with the aid of an antecedent grasp of the concept to be operationally defined. This is an instance of a logical empiricist epistemological procedure outrunning its posited resources.

The source of this difficulty is the belief that a specific range of sensory experience exhausts the meaning of a term. Once we need to use theory (as turns out to be the case in foundational justification) to *select* the relevant sensory experiences, we end up blurring the distinction between observation and theory. Since the portions of theory being used are assumed to be cognitively significant, in our view it is more reasonable to suppose that the significance of more theoretical terms like 'electron' or 'quantum' resides in their *conceptual role* within the theory rather than in any immediate connections with experience.

The view we wish to defend is that moral terms like 'good' or 'right' are significant in the same way that the most theoretical terms of science are. In realist fashion, we assume the unobservables of science exist because they are posited by the most coherent global account we can give of our interpreted experience (Quine 1960). Similarly, we suppose a moral theory and its associated judgments to be warranted to the extent that they also are part of the same global theory. On a coherentist approach to scientific knowledge there is therefore no sharp epistemological boundary to be drawn between administrative theory on the one hand and a large class of naturalistic moral theories and their normative claims on the other.

Critical Theory and Administration

Among the many critics of traditional science of administration are those who have been influenced by the writings of Jürgen Habermas. There is now a considerable body of literature in educational administration that might be regarded as falling within the critical theory perspective (for an overview see Foster 1986). Although critical theory approaches to administration are complex and multi-faceted, covering ethical, political, social, linguistic and personal dimensions, at least one strand of Habermas's thought that has been developed and applied to administrative contexts is uncompromisingly epistemological and lends itself readily to some brief coherentist remarks here.

We have in mind Richard Bates's thesis that a science of administration is essentially manipulative and concerned with social control (Bates 1980, 1983). In developing this claim Bates draws on a reading of the early work of Habermas for an understanding of science; particularly the epistemological theses of *Knowledge and Human Interests* and the 'General Perspective' lecture published as the Appendix to the English translation (Habermas 1972). In this work, Habermas (1972, p. 308) identifies three 'knowledge-constitutive' interests: the *technical* presumed by the empirical analytic sciences; the *practical* underlying the historical–hermeneutic sciences, and the *emancipatory* represented by critical social science. Traditional (so called positivist) science, which Bates, following

Habermas, identifies with empirical–analytic science, is seen as hypothetico-deductive after the Nagel/Hempel empiricist model, with predictive success a measure of technical exploitability. For Habermas (1972, p. 309)

> ... theories of the empirical sciences disclose reality subject to the constitutive interest in the possible securing and expansion, through information, of feedback-monitored action. This is the cognitive interest in technical control over objectified processes.

In Bates's view, the technical scientific definitions of knowledge and rationality are far too narrow for social science and need to be supplemented by critical discourse. A suitable broadening, he suggests,

> ... is argued at length by Habermas, who contends that the annexation of rationality by dominant scientific, technical, manipulative interests has prevented the continuation of an historical discourse directed towards a rational administration of the world (Bates 1980, p. 68).

He goes further, asserting that 'as currently conceived by professor and professional alike, educational administration is a technology of control' (Bates 1983, p. 46). And finally, in summarizing a robust and systematic indictment of poor philosophy for this state of affairs, he declares:

> The inadequacies of the hypothetico-deductive model of positivist science and the positivist, apolitical model of society were argued to be intellectual products that provided the illusions necessary for the continued employment of techniques of hierarchical administrative control that perpetuate the injustices of an unequal society (Bates 1983, p. 30).

There are a number of things that are puzzling about this account of science and administration, especially in view of the fact that Bates thinks traditional empiricist accounts of the practice and conduct of science are mistaken. For if the traditional view of science is wrong, and we know that it is thanks to the work of Quine, Kuhn, Feyerabend, Hesse and others, then the story Habermas tells of empirical science being *constituted* by technical interests of control and manipulation is also wrong. This is because Habermas's account of empirical–analytic science is as much dependent on traditional empiricist theories of science as the traditional science of administration that Bates is using Habermas's machinery to criticise. In more recent work Habermas recognises some of these difficulties (see Hesse 1982).

One attempt at avoiding the major incoherence threatening knowledge constitutive interests is worth briefly noting. The key move would involve distinguishing between traditional accounts of science being wrong on the one hand and people *acting* as though these wrong accounts are true on the other. A revised Habermasian argument might then run as follows:

> If traditional views of science (positivism, logical empiricism and the like) were true then technical control and manipulation would occur. Therefore if everyone (professor and practitioner alike) acted as though they were true then technical control and manipulation would occur.

The missing premise in this argument is a subjectivist claim to the effect that my having a particular theory of the world somehow *makes* the world that way, or brings it into line with my theory. This is perhaps an extravagant extension of the reasonable epistemological thesis that all observation is theory-laden. To see the limits of the thesis, however, consider another example. Suppose, for the moment, that the dominant orthodoxy concerning water says that it flows uphill. We know that the orthodoxy is wrong but we also know that if water did flow uphill it would require a special form of technical handling. It does not follow that if everyone *acted* as though water flowed uphill it would *require* a special form of technical handling. Presumably, water would continue to defy orthodoxy in a range of ways.

Critical theorists may be sympathetic to certain subjectivist theses since bad theory may influence humans more than it influences water. But they cannot be too sympathetic to this one without undermining the reality of human suffering and injustice, or the objectivity of the class and political analyses that underwrite their approaches to human emancipation.

Administrative Decision-Making

The last two examples of epistemological critique were directed mainly at views of the content and structure of administrative theorizing. Our final example will draw attention to the importance of epistemological views for organizational design.

The classic work in the field of administration is undoubtedly Herbert Simon's *Administrative Behavior*, first published in 1945. In that work Simon identifies rational decision-making as the locus of administrative theory. As against the prescription to make *optimal* decisions, the cornerstone of Simon's theory of decision-making is the *bounded* or limited nature of human rationality. We satisfice rather than optimize. He identifies three sources of limitation that organizational structures would need to address to enhance decision-making. First, an individual is limited in skills: dexterity, reaction times, powers of computation, thought, and understanding. A second limitation concerns individual values and the understanding of organizational values and goals.

Finally, there are limits to relevant knowledge — both knowledge of theory and knowledge of all the conditions that must obtain for a sound application of theory (Simon 1976, pp. 34–41).

Simon is reluctant to endorse any 'principles of administration' for enhancing administrative efficiency in advance of specific analyses of case by case administrative arrangements for reducing these limitations. Nevertheless, a particular approach to administrative reform is suggested by his theory. For example, if the growth of knowledge is a matter of *accumulating* more and more information, as the empiricism behind *Administrative Behavior* implies, a satisficing strategy will have a characteristic emphasis. For if optimal decision-making requires optimal initial information inputs to best theory, a less than optimal or second-best approach will involve not a difference in kind but a difference in degree of ambition. Resulting administrative arrangements will place a premium on ensuring the highest practicable quality of *initial input* into the actual point of decision-making. Depending on cases, reforms may focus on ensuring suitable concentrations of expertise, communications structures aimed at enhancing the availability of that expertise, and so on.

But on a coherentist epistemology, very large changes in knowledge are seen to occur through the promotion of a systematic virtue like simplicity in a theory network that includes among its statements a number of theory-laden contrary observations. For the big gains in knowledge appear to flow more from the theoretical resolution of error than the incremental accumulation of data. Given that the existence of limitations to our knowledge is likely to promote the occurrence of *error*, a case by case analysis of decision-making may show greater gains to be had by the promotion of *error correction* at the expense — given only finite resources — of extensive attention to *error prevention*. Of course, in any administrative design for sound long term decision-making there is always some trade-off, in the allocation of resources, between error prevention and error correction. But a theory of learning, the core of which conceives knowledge as growing through a process of conjecture and refutation, is more sensitive to the possibility of learning through mistakes. The option of securing efficiencies in decision-making through the rapid correction of error becomes, on this approach, a more explicit methodological guide to defining a suitable prevention/correction trade-off. Empirical studies by Chris Argyris and co-workers (Argyris 1982, Argyris and Schon 1978) show some of the conditions under which error correction by administrative feedback loop structures is more valuable. Not surprisingly, these are where the organizational environment is unstable or undergoing rapid change; where organizational knowledge and expectations are most likely to be falsified, and where there is a greater premium on more rapid acquisition or growth of knowledge.

This epistemological consideration suggests that theories of organizational *learning* can impose important constraints on the administrative structures of decision-making. Simon's position in *Administrative Behavior* does not deny this. However, in noting the relevance of epistemology for administrative theory it is sufficient to observe, for our purposes, that significant differences in organizational consequences can flow from adopting divergent theories of human knowledge acquisition.

Summing up, in applying coherentist considerations very briefly to some issues concerning the Theory Movement, critical theory, and a view of decision-making, we suggested a number of conclusions. Since our holistic epistemology places severe limits on attempts to partition knowledge into different compartments, we challenged both the fact/value distinction posited by logical empiricism and the three-fold division in knowledge posited by Bates's use of the earlier work of Habermas. Both challenges have extensive consequences for the administrative theories that employ these partitions in knowledge. Finally, we explored some organizational consequences of different views of the growth of knowledge on the question of enhancing structures for decision-making.

References

Achinstein P. and Barker F. (1969). (eds.) *The Legacy of Logical Positivism*. (Baltimore: Johns Hopkins Press).
Argyris C. (1982). *Reasoning, Learning and Action*. (San Francisco: Jossey-Bass).
Argyris C. and Schon D. (1978). *Organisational Learning: A Theory of Action Perspective*. (Menlo Park: Addison-Wesley).
Bates R.J. (1980). New developments in the new sociology of education, *British Journal of Sociology of Education*, 1(1), pp. 67–79.
Bates R.J. (1983). *Educational Administration and the Management of Knowledge*. (Geelong: Deakin University Press).
BonJour L. (1985). *The Structure of Empirical Knowledge*. (Cambridge, Mass.: Harvard University Press).
Churchland P.M. (1985). The ontological statues of observables: in praise of the superempirical virtues, in P.M. Churchland and C.A. Hooker (eds.) *Images of Science*. (Chicago: University of Chicago Press).
Churchland P.S. (1986). *Neurophilosophy: Towards a Unified Science of the Mind-Brain*. (Cambridge, Mass: MIT Press).
Churchland P.S. (1987). Epistemology in the age of neuroscience, *Journal of Philosophy*, 84(10), pp. 544–553.
Culbertson J.A. (1981). Antecedents of the Theory Movement, *Educational Administration Quarterly*, 17(1), pp. 25–47.
Evers C.W. (1988). Educational administration and the new philosophy of science, *Journal of Educational Administration*, 26(1), pp. 3–22.
Feigl H. (1953). The scientific outlook: naturalism and humanism, in H. Feigl and M. Brodbeck (eds.) *Readings in the Philosophy of Science*. (New York: Appleton-Century-Crofts). Also reprinted in Feigl (1974).
Feigl H. (1974). *Inquiries and Provocations: Selected Writings 1929–1974*. (Boston: Reidel).
Feyerabend P.K. (1981). *Philosophical Papers*, Vols. 1 and 2. (Cambridge: Cambridge University Press).
Foster W. (1986). *Paradigms and Promises*. (New York: Prometheus Books).
Garson G.D. (1986). From policy science to policy analyses: a quarter century of progress, in W.M. Dunn (ed.) *Policy Analysis: Perspectives, Concepts and Methods*. (London: JAI Press Inc.).
Giddens A. (1982). *Sociology: A Brief but Critical Introduction*. (London: Macmillan).
Greenfield T.B. (1975). Theory about organization: a new perspective for schools, in M.G. Hughes (ed.) *Administering Education: International Challenge*. (London: Athlone Press).
Greenfield T.B. (1978). Reflections on organization theory and the truths of irreconcilable realities, *Educational Administration Quarterly*, 14(2), pp. 1–23.
Griffiths D.E. (1959). *Administrative Theory*. (New York: Appleton-Century-Crofts).

Griffiths D.E. (1979). Intellectual turmoil in educational administration, *Educational Administration Quarterly*, **15**(3), pp. 43–65.
Habermas J. (1972). *Knowledge and Human Interests*. (London: Heinemann).
Halpin A.W. (1958). The development of theory in educational administration, in A.W. Halpin (ed.) *Administrative Theory in Education*. (Chicago: Midwest Administration Centre).
Hempel C.G. (1965). *Aspects of Scientific Explanation*. (New York: The Free Press).
Hempel C.G. (1966). *Philosophy of Natural Science*. (Englewood Cliffs: Prentice-Hall).
Hesse M. (1970). Duhem, Quine and a new empiricism, in G.N.A. Vesey (ed.) *Knowledge and Necessity*. (London: Macmillan).
Hesse M. (1982). Science and objectivity, in J.B. Thompson and D. Held (eds.) *Habermas: Critical Debates*. (London: Macmillan).
Hooker C.A (1975). Philosophy and meta-philosophy of science: empiricism, popperianism and realism, *Synthese*, **32**, pp. 177–231.
Kuhn T. (1962). *The Structure of Scientific Revolutions*. (Chicago: University of Chicago Press).
Lycan W.G. (1988). *Judgement and Justification*. (Cambridge: Cambridge University Press).
Moore H.A. (1964). The ferment in school administration, in D.E. Griffiths (ed.) *Behavioral Science and Educational Administration*. (Chicago: University of Chicago Press).
Phillips D.C. (1987). *Philosophy, Science and Social Inquiry*. (Oxford: Pergamon Press).
Popper K.R. (1959). *The Logic of Scientific Discovery*. (London: Hutchinson).
Popper K.R. (1963). *Conjectures and Refutations*. (London: Routledge and Kegan Paul).
Quine W.V. (1951). Two dogmas of empiricism, *Philosophical Review*, **60**, pp. 20–43. Cited as reprinted in W.V. Quine (1961). *From a Logical Point of View*. (Cambridge, Mass.: Harvard University Press).
Quine W.V. (1960). Posits and reality, in S. Uyeda (ed.). *Basis of the Contemporary Philosophy*, Vol. 5. (Tokyo: Waseda University Press). Reprinted in W.V. Quine (1976). *The Ways of Paradox and Other Essays*. (Cambridge, Mass.: Harvard University Press, revised and enlarged edition).
Quine W.V. (1969). Epistemology naturalised, in W.V. Quine *Ontological Relativity and Other Essays*. (New York: Columbia University Press).
Quine W.V. and Ullian J.S. (1978). *The Web of Belief*. (New York: Random House, second edition).
Simon H.A. (1976). *Administrative Behavior*. (New York: The Free Press, third edition, revised and enlarged).
Williams M. (1977). *Groundless Belief: An Essay on the Possibility of Epistemology*. (Oxford: Blackwell).
Willower D.J. (1981). Educational administration: some philosophical and other considerations, *Journal of Educational Administration*, **19**(2), pp. 115–139.
Willower D.J. (1988). Synthesis and projection, in N.J. Boyan (ed.) *Handbook of Research on Educational Administration*. (New York: Longman).

3
Towards Coherence in Administrative Theory

Much of the theoretical debate that has characterized recent educational administration has concerned the adequacy of traditional conceptions of science to provide a systematic and plausible science of administration. Thus, while traditional science focusses on the objectivity of observable phenomena, such as behaviours, subjectivist critics draw attention to the importance of human intentions, purposes, motives, and beliefs — to the dynamics of a hidden, inner, mental life. The study of organizational culture, in particular, seems to draw heavily on a repertoire of such hidden phenomena, with its typical appeals to interpretations, meanings, and understandings.

As we have seen in earlier chapters, other critics lament the failure of science to provide ethical guidance, to offer suggestions on what ought to be done beyond, say, the choice of efficient means to achieve morally unanalysed goals (although the appeal to efficiency itself has moral consequences). Some writers, for example Hodgkinson (1991), think that science is in principle unable to offer administrators ethical guidance because it can deal only with facts and there is thought to be a fundamental distinction between facts and values; ethical knowledge is simply different from factual knowledge. This alleged distinction is reinforced by a tradition in critical theory that posits a distinction among three types of knowledge constitutive interests. On this taxonomy, the scientific is primarily concerned with meeting a human interest in manipulating and controlling the physical world. According to critical theorists, it is due to the dominance of science-influenced conceptions of reason and knowledge that one type of knowledge, the scientific, has come to serve as a model for all knowledge.

Further criticisms of science allege its inability to theorize gender issues, to support political analyses, or to adjudicate conflicts of interest (Griffiths 1979). And if success in all of these tasks were to be achieved, it might still be argued that the dominance of science as a 'master narrative' has now been eclipsed in this post-modern age.

In our view, most of these criticisms of traditional science of educational administration are sound. However, the response we recommend is not to look for alternatives to administrative science, or even to augment it with non-scientific knowledge. Rather, our response is to urge the development of a better, broader, and more inclusive account of science; one which is able to accommodate, for example, human subjectivity and ethics.

In what follows, we sketch some of the philosophical background to a number of theoretical disputes in educational administration, and provide a very brief account of the framework for our own coherentist position. We then offer three examples where our coherentism provides a number of fresh insights into issues in educational administration. These discussions are rather speculative, very much a reflection of work in progress, and of course controversial. We hope, nevertheless, that they show something of the scope and explanatory promise of this research programme. (For more on the programme see Walker and Evers 1984, Evers 1987, Evers and Lakomski 1991.)

Philosophical Background

Philosophical assumptions have always had a significant coordinating function on the development of distinctive positions in educational administration. For example, assumptions about the nature of theory have determined what is included and what is omitted from the scope of administrative theory. Similarly, assumptions about the nature of ethics help determine whether the moral evaluation of organizational goals can be considered part of administrative theory or not. However, for us, the most important general assumptions are about the nature and justification of knowledge, dealt with most systematically in that branch of philosophy known as epistemology. Where some kind of justification is demanded for administrative choices, decisions, practices, and theoretical commitments, the relevant background epistemology, or theory of knowledge and its justification, will be significant in determining the structure and content of administrative theories.

Content is shaped by the common practice of omitting from theory claims that are in principle unknowable. And structure is shaped by characteristic patterns of justification. To see how these points operate, consider the case of traditional science of administration with its background epistemology of logical empiricism. According to logical empiricist canons, claims are justified by appealing to some empirical foundation for knowledge, usually observation reports that can in principle be intersubjectively made by several observers, in order to secure objectivity (Feigl 1974). Since a person's inner mental life cannot be directly observed by others, subjective experience drops out as objectively unknowable. Similarly, if it is agreed that all observations are of facts, and facts are separate and distinct from values, then value claims will not admit of justification either and so can be omitted from theory without loss.

The matter of structure is a bit more complex and involves spelling out the nature of empiricist appeals to foundations. Roughly speaking empirical foundations provide knowledge whereby theories are *tested against experience*. If observations match what the theory predicts, the theory is said to be confirmed. If contrary experience is reported, the theory is disconfirmed and in need of revision. On this epistemology, theories that are extensively confirmed and have no disconfirmations are more justified than those which that only a few confirming instances. To meet the demands of testability, theories are structured in a certain way. Since the aim is to test deduced observable consequences against experience, justification imposes a hierarchy on the sentences used to express a theory. At the top of the hierarchy are the theory's most general empirical claims, because these are needed to deduce more particular claims. As we go down the hierarchy we go from the most general universal law-like claims to the most particular singular observational claims. The theorists of the Theory Movement, both early and late, clearly regarded administrative theory as possessing this kind of pyramidal hypothetico-deductive structure (as we saw in Chapters 1 and 2).

Weaknesses in empiricist epistemology have been known and discussed for a long time in educational studies. Perhaps the best known critic is John Dewey, who argued against the whole idea of justifying knowledge by appeal to foundations, whether empirical or otherwise (Dewey 1929). However, we think that the current flourishing of alternatives in educational administration is due primarily to the influence of Thomas Kuhn's (1970) attack on empiricism and the widespread acceptance of his paradigms perspective on theoretical diversity. In the late 1950s and early 1960s Kuhn (among others) was able to show that the results of testability were never sufficient for rationally choosing among competing large-scale theories, or paradigms. Rather, such theories have their own internal criteria for interpreting, evaluating, and responding to experience. The business of justification was therefore claimed to be *paradigm-relative*.

Thomas Greenfield harnessed these arguments to great effect in his critique of administrative science, and so did many other critics. Moreover, by the early 1970s, the paradigms perspective was becoming orthodoxy in educational research methodology (Walker and Evers 1988). This coincidence was mutually reinforcing. Subjectivism in administration meshed with a range of qualitative methodologies, critical theory with, for example, action research, and cultural theory with interpretative research, each having its own epistemology. To the extent that it was seen to be valid, science was displaced to become just one more paradigm, though perhaps suitable for theorizing about only a fairly narrow range of phenomena where quantitative methods were useful (but certainly not about something as broad and complex as administration).

Although the above sketch glosses much relevant and controversial detail, two features are worthy of comment. First, traditional science approaches to administration are methodologically unable to exploit many valuable sources of knowledge, particularly those accessible to qualitative modes of inquiry. In education especially, this represents a major loss for policy analysts, decision-makers, and practitioners who as a matter of course have to extract knowledge from, and act upon, scant or partial information from complex changing scenes.

Second, however, because alternatives fostered by the paradigms framework were developed in opposition to traditional science, there is the opposite risk of scientifically accessible knowledge being discounted. An attack on the pretensions of science quickly becomes an argument for a dichotomy, or partition, between natural science and social science. Our general concern here is that partitionist epistemologies actually lose already scarce information.

The detail and defence of our version of a post-positivist science is complex and has been given elsewhere (see Evers and Lakomski 1991). However, the key moves can be sketched briefly and informally. Essentially, we see the successful attacks on logical empiricist epistemology as a case for asserting the narrowness of that epistemology's standards of evidence. In short, there is more to justification than appeal to some foundation of public, sensory evidence. Administrative theories need also to be consistent, as a contradiction sanctions any inference whatsoever. Simplicity is a further theoretical virtue since without it any event can end up being explained by just treating a description of the event as primitive and adding it to a theory. We want to explain a large class of things by invoking a small class of principles. Comprehensiveness is another virtue. A slight addition to a theory may be warranted if it makes for a big increase in explanatory power. And there is a premium on coherence in the special sense that all parts of a theory should fit together, or consistently share the same explanatory resources, or enjoy explanatory unity. Except for contradiction, these canons of *coherentist* justification come in degrees. A further constraint on theories is that they should be learnable; that is, they should cohere with accounts of how knowledge of them may be acquired. We conjecture that the best accounts of learning are provided by natural science, which means that to that extent administrative theories should cohere with natural science. In this sense our post-positivist theory is a species of scientific naturalism justified on coherentist principles.

This rather abstract account of theory and knowledge needs to be interpreted within the context of application. The following three examples illustrate how the impact of this epistemology on the content of administrative theory can be progressively sharpened and refined. Our first example draws attention to some modest advantages of coherentism's methodological requirement to mesh administrative and educational theory in educational administration. Our second example uses some assumptions of our preferred learning theory to reshape how the relationship between theory and practice might be more fruitfully construed. Finally, we suggest a relationship between organizational learning and ethics that is rich enough to provide a moral framework for educational organizations.

Administrative and Educational Theory

For the writers of the Theory Movement, educational administration owes much more to administrative studies than to educational theory. Educational content is generated primarily by the nature of the organizations under study; namely organizations such as schools, universities, and government and non-government educational bureaucracies (or networks) and systems. There are at least two reasons for this state of affairs. In the first place, educational organizations are regarded as *particular* kinds of organizations. If theory has the pyramidal structure suggested by logical empiricism then what occurs in educational organizations will admit of deeper explanation by appeal to the features of organizations in general. That is, a comprehensive traditional administrative science will subsume the particular under the law-like, or general.

Secondly, because of the dominance of decision-making models in administration, logical empiricism's means/ends account of rationality tends to detach consideration of organizational goals and purposes from accounts of organizational life and structure. That is, goals and purposes appear as external givens, with the main emphasis of theory being to determine optimal or satisfactory means. One consequence of combining this view of rationality with the quest for empirical generality is that the resulting administrative theory will come up with similar prescriptions for organizations with widely differing aims. Hence the current use of private sector corporate management models as guides for the restructuring of schools and school systems (see Chapter 5). However, a failure of explicit congruence between educational theory and administrative theory can lead to a failure of educational organizations to achieve certain educational outcomes.

Consider a school that is concerned to prepare students in three ways — to provide vocationally relevant skills, to socialize students into the surrounding culture(s), and to develop each student as an autonomous agent — though with a special emphasis on the third (Hodgkinson 1991). The school's educational theory specifies certain conditions under which student autonomy flourishes. Let us suppose that in keeping with a broadly progressive tradition, these conditions for developing autonomy are the same as the conditions for exercising autonomy (Walker 1981). So this development/exercise equivalence means that the social relations for becoming autonomous are the same as those required for a person to act autonomously. Now let us suppose further that however else these social relations are characterized they at least involve respecting a person's freedom to choose within limits that are equally applicable to all. A systematic expression of this account of personal autonomy usually implies that the social relations of schooling are best instantiated in an egalitarian, participatory, and democratic organization. Indeed, it is arguable that the school's educational aims could not be achieved in any other way. Requiring educational ends to cohere with organizational means for their realization thus blurs the means/ends distinction (as Rizvi 1995 has argued in relation to ethics).

Where the administration of learning is not readily separable from the educational processes of learning, as is clearly the case on some accounts of learning to be autonomous, it is especially important that administrative and educational theories be known to cohere. Of course, ignoring a demand for coherence does not prove that the two can be detached in fact. Rather, it means that some different view of education is likely to be implemented, or instantiated organizationally, *though without educational scrutiny or evaluation*. Much of the literature on the 'hidden curriculum' posits a causal role for the organizational framing of learning on what is actually learnt. Similarly, the use of benchmarking as a decision tool for the allocation of educational resources will *de facto* favour the implementation of one set of educational aims and objectives over another where there are differing associated costs. Indeed, the adoption of traditional administrative science approaches to educational administration will in general favour some educational programmes over others. For example, applying economic interpretations of efficiency and effectiveness to schools (perhaps even in the name of achieving theoretical generality across organizations) will place a premium on vocational construals of educational purposes, since the corresponding criteria for justification will be economic.

We see two main advantages in explicitly meshing administrative and educational theory into a coherent theoretical package. First, since the administration of education brings about specific educational practices, the effectiveness of educational administration can be evaluated by appeal to an entirely relevant body of knowledge — educational theory. A partitionist approach thwarts this possibility. Second, it provides a framework for adjudicating educational priorities against administrative priorities. This raises the option of permitting the main educational purposes of an organization a major say in how it is to operate, rather than using administrative principles to determine what is educationally possible. These advantages may seem modest, but they do have some purchase by raising the matter of proof against the prevailing fashion for generic administration.

Theory and Practice

Discussions of administrative theory usually begin with a recounting of the advantages of theory for practice (Halpin 1957). Thus good theory is said to provide useful knowledge that may lie beyond what an administrator has learned from experience. Or it may permit the rigorous development of further knowledge to guide practice, or the anticipation of unseen problems. And all these claims can be reasonable. However, behind the benign formulations lurks a deep and difficult issue generated by the common acceptance of a fundamental dichotomy concerning how knowledge is to be represented. A long tradition of conceiving knowledge as propositions represented by symbols, usually in the form of sentences of a natural language, implies an obvious solution to the representation problem for theories; theories can be represented explicitly by some symbolic formulation. However, there is no clear account of how knowledge of practice can be represented. Instead, we have some familiar dualisms that correspond roughly to the theory/practice split: for example, mind/brain, mental/manual, reason/cause, mainstream education/special education, knowing that/knowing how, propositions/skills, and so on.

Difficulties arise because the representation of theoretical knowledge highlights logical and quasi-logical relations among semantically significant symbols, whereas the activities of practice flow from being enmeshed in a causal field. The problem is most acute in two areas of human activity. The first is learning, where we may plausibly say that the best accounts portray learning as a complex set of processes that produce certain changes to the central nervous system. That is, human reasoning is a causal process. And the second is motivation, where we need a causal story which connects up the semantical decoding of symbolic tokens with desired responses. Until we can find some way of making a reasons based account of theory cohere with a causal account of practice, the theory/practice nexus will continue to generate serious problems. This will be especially the case in administration, which has traditionally acknowledged the importance of theorizing decision-making which, if seen as the business of making reasoned choices among desired alternatives, is the point where both reason and motivation converge.

Since we plainly learn from both theory and practice, our proposal is to approach the symbolic representation issue from the causal learning side rather than approaching the learning issue armed with the usual notions of representation. We therefore have to tackle the question of knowledge representation from the causal side as well, from the way the brain represents knowledge. Fortunately, advances in this area have been very impressive over the last ten years, to the point where researchers in cognitive science can now build plausible models of neural information processing — called 'neural nets' — which are capable of quite powerful feats of learning (for an introduction see Bechtel and Abrahamsen 1991; for the most influential text see Rumelhart and McClelland 1986; and for applications to educational studies see Chapters 8 and 9).

Consider the sort of neural net displayed in Figure 3.1. Although biological realism has been sacrificed for simplicity, this net still has considerable learning powers. Perhaps the most startling example of network learning discussed in the early literature is NETtalk, a network that learns to pronounce English text (Sejnowski and Rosenberg 1987). Its architecture consists of 309 'neurons' arranged in three layers: a 203-node input layer (29 groups of 7), 80 hidden nodes, and 26 output units, with each neuron in one layer connected to every neuron in the next layer. Each letter of text is presented to the appropriate part of the input layer as a seven-place vector consisting of that letter plus three letters on either side (or more precisely, letters, punctuation, and word boundary markers). Each letter thus has a small window of context. Inputs are then multiplied by numerical weights between layers, summed and checked against a set threshold value, and then passed from the output layer to a speech synthesiser for pronunciation. Differences between output and correct pronunciation are fed back into the network by a formula which adjusts weights to minimise error. At first the network just produces a babbling noise. After a short while it recognizes discrete words. Gradually, under the steady pressure of weight adjustment it produces clearly recognizable speech, being correct about 95% of the time on the 1024-word training text.

Note that no *rules* for text-to-speech mappings have been coded into the

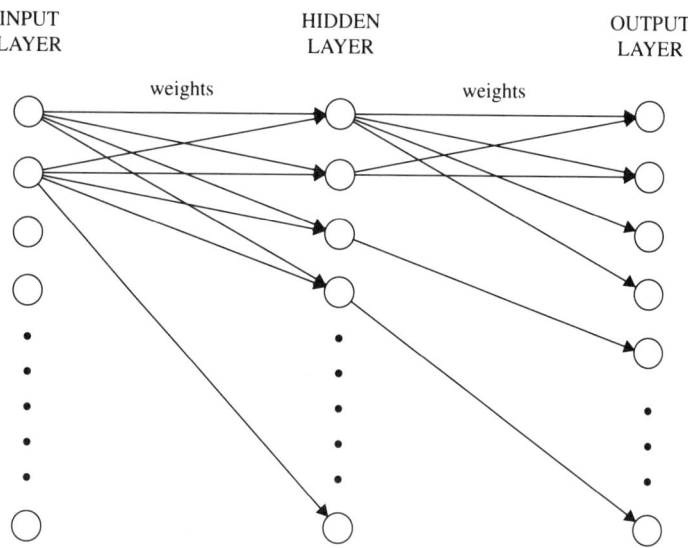

Figure 3.1 A three-layer net with some connection shown. Each connection has an associated weight for changing the signal transmitted from one layer to the next.

network. Rather, it is the case of an efficient learning device extracting the relevant regularities from experience. The network's knowledge, or *theory*, of these regularities is located in, and distributed across, the 18,000 plus weights in this configuration. Not having a symbolic theory formulation of text to speech mappings is no barrier to acquiring a powerful 'neural' theory of these mappings. Much human knowledge from experience must be along these lines. Under the steady pressure of feedback from practice we build up non-symbolic internal representations of the main regularities we find useful in our interactions with the world. Now in educational administration, where current symbolic theory formulations are of only modest power, we would seem to have little alternative but to regard a considerable amount of day-to-day knowledge as acquired and represented in this fashion. Teaching may be another example, and Paul Churchland (1989, p. 133) applies the same reasoning to culture:

> The set of weights that constitutes a child's developing consciousness is continually being shaped by the linguistic, conceptual, and social surround. The developing brain comes to reflect the elements and structure of that surround in great detail. For that is what networks do. What shapes them is the stimuli they typically receive, and the subsequent corrections in their responses to which they are typically subject. Small wonder we become attuned to the categories of the culture that raises us.

Just as the notion of theory can be extended to admit theory formulations in neural stuff, so the notion of practice can be extended to include the processing of symbols. These extensions have the effect of making any so-called divisions between theory and practice matters of pragmatic emphasis over promoting learning, rather than principled bifurcations. For example, symbols have the advantage of being public, though causally more remote as a source of behaviour. Networks effortlessly represent exceptions as well as regularities, whereas symbolic formulations fare best on exceptionless rules.

On this analysis it is tempting to conclude that the training of administrators can occur entirely by engaging in a class of practices that excludes dealing with symbolic theory formulations altogether: one becomes an administrator by doing the things an administrator does, or a teacher by getting out there in the classroom. (This is a version of the development/exercise equivalence we canvassed earlier in relation to autonomy.) This is tempting but wrong. For whereas networks represent and store learning from past experience, and permit extrapolations, the manipulation of sentences permits inferences to be made about matters that are counterfactual, or beyond the range of experience. Knowledge can be extended in some cases simply by the sound (neural) processing of a good representational structure. Clearly, it is important to maintain all sorts of possibilities for learning where information is scarce and situations are complex and changing.

The connection between symbols, cognition, and organizational design has been the major preoccupation of Herbert Simon (1976, 1983), without doubt the most influential theorist of administrative studies during the last 50 years. In Simon's work, understanding administratively relevant human subjectivity is a matter of developing a psychologically realistic account of human rationality. However, Simon's models have all been within the symbolic paradigm and hence within a correspondingly narrow view of rationality. We think that the recent arrival of neural network models of thought will provide vastly greater scope for meshing natural science with a view of human subjectivity. The use of terms such as 'intention', 'meaning', 'understanding', 'interpretation', 'belief', 'desire' and 'deciding' capture familiar categories for describing inner mental processes. But these are all terms embedded within our commonsense theory of folk psychology. Just as science has revolutionised our understanding of the external world over the last 400 years, so we expect the new cognitive science to provide the beginnings of a revolution on our inner world. Our sketch of how the terms 'theory' and 'practice' might be embedded within the less familiar categories of this new approach gives some indication of how that revolution might proceed.

Ethics in Educational Administration

Different epistemologies can generate quite different accounts of ethical knowledge. For Plato, moral knowledge depends on one's intellectual apprehension of abstract objects called 'forms'. Being aware of the form of the Good thus provides knowledge for distinguishing good from evil. Philosophers possess this kind of awareness beyond the common measure, so according to Plato, in the *Republic*, a just society will be one ruled by philosophers, or philosopher-kings. Here, in a systematic educational treatise, we find defended a meritocratic distribution of administrative and political power: an epistemically based hierarchy of access to knowledge, especially moral knowledge, is isomorphic to an organizational hierarchy of authority. Christopher Hodgkinson (1991) defends a similar link between knowledge and power, though one based on the requirement that leadership and moral insight intersect.

The place of ethics in traditional science of administration has been shaped by its long association with empiricism which tends to identify knowledge with what corresponds to observable facts, and ethics as claims concerning one's subjective reactions to those facts. Subjectivist critics such as Greenfield have challenged the exclusion of ethics from administrative science by attacking science's profession to objectivity (see Gronn 1995). Our response has been to defend the unity of knowledge and the objectivity of ethics (Evers and Lakomski 1991, pp. 166–191).

We begin with the familiar Deweyan point that all experience is interpreted, or filtered, by the brain's past learning (or initial configuration of material dispositions that we now call weights). In the terminology of philosophy of science, all observation is laden with theory. Some interpretations, or theories are better than others, however: they lead to more successful predictions, or explain more phenomena, or manage to do both while being simpler than alternatives, or cohere better with what is known about human learning and other bodies of (provisionally) accepted knowledge. This approach to knowledge helps explain why, in science, we might reasonably suppose that electrons exist, or atoms, or molecules, even though they are not directly observed by the senses. It is partly because their supposition leads to simpler, more comprehensive, accounts of otherwise disparate, disconnected phenomena. Chairs and tables are also known inferentially, as posits, to integrate and simplify accounts of experience—for example, we suppose they continue to exist when observed by no one. The difference between these two sorts of posits is one of degree, not kind. Imagine, to use Quine's metaphor, knowledge as a web of belief (Quine and Ullian 1978). At the centre of the web can be found the most theoretical parts of our knowledge (atoms and molecules); towards the periphery, closer to observation, the less theoretical parts (chairs and tables).

Now if we suppose, with Churchland, that there is no special or additional faculty for acquiring moral knowledge, that it is learned by the same cognitive machinery responsible for everything else that we know, then it is simpler to conclude that the apparent distance of values from observations merely reflects their theoreticity. That is, value claims are part of the one continuous fabric of knowledge, to be justified by their contribution to fabric's overall coherence in relation to systematic alternatives.

Controversy over values often reflects the sheer complexity of many moral issues, especially in applied fields such as medicine, law, education, or administration. With the abandonment of any sure or certain foundation for knowledge, the resulting fallibility of all knowledge also contributes to controversy. As fallibilists in epistemology, we therefore see the business of justification as merging with the task of improving and revising existing, provisionally accepted claims, including moral claims; of promoting, as Dewey saw, the growth of knowledge. Whatever disagreement there might be about value positions, our theory of knowledge requires the maintenance of a social and ethical infrastructure that supports the further growth of knowledge and that can lead to the revision of all value positions (including those assumed necessary for sustaining inquiry).

For us, this social and ethical infrastructure coheres best with the organizational learning tradition of administration (Argyris and Schön 1978). Politically and ethically it has much in common with Dewey's liberal democratic values, or Karl Popper's (1945) 'Open Society', championing respect for persons and differing points of view, tolerance, the value of criticism and critical feedback, education and learning, and equitable participation in the epistemic community.

Although compatible with a wide range of organizational designs, there is a premium on developing feedback loops for correcting and improving decisions. The emphasis is not so much on the soundness of visionary and authoritative leadership, which always contains the possibility of error, but on the mechanisms of quality assurance: the steady pressure of regular checks of expectations against outcomes, goals against performance, and the coherent adjustment of both against the total fabric of organizational purposes and possibilities. We therefore favour an *educative* model of leadership; that is, one where leadership is concerned with promoting individual and organizational learning (Duignan and Macpherson 1992). In educational organizations where the process of feedback requires acquaintance with a broad and diffuse set of community expectations, more democratic forms of community participation can be defended, perhaps by invoking the notion of quality improvement through stakeholder participation (see Chapter 5).

After investigation, it is a fairly commonplace result to find that proposed structures of human knowledge acquisition and cognition are isomorphic to proposed organizational structures. Our epistemology is certainly no exception in sustaining this consequence. However, because our coherentism justifies the claims of administrative theory by assessing their contribution to the overall coherence of a more global account of experience, the outcome is a much closer integration of administrative concerns with the relevant ethical, educational, scientific, and social theories that bear on administrative contexts (Evers 1987). And because the resulting administrative theory both coheres with natural science and employs the same pattern of coherentist justification, the aim of our research programme may be regarded as the development of a new post-positivist science of administration. The above background and examples should give some indication of how we see that development progressing.

References

Argyris C. and Schön D. (1978). *Organisational Learning: A Theory of Action Perspective.* (Menlo Park: Addison-Wesley).
Bechtel W. and Abrahamsen A. (1991). *Connectionism and the Mind.* (Oxford: Blackwell).
Churchland P.M. (1989). *A Neurocomputational Perspective: The Nature of Mind and the Structure of Science.* (Cambridge, Mass.: MIT Press).
Dewey J. (1929). *Experience and Nature.* (New York: Dover).
Duignan P.A. and Macpherson R.J.S. (1992). (eds.) *Educative Leadership.* (London: Falmer Press).
Evers C.W. (1987). Naturalism and philosophy of education, *Educational Philosophy and Theory*, 19(2), pp. 11–21.
Evers C.W. and Lakomski G. (1991). *Knowing Educational Administration: Comtemporary Methodological Controversies in Educational Administration Research.* (Oxford: Pergamon Press).
Feigl H. (1974). *Inquiries and Provocations: Selected Writings.* (Boston: Reidel).
Griffiths D.E. (1979). Intellectual turmoil in educational administration, *Educational Administration Quarterly*, 15(3), pp. 43–65.
Gronn P.C. (1995). Subjectivity and the creation of organizations, in C.W. Evers and J.D. Chapman (eds.) *Educational Administration — An Australian Perspective.* (Sydney: Allen and Unwin).
Halpin A.W. (1957). A paradigm for research on administrator behavior, in R.F. Campbell and R.T. Gregg (eds.) *Administrative Behavior in Education.* (New York: Harper and Bros).
Hodgkinson C. (1986). Beyond pragmatism and positivism, *Educational Administration Quarterly*, 22(2), pp. 5–21.
Hodgkinson C. (1991). *Educational Leadership: The Moral Art.* (Albany: SUNY Press).
Kuhn T. (1970). *The Structure of Scientific Revolutions.* (Chicago: University of Chicago Press, second edition).
Popper K.R. (1945). *The Open Society and its Enemies*, Vols 1 and 2. (London: George Routledge).
Quine W.V. and Ullian J.S. (1978). *The Web of Belief.* (New York: Random House, second edition).
Rizvi F. (1995). Ethics in educational administration, in C.W. Evers and J.D. Chapman (eds.) *Educational Administration — An Australian Perspective.* (Sydney: Allen and Unwin).
Rumelhart D.E. and McClelland J.L. (1986). (eds.) *Parallel Distributed Processing*, Vols. 1 and 2. (Cambridge, Mass.: MIT Press).
Sejnowski T.J. and Rosenberg C.R. (1987). Parallel networks that learn to pronounce English text, *Complex Systems*, 1, pp. 145–168.
Simon H.A. (1976). *Administrative Behavior.* (London: Macmillan, third revised edition).
Simon H.A. (1983). *Reason in Human Affairs.* (Stanford: Stanford University Press).
Walker J.C. (1981). Two competing theories of personal autonomy: a critique of the liberal rationalist attack on progressivism, *Educational Theory*, 31(3/4), pp. 285–306.
Walker J.C. and Evers C.W. (1984). Towards a materialist pragmatist philosophy of education, *Education Research and Perspectives*, 2(1), pp. 23–33.
Walker J.C. and Evers C.W. (1988). The epistemological unity of education research, in J.P. Keeves (ed.) *Educational Research, Methodology, and Measurement: An International Handbook.* (Oxford: Pergamon Press).

4
Educational Organisations as Systems

The scientific study of schools as formal organizations has traditionally relied on theoretical tools and frameworks derived from the study of organizations in general. Organization theory evolved from the *classical* doctrine, associated with the work of Frederick Taylor (1947), which was largely concerned with the internal structure, or *anatomy*, of formal organization (see Scott 1983). The anatomy of an organization was seen to consist primarily of four features:

(1) the division of labour;
(2) the scalar and functional processes;
(3) structure; and
(4) span of control.

These concepts have become the stock-in-trade of organization studies but, although still useful, have come to be seen as presenting too narrow a focus for understanding the nature and functioning of an organization. What was missing was any acknowledgement of the fact that it is humans who people organizations and that the most important aspect aiding our understanding of organizations is understanding the complex interplay of individuals who work in and shape the organization (for example, in the famous Hawthorne studies, see Roethlisberger and Dickson 1939). However, it needs to be stressed that it was the *neo-classical* school of organization theory, commonly identified with the *Human Relations Movement* and the name of Chester Barnard (1938), that first focussed on the human element. It is an important feature of the human relations school that it drew on the behavioural social sciences, which presented theories explaining various aspects of human behaviour. This, of course, was invaluable for a discipline centrally interested in understanding how it is that humans act in and interact with formal organizational structures. As a consequence of this influence, neo-classical organization theory, whether in industrial or educational contexts, could attend to such issues as communication, motivation, climate, change, leadership, and conflict — concerns that had not come into the purview of its classical ancestor. It was the Human Relations Movement, in particular, that provided the framework for the behavioural science of administration.

Both the classical and neo-classical views are so complex that they cannot be covered in the present chapter (for good discussions and overviews of the history of organization theory see for example Hoy and Miskel 1991, Ogawa 1985, Ecker 1985, Greenfield 1985, Campbell 1987, Harmon and Mayer 1986, Perrow 1983, Burrell and Morgan 1982, Gronn 1982). Nevertheless, it would be fair to say that the overriding difference between the two schools is their respective emphasis on *formal* versus *informal* organizational features. The study of the informal leader, for example, is a most important research issue generated by the neo-classical tradition. And this topic could only become an issue because the interactions and interrelationships between the formal and the informal organizations were considered of prime importance by the neo-classical school. The major emphasis signalled by the shift from classical to neo-classical thinking is therefore the move away from the study of the components of the organization to the study of the relationships among its parts. This reorientation had become possible, in part, because of an accompanying shift of viewing organizations conceptually as *social systems*, which, as will be explained in the following, are by their nature open.

Although already embedded in the neo-classical view, the idea that organizations are best viewed as *social* systems is the main characteristic of modern mainstream organization theory. This also holds for educational administration. Hoy and Miskel's (1991) text *Educational Administration: Theory, Research and Practice* is arguably the most prominent representative of the social systems approach. While advocating an open rather than closed systems perspective (the latter being associated with the Theory Movement), Hoy and Miskel nevertheless align themselves theoretically with the behavioural science approach. Arguing for the scientific nature of educational administration, and maintaining an empiricist definition of science (albeit in modified form) they are the modern descendents of the Theory Movement. There is little indication in the fourth edition of their widely used text that they have rethought their epistemological views. Thus they continue to be important representatives of what Evers and Lakomski (1991, p.60) have termed the 'new orthodoxy' in educational administration. It is therefore worthwhile giving their social systems theory of educational administration a closer inspection.

While the idea of treating educational organizations as social systems appears plausible and may be helpful as a heuristic device, we still need to ask whether the systems metaphor enables us to understand organizations better. In other words, does the systems' view have *explanatory* value? There is reason to believe that it has important limitations. For example, the generality of the systems metaphor, advocated as its great strength, is also its Achilles heel because great generality can also signify that a claim is compatible with almost any state of affairs and hence is empirically vacuous. This consequence has to do with problems of functionalism and functionalist explanation, which are inherent in the open systems perspective. But before we can address the issue of its explanatory nature, it is necessary to consider some of the major features, claims and assumptions of the systems perspective itself, because it is these that also shape educational organization theory. The best way to do this is by beginning with a brief outline of General Systems Theory as the parent discipline, and considering its relationship to modern organization theory. In the following, then, let us explore what it means to speak of a social organization as an *open system* and what the differences are between a closed system and an open system. (For a fuller treatment, including explicitly rational and natural systems, see Scott 1981. On loosely coupled systems see Cohen and March 1974, Weick 1969, 1976, 1982.)

General Systems Theory and Modern Organization

In his characterization of the relationship between modern organization theory and general systems theory, Scott (1983, pp. 60–61) points out that modern organization theory is primarily defined in terms of its 'conceptual–analytical base, its reliance on empirical research data and, above all, its integrating nature. These qualities are framed in a philosophy which accepts the premise that the only meaningful way to study organization is to study it as a system.' This view rose to prominence and become the conceptual foundation of contemporary organization theory primarily because of the pioneering work of Ludwig von Bertalanffy (1973), the founder of general systems theory. His work is of such importance that some of the major features of his general systems thinking are sketched below. This also allows us to trace the origins of theoretical problems that continue to plague contemporary advocates of the systems perspective in the field of educational administration (Hoy and Miskel for example).

General open systems theory was designed to explain no less than the 'the world as organization'. Von Bertalanffy conceived it in the hope of 'attempting scientific interpretation and theory where previously there was none, and [providing] higher generality than that in the special sciences' (von Bertalanffy 1973, p.14). Motivated by his observation of the inadequacies of the 'mechanistic approach' then employed in classical physics, especially when applied to solving theoretical problems in the biosocial sciences, and struck by structural similarities (and their isomorphism) in different fields, von Bertalanffy believed that a 'general system theory', as a general theory of organization, would be able to provide an overarching framework for all the sciences. A general system theory was to provide precisely what its name denotes, '... the formulation and derivation of those principles which are valid for "systems" in general' (von Bertalanffy 1973, p. 32). More specifically, general systems theory, as outlined in the program of its society founded in 1954, was expected to perform the following functions:

(1) investigate the isomorphy of concepts, laws, and models in various fields, and to help in useful transfers from one field to another; (2) encourage the development of adequate theoretical models in the fields which lack them; (3) minimize the duplication of theoretical effort in different fields; (4)

promote the unity of science through improving communication among specialists (von Bertalanffy 1973, p.15).

These functions are underwritten by a 'systems philosophy' that von Bertalanffy outlines in the Preface to the revised edition (1973). Its most important features for us in the present context are its ontological and epistemological elements. While we may agree that systems of various kinds can be identified by direct observation, such as 'a galaxy, a dog, a cell and an atom', there are also other systems such as conceptual and abstracted ones not directly observable but equally real — that is, 'corresponding to reality'. But, von Bertalanffy warns, 'the distinction between "real" objects and systems as given in observation and "conceptual" constructs and systems cannot be drawn in any commonsense way' (von Bertalanffy 1973, p. xxi). This difficulty poses 'deep problems', which are not addressed by him in the present context. A definition of systems is given as 'sets of elements standing in interrelation' (von Bertalanffy 1973, p. 38), a definition that, he states, acquires more precise meaning when we consider that 'systems can be defined by certain families of differential equations' — that is, mathematically.

But, how do we get to know systems of either kind? Von Bertalanffy stresses that his systems epistemology differs quite markedly from both logical positivism and empiricism, while retaining their scientific attitudes. Rejecting the physicalism and reductionism of both forms of positivism, and believing knowledge to be 'an interaction between knower and known', he opts for what he terms a 'perspective' philosophy in which science is 'one of the "perspectives" man with his biological, cultural and linguistic endowment and bondage, has created to deal with the universe he is "thrown in", or rather to which he is adapted owing to evolution and history' (von Bertalanffy 1973, p. xxii). Finally, now that the closed system of classical physics has been replaced with its equally confined image of humans existing in a world 'of physical particles governed by chance events as ultimate and only "true" reality', von Bertalanffy (1973, pp. xxii–xxiii) believes that 'the world of symbols, values, social entities and cultures' can also be considered as real, seeing that all reality is 'a hierarchy of organized wholes'. Thus, general systems theory, according to von Bertalanffy, has a humanistic orientation that is directly opposed to the narrow vision engendered by classical systems theorists, which, he warns, runs the danger of mechanizing society.

Even in its early days general systems theory was criticized for its lack of explanatory value, among other things. Von Bertalanffy acknowledges the difficulty but also points out that at this stage of theory development, 'explanation in principle', an expression borrowed from the economist Hayek, has to suffice (von Bertalanffy 1973, p. 36). Just as economists are not able to predict fluctuations in the stock market accurately while explaining general economic phenomena well enough, so system-theoretical explanation has to await further developments. 'Explanation in principle . . .', he notes, 'is better than none at all' (von Bertalanffy 1973, p. 36). We shall return to this point later.

Heavily influenced by biology, his home discipline, von Bertalanffy was quite careful, however, to note the limitations of the (biological–organizmic) systems metaphor in any attempt to explain *social* activity. When looking for similarities between very different kinds of systems, von Bertalanffy (1973, pp. 35–36) warns, one always also finds important dissimilarities:

> Analogies as such are of little value since besides similarities between phenomena, dissimilarities can always be found as well. The isomorphism under discussion is more than a mere analogy. It is a consequence of the fact that, in certain respects, corresponding abstractions and conceptual models can be applied to different phenomena. Only in view of these aspects will system laws apply. This is not different from the general procedures in science. It is the same situation as when the law of gravitation applies to Newton's apple, the planetary system, and tidal phenomenon. This means that in view of certain limited aspects a theoretical system, that of mechanics, holds true; it does not mean that there is a particular resemblance between apples, planets, and oceans in a great number of other aspects.

Despite such delimiting comments, von Bertalanffy believed that his general systems theory, because it was truly interdisciplinary, had the potential to contribute to the unification of all science. Eschewing reductionism, he describes his philosophy of perspectivism in the following way:

> The world is, as Aldous Huxley once put it, like Neapolitan ice cream cake where the levels —the physical, the biological, the social and the moral universe— represent the chocolate, strawberry, and vanilla layers. We cannot reduce strawberry to chocolate — the most we can say is that possibly in the last resort, all is vanilla, all mind or spirit. *The unifying principle is that we find organization at all levels* (emphases added; von Bertalanffy 1973, p. 49).

Finally, there are also some educational implications deriving from the general systems theory. On the one hand, von Bertalanffy argues that general systems theory helps to train scientific generalists who are able to work in an interdisciplinary way, and on the other, he believes that his theory also helps in regard to developing ethical values. His solution is not to advocate straight-out scientific control of society, which, in his words, '... is no highway to Utopia' (von Bertalanffy 1973, p. 52), but to pay attention to the individual, a humanistic concern that, interestingly, he shares with Greenfield, although of course he would disagree with Greenfield on other matters.

Having thus traced some of the more important features of general system theory, and identified some difficulties, let us note for the moment that both general systems theory and contemporary organization theory consider organizations as 'wholes'. The major difference between them is that general systems theory, as the more comprehensive framework, is concerned with examining every level of system, whereas organization theory restricts itself to human organizations in particular (Scott 1983, p. 65, provides a useful classification of systems, following the economist K. Boulding). Perhaps another way of expressing the relationship between the two is to say that modern organization theory, in terms of its theoretical structure, is a specific social science application of general systems theory. This also means that problems inherent in one are likely to show up in the other, as we shall see.

Characteristics of General and Open Systems

The discussion of general systems theory and system features does suggest that the theorists employing this concept know and can define what makes up a 'system'. Systems are generally defined as consisting of interrelated objects, attributes and events (see Litterer 1969). This is a very wide definition, but, as Burrell and Morgan note, it appears to be all that is available. They observe that 'Despite its popularity ... the notion of "system" is an elusive one. Many books on systems theory do not offer a formal definition ... and where a definition is attempted, it is usually one of considerable generality' (Burrell and Morgan 1982, p. 57). Given the brief exposition of general systems theory and von Bertalanffy's own definition, Burrell and Morgan's concerns are not surprising. Bearing in mind the generality of definition, it appears that *interrelatedness* is a system's most important characteristic. This is followed by *holism* as the second most fundamental characteristic. We can already note some interesting methodological problems here. For example, changes in any one of the interrelated objects, attributes or events will result in changes or adjustments across the whole system. This means that it is very difficult, or near impossible, to assess what affected what, and to determine exactly what accounts for the changes in the system. Also, the second important characteristic, *holism*, brings its own difficulties when examining a system. Often expressed in the phrase 'the whole is bigger than the sum of its parts', the problem, as Litterer (1969, p. 4) acknowledges, is the following:

> Is a set of bricks just a wall, or is it a part of a building, is the building part of a city, is the city part of a nation? Any element contributes to a system, but all too typically that system is part of another system, and the question then is, 'What is the system under study?'

Litterer answers his own question, 'This depends to a considerable degree upon the interests of the observer or person who is concerned with defining a system' (1969, p. 4). Here let us note simply that because different observers bring different interests to the study of social phenomena, the question of determining 'the system under study' becomes very elusive indeed and, on this reading, is one that cannot be answered decisively by system theorists, as Litterer (1969, p. 4) appears to concede when he notes that 'The definition of system ... is somewhat arbitrary'. Despite the uncertainties of definition, it is nevertheless true that complex systems such as banks, automobile companies, and governments have stability and continue over time. This third characteristic is described in terms of open systems being *goal-oriented*. What is meant by this is that all systems, open or closed, tend to return to a position of equilibrium after they have been disturbed. The return to equilibrium is interpreted as the system's seeking to achieve this goal. There may be more than one point of equilibrium in complex systems, meaning that the system may have multiple goals, which, however, cannot all be attained simultaneously, and this indicates that the system may be in conflict.

A fourth characteristic of general systems is that they are self-regulating, and *regulation* includes three different kinds: adjustment, control (which includes feedback) and learning. Fifth, open systems, as Litterer (1969, p. 5) points out, are almost by definition characterized by their acceptance of *inputs* from the environment, usually described in terms of energy and information, as well as by the *outputs* they give back to their environment. A sixth characteristic, *transformation*, indicates that what systems deliver back to their environment is not what they received, because transformation has taken place. Seventh, what is termed 'hierarchy' is a system characteristic that suggests that complex systems may contain simpler and smaller ones. The study of the latter is one of the main tasks of general systems theory. The eighth characteristic is called *entropy*. Following Katz and Kahn's (1983) characterization, it is the concepts of *entropy* and the *second law of thermodynamics* in particular that demarcate open (social) systems from closed ones. Expressed simply, the second law of thermodynamics means that systems tend to run down (see Harmon and Mayer 1986, pp. 162–163). Katz and Kahn (1983, p. 100) provide the example of an iron bar that is heated at one end by a blowtorch. The iron bar is now in an unstable state, with fast molecules in the end being heated and slow molecules in the other, the cool end. After the blowtorch is switched off, a steady state is eventually reached (in that the distribution of fast/slow molecules in the iron bar becomes random) and the bar reaches the same temperature overall. The iron bar cools, until its temperature approaches that of the room in which it is located, when this has happened the system has returned to equilibrium. However, unlike closed physical system, social systems tend to become more complex over time rather than simpler. This characteristic is called 'differentiation'. An open system does not run down, because it imports energy from its environment. This constitutes the major difference between the two types of system. Because of this specific characteristic, open systems experience what is called *negative entropy*. As Scott (1981, pp. 109–110) explains it, 'By acquiring inputs of greater complexity than their outputs, open systems restore their own energy and repair breakdowns in their own organization'. Finally, given that open systems are goal-seeking entities, they are also characterized by the principle of *equifinality*. As Harmon and Mayer (1986, p. 164) put it:

The idea is that an end state can be reached by a variety of paths and from widely different initial conditions. Essentially this [equifinality] is a notion of causality, but one that differs radically from that embodied in, for instance, the work of Frederick Taylor ... From the systems perspective, 'the one best way' does not exist.

Now that we have briefly outlined some of the major features of general and open systems, and also considered some differences between closed and open systems, let us see how the open systems perspective works in educational administration. This is best done by considering first why theorists objected to the closed systems view when applied to organizations such as schools.

The Systems Perspective in Educational Administration

Not surprisingly, the closed systems perspective was never popular in organization studies because of its many perceived limitations. It has mainly been applied to theorists such as Max Weber, Frederick Taylor, and Herbert Simon because their work emphasized the internal structure and functioning of bureaucracies and/or (industrial) organizations and did not pay sufficient attention to the influence of the external environment. In fact, the criticism levelled against these theorists, commonly described as advocates of the *rational model of organizations*, was not expressed directly in terms of closed versus open systems models, but in terms of their narrow *scientific* approach to the study of organizations. As a result of these criticisms, some writers, such as Griffiths, argued for a more flexible scientific approach in educational administration that would expand the traditional empiricism of the Theory Movement, while others such as Greenfield, Hodgkinson, Foster, and Bates reasoned that administration is not to be understood as a science at all, but ought to be seen as a humanism or an art. So when Griffiths (1979) took stock of what the traditional science of administration had accomplished, he noted that one of its major shortcomings was that it had not included discussion of women, unions, and racial minorities in administration. This is another way of saying that orthodox administration science's primary focus on rational-bureaucratic features excluded the external environment from consideration (for a first major study on the importance of the environment see Lawrence and Lorsch 1967). The closed systems model therefore sanctioned only very limited *empirical* research. The dynamic interplay between environment and organization could not be studied as the major factor of organizational change. This was acknowledged by Bidwell (1979, p. 111), who in a recanting of his earlier, closed systems, views (for example Bidwell 1965), describes organizations as follows:

> Closed-systems theories have approached organizations as if they were machines. The organization-as-machine is a system that remains

undisturbed by events outside its boundary, unless a prime mover of some kind — most often in these theories either an entrepreneur or top-level administrator — intervenes to change parts of the system or change the ways existing parts act on one another. Moreover, the action of such a prime mover is used to account for the machine's existence in the first place.

The features Bidwell alludes to here, the system's static and apparently self-contained nature, the assumption of a 'creator' or 'designer' whose rational purposes determine the system's goals, and the exclusive focus on its internal workings, are the very features that are too narrow to encapsulate real-life social organizations. As Bidwell (1979, p. 111) notes in relation to schools, the closed system idea could not account for change, for school productivity, and for 'the connectedness of schools and their environments'. From a closed systems perspective, the influence of the economy on organizations, for example, would not be an issue which could be raised. If we assumed a closed system model, Victoria's 'Schools of the Future' project would be inexplicable since Australian economic rationalism, the relevant 'environment', could not be drawn on as an important source of 'input'. The kind of study possible from a closed perspective would emphasize how well or how badly schools manage to restructure themselves internally — that is, how efficiently or inefficiently they manage their human and other resources. While this is, of course, one way of saying that the environment is disregarded, there is another added cost. Internal restructuring takes place without regard for the effects it might have on the environment. Katz and Kahn (1983, p. 101) spell out in more detail what the consequences are:

> The effects of such [internal] moves on the maintenance inputs of motivation and morale tend not to be adequately considered. Stability may be sought through tighter integration and coordination, when flexibility may be the more important requirement. Coordination and control become ends in themselves, desirable states within a closed system rather than means of attaining an adjustment between the system and its environment. Attempts to introduce coordination in kind and degree not functionally required tend to produce new internal problems.

In addition to the preoccupation with internal functioning and its resultant consequences, there are two more features of closed systems thinking that are particularly detrimental, 'the neglect of equifinality and the treatment of disruptive external events as error variance' (Katz and Kahn 1983, p. 101). Equifinality means that there are many ways of reaching an objective; there is no single best way. Insofar as there might be social conditions that are fixed and known, there may well be one best way, and Katz and Kahn (1983, p. 101) mention the coaching of baseball players as an example. However, this is the exception rather than the rule in social organizations, as was indeed recognized by Simon (1976) and expressed in his well known phrase of 'satisficing' (this concept makes him rather less of a closed systems thinker than is commonly assumed).

The second detrimental feature mentioned by Katz and Kahn, that of treating disruptive external events as error variances, means that they are controlled out of studies of organizations because they are considered to be irrelevant from the organization's point of view. This means that troublesome external factors are simply bracketed out, defended against, or ignored. In the open systems perspective, because environmental inputs of any kind are integral to system functioning and maintenance, disruptive factors are also included as an important source of input.

A final, and most important worry for closed system thinking is that it has a very limited view of organizational learning. What is meant by this is captured in the feature of regulation, particularly the feedback function which in closed systems is a single feedback loop. An organization's feedback function is important because it delivers information about external changes back to the organization so that it can react or change as required. This learning function has not been developed in closed system thinking, but is of central importance in the open systems view.

Despite the conceptual advantages seen in the open systems metaphor, even Katz and Kahn, whose book *The Social Psychology of Organizations*, initially published in 1966, can be described as the classic open systems primer for organization theory, note that organizational openness must still be developed:

> Open is not a magic word, and pronouncing it is not enough to reveal what has been hidden in the organizational cave. We have begun the process of specification by discussing properties shared by all open systems... (Katz and Kahn 1978, p. 33).

This warning notwithstanding, the general consensus in organization theory is that the debate between the two versions of systems theory is, as Meyer (1978, p. 18) expresses it, closed 'on the side of openness'. This shift is considered quite significant because, as Katz and Kahn (1983, p. 100) observe, it allowed a refocussing of the *scientific* study of organizations. That is, leaving behind the laws of traditional physics that sanctioned the closed systems view, they believe, did not mean giving up the scientific study of organizations. Rather, the turn to the biological sciences with their emphasis on living open systems enabled a new and more promising scientific way of studying social organizations, which, after all, are also living entities. Hence, the open systems metaphor appeared to be more appropriate for all manner of organizations, including schools. As Hoy and Miskel (1991, p. 21) put it:

> ... schools are social systems that take resources such as labor, students, and money from the environment and subject these inputs to an educational transformation process to produce literate, educated students and graduates.

As was the case in the (1987) third edition of their text *Educational Administration*, Hoy and Miskel, in their fourth (1991) edition, again make clear that they consider their theory of educational administration as a continuation of the behavioural science approach, with the open systems perspective providing the relevant conceptual framework (Hoy and Miskel 1991, pp. 23, 25). As well as reconfirming their commitment to this perspective, they also remain loyal to the view of theory they advocated previously. While noting that there was initial agreement in the field that Feigl's definition of theory 'was an adequate starting point', they also cite Willower's criticism that Feigl's view 'is too rigorous as to exclude most of theory in educational administration' (Hoy and Miskel 1991, p. 2). They settle, as they did before, for Kerlinger's definition of theory as the most useful:

> Theory is a set of interrelated concepts, assumptions, and generalizations that systematically describes and explains regularities in behavior in educational organizations. Moreover, hypotheses may be derived from the theory to predict additional relationships among concepts in the system (Hoy and Miskel 1991, p. 2).

Hoy and Miskel believe, as did Griffiths, and others before them, that administrative theory guides action, that it is in the business of explaining the nature of educational administration, and that it contributes to the solving of administrative problems. Their major departure from the Theory Movement does not therefore involve abandoning its philosophical assumptions, but rather adopting an open systems view, of which they have high hopes. This is indicated in the introduction to their second chapter, 'The School as a Social System', where they cite Getzels, Lipham, and Campbell approvingly: 'There is little point in general models if they do not give rise to specific conceptual derivations and empirical applications which illuminate, in however modest a degree, significant day-to-day practices...' (Hoy and Miskel 1991, p. 28). Adopting the social system perspective is considered to be the appropriate way to achieve these aims. Also, the system perspective's utility is supported by their belief that the challenges faced by the behavioural science approach in the 1990s require that 'theory and research will have to become more refined, useful and situationally oriented', to which they add in the fourth edition, 'and will need to address emerging gender issues' (Hoy and Miskel 1991, p. 25). This is a welcome addition, yet one may wonder in passing why, at a time of extraordinary theoretical and philosophical debate, there is scant attention to values and cultural concerns in this edition, and still no discussion of Greenfield's or Hodgkinson's contribution to administration theory.

In keeping with general systems thinking, Hoy and Miskel (1991, p. 28) define a social system as follows:

> It is a model of organization that possesses a distinctive total unity

(creativity) beyond its component parts; it is distinguished from its environment by a clearly defined boundary; it is composed of subunits, elements, and subsystems that are interrelated within relatively stable patterns (equilibria) of social order (see also Thelen and Getzels 1957, p. 351).

In addition to drawing on the work of Litterer (1969), Getzels and Guba (1957), and Bidwell (1965), the authors also draw on the classic systems theory work of Getzels, Lipham, and Campbell (1968) in the fourth edition. This means, then, that for Hoy and Miskel the basic two elements of a social system are also (a) the institutional and (b) the individual (This is represented graphically on p. 29 of their text). Getzels *et al.* describe administration 'structurally ... as the hierarchy of superordinate–subordinate relationships within a social system ... Functionally, this hierarchy of relationships is the locus for allocating and integrating roles and facilities in order to achieve the goals of the system'. Hoy and Miskel are also concerned to explain organizational behaviour in terms of relating institutional elements — that is, roles and expectations (the nomothetic) — with individual elements, such as different personalities and need dispositions (the idiographic) found in any organization. These two elements together 'provide the basis of a social-psychological theory of group behavior in which a dynamic transaction between roles and personality interacts' (Hoy and Miskel 1991, p. 35). Thus the summary of their basic model is as follows:

> Behavior (B) in the system is explained in terms of the interaction between role (R), defined by expectations, and personality (P), the internal needs structure of an individual; that is, $B = f(R\ P)$ (Hoy and Miskel 1991, p. 35).

Although there may be a balance between role and personality factors in some systems, it is to be expected that the proportion of one to the other differs depending on the type of organization. The authors suggest that highly bureaucratic organizations are characterized more by role factors and that less highly structured ones display more personality attributes. Behaviour in all social systems, nevertheless, is determined by both institutional and personal needs. Hoy and Miskel refine their basic model, adding the concept of the work group. In fact, in formal organizations 'the work group is the mechanism by which bureaucratic expectations and individual needs interact and modify each other' (Hoy and Miskel 1991, p. 38). In the school context, for example, teachers might have developed informal procedures for disciplining students; these informal norms might then, in turn, become the hallmarks for judging good teaching, which is equated with keeping good control. Now since a social system is an open system, determining boundaries is difficult because the environment encroaches on what happens inside the organization. This raises an important question: 'Which features of the environment are most salient for constraining behavior in schools?' Hoy and Miskel (1991, p. 40) answer their own question by admitting that 'There is no quick or simple answer' because all kinds of external factors impinge on the operations of schools. The most useful way to predict the behaviour in schools is to study the interactions of the three elements (institutional, work and individual) in terms of their consistency:

> We posit a congruence postulate, *Other things being equal, the greater the total degree of congruence among the elements of the system, the more effective the system* (Hoy and Miskel 1991, p. 41).

An effective school, in their view, is one in which expected and actual performance are consistent (Hoy and Miskel 1991, p. 51). Although, in their second Chapter, the authors no longer speak in terms of goals, a section deleted for reasons of 'parsimony' (Hoy and Miskel 1991, p. 53, footnote 2), the goal model of organizational effectiveness is still maintained. Thus, in their Chapter 12, Hoy and Miskel (1991, p.375) continue to accept Etzioni's definition, 'An organization is effective if the outcomes of its activities meet or exceed organizational goals'.

With regard to the congruence model, Hoy and Miskel also suggest that it provides a good basis for organizational analysis and problem solving. In order to improve school outcomes, school decision makers

> gather information on the performance levels of their schools, compare the information with the desired performance levels, identify discrepancies and difficulties, search for causes of the problem, develop and select a plan to alleviate the problems, and implement and evaluate the plan ... The model is particularly useful in diagnosing conflict or lack of congruence among the key elements of the system (Hoy and Miskel 1991, p. 51).

Determining or improving congruence is not just a matter of intuition, according to Hoy and Miskel, who suggest that 'the theory and research in the remainder of this book should be extremely useful in this regard' (Hoy and Miskel 1991, p. 51).

Although this is only a brief outline of a more complex approach, the essential elements of the model are clear enough in order to make some critical observations. In the final section, then, let us address the issue of the systems theory's explanatory utility and indentify other difficulties for this perspective in educational administration.

The Utility of the (Open) Systems Metaphor

Because the conceptual framework for Hoy and Miskel's theory of educational administration is that of general systems theory, it is to be expected that problems inherent in the general theory will also show up in its organizational application. Furthermore, the fact that the writers still subscribe to a version (albeit modified) of a logical empiricist account of science has its own unhappy consequences. But let us begin by noting the following problem. In borrowing from Getzels *et al.*, Hoy and Miskel adopt a structuralist–functionalist framework for understanding schools, although they add a third element, that of the work group. Hence the difficulty of explaining observed behaviour found in the former reappears in Hoy and Miskel's account. To begin with, one cannot observe a role or role expectation; it has to be inferred. Furthermore, how a person perceives his or her role depends on how he or she *interprets* what needs to done in a given situation. The point is an unexceptional one; it is not a role but a human being who acts on the basis of certain beliefs or values. Because filling a role is essentially the interpretive act of an individual, the distinctiveness of roles and role expectations as categorically separate from the personal, or idiographic dimension, disappears. This means for Hoy and Miskel's model that explaining behaviour as a function of the interaction between the institutional and the personal dimensions is problematic because the institutional is laden with the personal. The result is that the generality of systems theory is compromised by the particularity of its major categories.

This tension between generality demanded by traditional science, and the particularity demanded by the conditions of applicability, appears in several other guises. Take, for example, the definition of 'system'. Given von Bertalanffy's definition of systems as 'sets of elements standing in interrelation', everything and anything would appear to qualify as a system *in the absence of specifying the nature of the relations in question*. But once we begin to specify the relation by stipulating, for example, that collections of objects have to stand in a *causal* relation to each other, many of sets of elements would not qualify as systems.

This difficulty with definition spills over into von Bertalanffy's point about isomorphism across different fields. Because relations between sets of elements are initially not specified, the fundamental question is, on the basis of which assumptions can they be seen *as* isomorphic? How does von Bertalanffy (and other systems theorists) *know* which phenomena to count as 'similar' and which as 'dissimilar'? This is the crucial epistemological point he attempts to answer by talking about a 'perspective' philosophy, which, according to Litterer, amounts to saying that it is the interest of the observer that ultimately defines what the system under discussion is. Such interests are, of course, theoretically motivated (see Lakomski 1986). For a trained biological scientist, what systems count as isomorphic would be determined by the relevant concepts, theories and models of that discipline. For a scientist from the behavioural sciences, psychological theories and constructs would be the relevant markers as they are for Hoy and Miskel. Or, to use von Bertalanffy's example, what is considered as the 'relevant' aspect that ties together Newton's apple, the planetary system and tidal phenomenon is 'relevant' in terms of the relations stipulated to hold by the law of gravitation. However, notice now that the plausibility of systems theory as a heuristic device depends on the integrity of the background substantive theory used to sort and classify phenomena (for more detailed criticism see Evers and Lakomski 1991, pp. 68–73; Chapter 13).

This is also the fundamental problem of functionalist explanation. Insofar as scientific knowledge and theory are implicitly drawn on to specify what are seen as relevant relations among phenomena or objects, systems theory (including its educational administration relative) piggybacks onto generalizations first developed elsewhere. Therefore talk of the respiratory system, or of the heart as an organ that pumps blood through the body, is of explanatory value precisely because the detailed scientific work has already been done. Useful talk of 'system' comes *after* rather than before empirical study. But now talking of schools as social systems, for example, presumes that the detailed work needed to explain the extraordinary complexity of human interaction and interrelations has already been done. In the end, our caution about the utility of the systems metaphor in administrative studies stems from our belief that this presumption is false. The upshot, ironically, is that the powerful generalities that systems theory was supposed to deliver will arrive only when the particularities have been given sufficiently detailed attention.

References

Barnard C. (1938). *The Functions of the Executive*. (Cambridge, Mass.: Harvard University Press).
Bidwell C.E. (1965). The school as a formal organization, in J. G. March (ed.) *Handbook of Organizations*. (Chicago: Rand McNally).
Bidwell C.E. (1979). The school as a formal organization: some new thoughts, in G.L. Immegart and W.L. Boyd (eds.) *Problem-Finding in Educational Administration*. (Lexington: DC Heath).
Burrell G. and Morgan G. (1982). *Sociological Paradigms and Organizational Analysis*. (Heinemann: London, reprinted).
Campbell R. (1987). *A History of Thought and Practice in Educational Administration*. (New York: Teachers College Press).
Cohen M.D. and March J.G. (1974). *Leadership and Ambiguity: The American College President*. (McGraw-Hill: New York).
Ecker G. (1985). Theories of educational organization: modern, in T. Husen and T.N. Postlethwaite (eds.) *International Encyclopedia of Education*. (Oxford: Pergamon Press).
Evers C.W. and Lakomski G. (1991). *Knowing Educational Administration: Contemporary Methodological Controversies in Educational Administration Research*. (Oxford: Pergamon Press).
Getzels J.W. and Guba E.G. (1957). Social behavior and the administrative process. *The School Review*, 65(4), pp. 423–441.
Getzels J.W., Lipham J.M., and Campbell R.F. (1968). *Educational Administration as a Social Process*. (New York, Evanston and London: Harper and Row).
Greenfield T.B. (1985) Theories of educational organization: a critical perspective, in T. Husen and T.N. Postlethwaite (eds.) *International Encyclopedia of Education*. (Oxford: Pergamon Press).
Griffiths D.E. (1979). Intellectual turmoil in educational administration, *Educational Administration Quarterly*, 15(3), pp. 43–65.
Gronn P.C. (1982). Neo-Taylorism in educational administration, *Educational Administration Quarterly*, 18(4), pp. 17–35.
Harmon M. M. and Mayer R. T. (1986). *Organization Theory for Public Administration*. (Boston, Toronto: Little, Brown).
Hoy W.K. and Miskel C.G. (1991). *Educational Administration: Theory, Research and Practice*. (New York: Random House, fourth edition).
Katz D. and Kahn R.L. (1978). *The Social Psychology of Organizations*. (New York: Wiley, second edition).
Katz D. and Kahn R.L. (1983). Organizations and the systems concept, in J.R. Hackman, E.E. Lawler III, and L.W. Porter (eds.) *Perspectives on Behavior in Organizations*. (New York: McGraw-Hill, reprinted).

Lakomski G. (1986). A meta-structuralist analysis of Palermo's structuralist analysis of 'Dewey's Impossible Dream', in D. Nyberg (ed.) *Philosophy of Education* 1985: *Proceedings of the Forty-First Annual Meeting of the Philosophy of Education Society, Normal, IL*, pp. 219–221.

Lawrence P.R. and Lorsch J.W. (1967). *Organization and Environment*. (Homewood, IL: Richard D. Irwin).

Litterer J. A. (1969). *Organizations, Systems, Control and Adaption*. Vol II. (New York: John Wiley).

Meyer M.W. (1978) (ed.) *Environments and Organizations*. (San Francisco, Washington, London: Jossey-Bass).

Ogawa R. (1985). Theories of educational organization: classical, in T. Husen and T.N. Postlethwaite (eds.) *International Encyclopedia of Education*. (Oxford: Pergamon Press).

Perrow C. (1983). The short and glorious history of organizational theory, in J.R. Hackman, E.E. Lawler III, and L.W. Porter (eds.) *Perspectives on Behavior in Organizations*. (New York: McGraw-Hill, reprinted).

Roethlisberger F.J. and Dickson W.J. (1939). *Management and the Worker*. (Cambridge: Mass.: Harvard University Press).

Scott W. R. (1978). Theoretical perspectives, in M. W. Meyer (ed.) *Environments and Organizations*. (San Francisco, Washington, London: Jossey-Bass).

Scott W. R. (1981). *Organizations*. (Englewood Cliffs: Prentice-Hall).

Scott W. G. (1983). Organization theory: An overview and an appraisal, in J.R. Hackman, E.E. Lawler III, and L.W. Porter (eds.) *Perspectives on Behavior in Organizations*. (New York: McGraw-Hill, reprinted).

Simon H.A. (1976). *Administrative Behavior*. (New York: The Free Press, third edition, revised and enlarged).

Taylor F.W. (1947). *Scientific Management*. (New York: Harper and Bros).

Thelen H.A. and Getzels J.W. (1957). The social sciences: Conceptual framework for education, *The School Review*, **LXV**(3), pp. 339–355.

von Bertalanffy L. (1956). General Systems Theory, *General Systems*, **1**, pp. 1–10.

von Bertalanffy L. (1973). *General Systems Theory*. (New York: George Braziller, revised edition).

Weick K.E. (1969). *The Social Psychology of Organizing*. (Reading, Mass.: Addison-Wesley).

Weick K.E. (1976). Educational organizations as loosely coupled systems, *Administrative Science Quarterly*, **21**(1), pp. 1–19.

Weick K.E. (1982). Administering education in loosely coupled schools, *Phi Delta Kappan*, pp. 673–676.

5
Schooling, Organizational Learning and Efficiency in the Growth of Knowledge

It has sometimes been supposed that the mode of organization most likely to promote efficient decision-making is a hierarchical, centralized decision structure. In what follows, we shall employ some arguments from theory of knowledge, control theory, and economics to suggest that this claim is, in general, untrue. In particular, we suggest that for a public good like education, where provided by government, there are good grounds for preferring less hierarchical and more decentralised decision structures. That is, under certain defensible conditions, efficiency in educational decision-making can be enhanced by reductions in the concentration of organizational control.

In summary, the main argument of the chapter runs as follows. Hierarchical organization, instantiated as a network of superordinate–subordinate authority relations promotes consistency or uniformity in the implementation of centrally produced decisions. At its best, it promotes the reliable transmission and diffusion of *directives*. Against this model, theory of knowledge nowadays stresses the *fallibility* of directives or decisions (Popper 1963). As a result, it urges that some attention be given to error *correction*, at the expense of attention to the *prevention* of error. This broadly Popperian view of the nature and growth of knowledge is, in general, isomorphic to the main structures of organizational control posited by control theory, which is concerned with the application of detected error to system management (Simon 1952, Beer 1966, Strank 1983). Combining these views of knowledge and control leads most naturally to an adaptive view of organizational learning (Argyris 1982, Argyris and Schön 1978). Where information about the outcomes of organizational decision-making can be produced and reliably transmitted to centrally located decision-makers, even hierarchical organizations can learn adaptively (from their mistakes, or better advice) to make better decision. Our claim is that if these organizations are educational systems, the information assumption is less likely to be true. Given an understanding of the measurement of efficiency for the control of public goods, and the epistemological advantages of adaptive learning strategies, we claim that educational organizations can make certain decisions more efficiently (and effectively) where some form of participation in a less centralized decision process exists, including participation at the school level.

Participative Decision-Making in Education: Some Empirical Considerations

There are many arguments in the literature for participative decision-making, some more easily adjudicated empirically than others. One large survey of studies (Conway 1984) usefully reflects on some of the methodological complexities here and the equivocal nature of much of the evidence. In assessing this evidence for the effects of participation, it is important to distinguish, for example, the external perspective: the involvement of parents and other lay citizens in the educational process; and the internal perspective: the involvement of teachers and students. In addition to the question of who participates, there is also the question of the qualities of the process. According to Conway (1984, p. 20) 'the set of qualities that appear to be most useful deal with the *degree* of participation, the *content* for decisions, and the *scope* of the participant powers or the stage of the decision process involved'. The degree of participation has to do with the amount of input a participant has in the decision process and the extent to which this input is heeded. Content concerns the type of decisions made: financial matters, curriculum content, personnel issues, the organization of the school, educational policy, and so on. Finally, the scope of participation refers to the stage of decision-making or problem solving at which participation occurs; that is, whether it occurs at all stages or only some.

Now if the justification of internal (teacher and student) participation in school-based decision-making depends on current evidence for the promotion of more effective change, satisfaction with job, or greater productivity then for Conway (1984, pp. 23–29), the support is at best modest. On change, there appears to be little evidence of a difference in effect between change initiated in collaboration with rank-and-file and change without such participation. On the relationship between job satisfaction and participation, Conway (1984, p. 25) notes an increase in job satisfaction observed in studies, but cautions that 'the ratio still shows about one of three investigations not confirming the relationship'. This is slightly more favourable than the 40% of studies not confirming the relationship reported in Locke and Schweiger's (1979) large survey of (mainly) non-educational organizations.

On the question of productivity, Locke and Schweiger (1979, p. 316) concluded 'that there is no evidence that participation in decision-making is superior to more directive methods in increasing productivity'. However, they made the important concession that a major contextual factor was *knowledge* possessed by participants. Although Conway (1984, p. 17) is cautious about making inferences from findings on non-educational organizations to educational organizations, as Chapman (1990, p. 19) notes, what amounts to the knowledge factor appears sufficiently important to suggest 'that in the longer term, at least, there may be a direct positive relationship between increased school based decision-making and improvement in educational outcomes'. Indeed, the relevance of participant knowledge may explain why, on even quite narrow productivity criteria of teacher teaching behaviours and student test scores, the cumulative evidence seems 'to indicate that mid-level participation is probably desirable for both effective teaching and student achievement' (Conway 1984, p. 29).

With respect to the external perspective of lay participation, for comparable indicators much the same pattern of evidence exists. In what follows, therefore, we propose to examine more explicitly the kind of organizational structures that permit a more focussed application of participants' knowledge and judgement to the decision process.

Knowledge, Complexity and Error

A one way directive view of decision-making, with instructions for implementation propagating down an organizational hierarchy, places a great premium on the initial correctness of decisions. Given the magnitude of the complexities that invest educational policy and decision-making, this top-down approach to decision can be seriously affected by error. It is worth considering at least three types of error and possible associated hierarchical resources for prevention.

Perhaps the simplest is error of calculation. All the knowledge required for a decision may be at hand, but an error has occurred somewhere in, say, the calculation of consequences. The simplest organizational hedge against any malfunction is to build in redundancy of function. Where the cost of malfunction is high, and its stochastic properties known, reliable calculation, and hence the quality of decision-making, will invariably depend on maintaining a certain amount of functional redundancy. Indeed, von Neumann has proved that given enough redundancy, it is possible to construct systems of arbitrarily high reliability from unreliable parts. (The massively parallel architecture of the human brain is a design instance of how long-term reliability is maintained against the ongoing degradation of components.)

Strictly speaking, duplication of function is a horizontal extension of organizational structure, and so involves a measure of decentralization. However, to provide information, independent calculators need to be coupled somewhere by a match/mismatch feedback loop which in turn may involve supervision (though not always, since further calculation can continue to be initiated automatically until matching occurs).

Errors in knowledge which are not simply the result of malfunction call for more drastic prevention strategies. When it comes to cataloguing the limits of human knowledge in policy analysis and decision-making, we think the arguments of the incrementalists (e.g. Braybrooke and Lindblom 1963) and the satisficers

(Simon 1976) are decisive (see Evers 1988). For example, in knowing all the relevant details required in deciding on, say, a major reallocation of funding resources to schools, or even the introduction of local selection of school principals, it would be useful to know the relevant regularities or laws that govern the behaviour of the system. But our current knowledge of predictively useful law-like generalisations in social science is modest indeed, though a little better in some quantitative disciplines like economics, provided the drastic simplifying assumptions about human (economic) behaviour hold up.

It may be thought that this failing, occurring as it does in the most theoretical reaches of theory, can be quarantined from the descriptive taxonomies used to make observations and gather data. We now know, however, that there is no sharp theoretical/observational distinction to be drawn in accounting for the structure of empirical theories (see Quine 1960, pp. 1–79). Since failure of generalization may reflect not so much a concession to real world complexity as a failure to carve the world in theoretically interesting ways, the infirmities of generalization can also be the infirmities of data gathering.

Theory, of course, also shapes perceptions of the *relevance* of information, including that which bears on the tracing of more attenuated causal chains. As an example of this, Herbert Simon (1976, p. 82) traces a devious causal link between the number of elderly English spinsters and the size of the clover crop in different English counties, inviting legislators on matters of marriage to beware! Knowledge for decision can thus fall short of what is required through weaknesses of theory, poverty of imagination when it comes to exploring and testing possible relevant information, and the limited resources of research. Since the consequences of decision alternatives can ramify in arbitrarily extensive ways without the simplifying assumptions and approximations of theory, the costs of research needed to minimize the occurrence of error will fairly quickly outrun the costs of detecting and correcting the consequences of error. It was partly for these reasons that Lindblom suggested a range of simplifying and approximating stratagems in the approach he called 'disjointed incrementalism': for example, limiting consideration to a few familiar alternatives; exploring only some of the (more important) possible consequences of an alternative; and, more controversially, dividing the work of analysis 'to many (partisan) participants in policy making' (Lindblom 1979, p. 517).

However, this approach does not so much solve the problem of error prevention in knowledge as give up on it. The obvious efficiency trade-off against the costs of increased lapses in prevention due to savings and real-time constraints on knowledge acquisition is the provision of a mechanism for error correction. Administratively, within a hierarchy, this amounts to the provision of a feedback loop between sampled outcomes (or perhaps localized expert knowledge) and centralised decision structures. Where knowledge is reckoned as fallible, there is nothing out of the ordinary in a design that confers, at its strongest, *de facto* veto over the outputs of centralised decision-making. As Popper (1968) has argued, this kind of design is ubiquitous in natural learning systems. For all learning, all growth in fallible knowledge, is really a matter of improving some pre-existing body of knowledge. There can be no learning without some prior (unlearned) knowledge or set of dispositions (see Popper 1963, pp. 43–48, Watanabe 1969, pp. 376–379, Evers 1985, 1988). And the principal mode of improvement is by resolving mismatches between prediction and theory-driven experience. One may expect, therefore, that organizations that instantiate in their structure a basic learning design would, over time, become more efficient decision-makers.

The last type of error we want to consider, error of judgement, is really in our view just an extension of the knowledge category. This claim requires some argument, however, since we use the term 'judgement' to include values or very broad goals, and these are sometimes supposed to be either other than knowledge or immune from the usual forms of refutation (for an extended argument see Evers 1985, 1987, 1988) and we do not propose to add to it here. The central claims are as follows. Our learning of values appears to be contiguous with our learning of everything else. The epistemological failure of empirical content to distribute unevenly over the sentences of a theory, which leads to a blurring of the theoretical/observational distinction, also appears to blur the fact/value distinction in the direction of ethical naturalism. This suggests that the same apparatus and details of theory revision that apply to the more central elements of a theory also apply to value claims. A merit of this conclusion is that it posits a needed mechanism for explaining the adjustments experience counsels we make to the *weightings* we are willing to accord decision alternatives. And in positing the same mechanism for the learning or purported means and ends, it permits the justification of trade-offs to be embedded in the same epistemological framework.

Inasmuch as values (and maybe other knowledge claims) tend to be sensitive to variations in a person's background, interest and experience, the administrative provision of communication feedback loops for the correction of errors of judgement may require a sampling of those most affected or likely to be most affected by decisions. We shall argue later that this provision will be crucial for efficient decision-making concerning the public good of education. Needless to say, arrangements in this direction will be towards decentralization and away from hierarchy, involving as they do a weakening of executive forms of decision control. For as Strank (1983, p. 55) has argued, 'the normal hierarchical form of organizational structure puts great barriers in the way of such horizontal communication'.

A Note on Control Theory

The use of error feedback loops to regulate both the growth of knowledge and organizational decision-making is an application of designs that are part of the usual subject matter of control theory. For in general, control theory deals with the regulation of systems through the use of feedback (see Wonham 1984,

Schoderbek *et al.* 1975, pp. 57–84). The most basic design for a feedback-regulated control system can be expressed diagrammatically as in Figure 5.1.

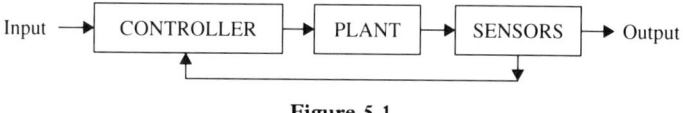

Figure 5.1

In this model, output from PLANT is regulated by the CONTROLLER-matching SENSED output from PLANT against input.

This simple design is sufficient to describe the operation of a thermostatically controlled heating system. To see this let input be Ti, the required room temperature. Let PLANT be the heater that is used to produce room temperature output To. And finally, let CONTROLLER be an error detector that measures the difference between Ti and To. The total system would then be as shown in Figure 5.2.

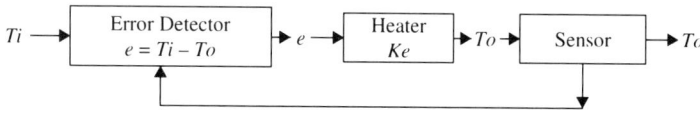

Figure 5.2

Although this model, as a first-order feedback system, is only capable of goal *attainment*, it is not a trivial design. For example, if we make the (reasonable) assumption that the energy consumed by the heater to raise room temperature is some multiple K of e, the model yields the equations

$e = Ti - To$
$To = Ke$

from which we can derive an expression for K $(= To/(Ti - To))$ to estimate the work done (and hence costs incurred) in heating the room to the required temperature (see Simon 1952).

A first-order design for decision-making would be slightly more complex and should contain at least the features shown in Figure 5.3, where d is decisions made, o is the outcomes of implementation or, perhaps, the results of localised expert review, and e the difference between the two. (We omit drawing in the sensors). Note that we have drawn Goals/Values/Some Information (i) apart from the Decision-Making box into which error feeds back and (ii) connected by a one-way arrow. This is to indicate that these items have been removed from feedback-driven critical scrutiny, a limitation that is a defining feature of first-order systems.

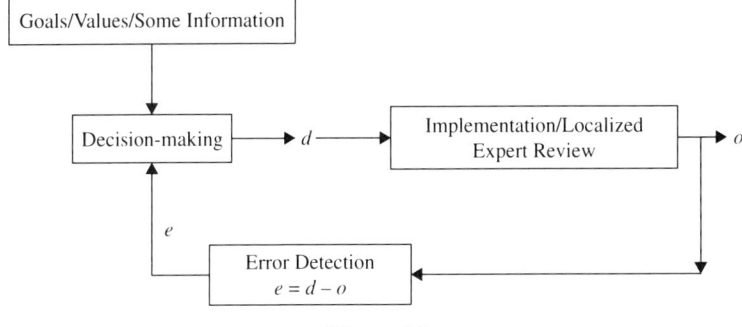

Figure 5.3

In second-order systems, however, goals, aims, objectives, values, in fact all knowledge that figures in the making of a decision, is subject to revision through error elimination. These systems are rather more complex to represent diagrammatically because goal evaluation will itself be a function of reviews of *past* successes and failures, thus implying additional feedback from a memory component. More extravagant reviews will keep track of the match/mismatch rate for whole decision strategies, a feature of third-order systems that provides feedback on the outcomes of different learning designs. Powerful heuristic problem solvers like the current best chess playing machines employ second- and third-order design features.

Some of these designs analysed by control theorists, especially second-order designs, have also been studied by organizational theorists interested in designing structures that promote organizational learning, and hence more efficient problem-solving and decision-making. In the next section we consider briefly the work of one such influential theorist who has studied both business and educational systems.

Organizational Learning

In a series of studies of a large range of organizations, Chris Argyris and co-workers have distinguished two main categories of organizational learning. In single-loop learning

> there is a single feedback loop which connects detected outcomes of action to organizational strategies and assumptions which are modified to keep organizational performance within the range set by organizational norms. The norms themselves — for product quality, sales, or task performance — remain unchanged... In order for *organisational* learning to occur, learning agents' discoveries, inventions and evaluations must be embedded in organisational memory (Argyris and Schön 1978, p. 19).

Organizational learning can occur by so structuring the possibility of individual learning that certain shared perceptions of the organization — the organization's theory-in-use — are changed.

Argyris regards single-loop learning, which in design corresponds to our earlier first-order systems, as being quite inadequate for the decisions and problems that most organizations face (for an extended analysis see Argyris 1982, pp. 3–155). For example, while such a design makes it possible for organizations to 'create continuity, consistency, and stability, and to maintain the status quo in order to achieve objectives within desired costs' (Argyris and Schön 1978, p. 123), it contains no provision for the correction of its goals and values, or the basis of weightings used to structure decision alternatives. So while participation in decision-making is enhanced through the process of organized feedback, the scope for error elimination is somewhat restricted. Single-loop organizational learning makes for 'dynamic conservatism', a condition inimical to long-term success or survival in a changing internal or external environment. It is a learning structure ill suited to coping with discontinuity, instability, and changes in the status quo.

The second type of learning, which Argyris regards as the most appropriate given the conditions under which most organizations exist, is called 'double-loop learning'. This applies to

> those sorts of organizational inquiry which resolve incompatible organizational norms by setting new priorities and weightings of norms, or by restructuring the norms themselves together with associated strategies and assumptions (Argyris and Schön 1978, p. 24).

This corresponds to our earlier second-order systems, and should also be regarded as containing the possibility for third-order learning — learning to learn more effectively — which Argyris calls 'deutero-learning' (Argyris and Schön 1978, p. 26).

What effect does the implementation of double-loop learning have on hierarchy in decision-making and how can it promote participation? One advantage of hierarchy that needs to be noted, in addition to its top-down control capacity, is its facility for resolving complexity. Complex tasks, when decomposed into more manageable specialised sub-tasks coordinated as part of the execution of organizational goals, admit readily of a hierarchical control structure. Indeed,

> the pyramidal pattern, which is so invariant that people often define organizations as pyramidal structures, derives from the principles of specialisation of work and of hierarchical control (Argyris and Schön 1978, p. 120).

A fundamental design problem for this type of structure consists in maintaining knowledgeable control in the face of expertise decentralised in order to provide high-level specialized function. Hierarchical control may be undermined for gains in efficient decision-making to the extent that decentralised error detection and corrective advice can be fed back to influence, perhaps decisively, all the relevant determinants of decision-making.

Whether or not these cybernetic participative designs do in fact make for more efficient decision-making or problem-solving is finally an empirical matter. Argyris's evidence suggests that they do, and this is a reasonable result given what we know of the epistemic virtues of these designs. However, there are some complexities in the measurement of efficiency that call for a distinction between the (mainly) business organizations that Argyris deals with and educational organizations that produce (at least to some extent) public goods. This distinction is also worth making when considering the participative perspective. For almost all of the evidence for second-loop organizational learning involves internal participation, whereas an important dimension in the argument over educational decision-making concerns external participation. Both of these distinctions have some bearing on the issue of efficiency.

Efficient Educational Organizations

'Efficiency' is well defined in engineering contexts, where the term has standard uses. Consider, for example, the mechanical system of a car engine. It converts the chemical energy contained in the fuel into mechanical energy for propelling the car. From an engineering perspective, the efficiency of the engine is the ratio of energy input, chemical energy, to desired or usable energy output, the mechanical energy available for car. Conservation of energy thus sets a maximum level of engineering efficiency for any physical system at 100%. (A good car engine is about 40% efficient in this sense.) When we move to consider organizational efficiency, some disanalogies appear. For example, there is no equivalent conservation principle. So the costs sustained in having, say, a fire department, the inputs, may very well be less that the costs incurred in not having one, the output being the consequent losses from fire (see Simon 1976, p. 181).

In the case of the car and the fire department the construction of an input/output ratio is possible because input and output are measured in the same units: energy for the car and dollars for the department. The possibility of a similar calculation for, say, the efficiency of an art gallery seems to be thwarted for want of a suitable identity of units of input and output. One solution often adopted is to shift to a measurement of relative efficiency. Instead of saying that a gallery, A, is efficient, we say that it is more efficient than some other gallery, B. If inputs be dollars and outputs be reckoned as some quantity, say 'arts', relative efficiency compares, legitimately, *ratios* of arts per dollar. Turning to educational systems, we can now say that a participative system is relatively more efficient that a non-participative system if it delivers more educational outputs per, say, dollar cost.

An alleged advantage of relative efficiency is that comparisons can be made independently of any adjudication on goals or values: 'the efficiency criterion is completely neutral as to what goals are to be attained' (Simon 1976, p. 14). This means that the efficiency formula can be stated in relation to any specified educational output regardless of the *merits* of that output. Goals are one thing, about which there can be disagreement and debate, efficiency in realising goals is something else again.

Some qualifications on this conclusion should be noted that imply that the matter of efficiency as a policy objective cannot be separated so readily from the matter of educational goals when these are considered as further policy objectives. The key difficulty emerges when one considers what we would call the *equivalence of complementary efficiencies*. To give a simple example, suppose system A produces knowledge more efficiently than B. Then if the complement of knowledge is ignorance, we can say that B produces ignorance correspondingly more efficiently than A. Since a complement can formally be defined for any arbitrary performance measure, it follows that for *unspecified* outputs any system is as efficient as any other system. (This counterintuitive result is related to the Popper/Quine/Watanabe claim, 'if X, Y and Z are distinct objects — say two swans and a duck — there are as many similarities between X and Y as there are between Y and Z', which provides the basis for an error elimination view of the growth of knowledge. The claim implies that logically, there is no such thing as a class of similar objects.)

The equivalence of complementary efficiencies means that for unspecified educational outcomes efficiency cannot function in a *normative* way in educational decision-making. So efficiency can be used normatively only given some *prior* weighting or selection of outputs. Unfortunately, this condition for normativeness does not decide in general on the desirability of efficiency, since presumably we would prefer a little less efficiency in the production of ignorance from a school. To secure desirability, we appear to need reliable knowledge of the sorts of outputs appropriate for an educational system. But the defensibility of sources here will presumably reflect judgements of expertise, knowledge of educational theory, relevant experience, and legitimate interests in the outputs of educational systems (on the last item see Walker 1987). The belief that all of this can be provided without recourse to (i) second-order feedback loops that permit the correction of educational aims and objectives and (ii) the provision of an external perspective is, if not clearly false, at least worth testing against the inputs from (i) and (ii). This suggests that our most reliable *knowledge* of some particular level of efficiency as a desirable goal for an educational system depends on the existence of some form of internal and external participative contribution to the broader range of educational policy and decision-making.

Developing an argument from Goodin and Wilenski (1984), we can now propose a considerable extension of external participation in educational decision-making. For once the virtue of efficiency is admitted to be relative to the satisfaction of other desirability conditions, the importance of competing desirable desiderata needs to be adjudicated. To see the consequences of this, consider a set W of wants, including the (first-order) want of efficiency. Let S be a system that satisfies these wants to such an extent that any other system always satisfies fewer wants. We can call S a Pareto optimal system. Now it is simply an open question whether optimal S implies a ranking of wants *within* S that places efficiency at the top. The (second-order) efficiency of S does not guarantee first-order efficiency *within* S. This is because first-order efficiency determines a ranking of other wants that can run contrary to a ranking determined by some other element of W, say some first-order want associated with a conception of justice. So if people have a very strong desire to live in a society governed by such a view of justice, increasing the second-order efficiency of S, that is optimizing the global satisfaction of wants, will entail reducing efficiency within S. Similarly, systematic views of educational goals — perhaps equity, democratic participation, and the growth of knowledge — imply preferred rankings of wants that can fail to coincide with a ranking that places the want of efficiency first. Of course, what people want in complex, continuously changing circumstances is not known a priori. But to ensure education contributes to a more efficient satisfaction of global wants — the good of the community — participative feedback from a broader community spectrum seems reasonable, though by now the amount of relevant input resulting from feedback output would require a more piecemeal approach to decision-making, with perhaps more decentralization.

Some of these consequences may tempt second thoughts about the move to relative efficiency which was motivated by unequal units of inputs and outputs. A sterner approach would seek to place a dollar value on outputs as well as inputs. The issues here are quite complex and we do not propose to discuss any of them at length. Instead, we shall briefly indicate the directions of a particular efficiency strategy based on certain results. One approach to input–output unit equivalence is to consider the constraints imposed on an educational system producing for a perfectly competitive market in equilibrium. Consider first, some industry under these conditions. To begin, we know that the market is Pareto efficient (for a proof see Sugden 1981, pp. 67–93). So for the industry it is not possible to increase production of goods without decreasing production in some other industry; the industry is (Pareto) production-efficient. It is not possible to reallocate the factors of production between industries to produce a different mix of goods without reducing consumer's utility, so it is product-mix-efficient. And so on. So one might suppose that educational systems can be reorganized more efficiently by requiring them to operate under conditions of perfect competition.

The trouble with this idea is that education is also a public good, since to some extent everyone gains some benefit from an educated community, and we know that there is no Pareto-efficient equilibrium for a market with public goods (see Sugden 1981, pp. 89–92). The central problem, then, is to devise reasonable efficiency rules for systems operating in a less than perfect market. Unfortunately, a result known as the 'general theory of second best' due to Lipsey and Lancaster (1956) implies that there are no precise rules to be had here. In achieving second best, global gains can sometimes be made by further system violations of efficiency, say of product, or mix, or exchange. As Mishan (1981, p. 223) notes in a review of the economic doctrines of the Chicago School,

> the traditional goal of promoting competition through the economy suffered a severe blow from ... the general theory of second best ... Although it is not possible to infer from the theorem that, in the absence of a first-best solution for the economy, any allocation was as good or as bad as any other, it certainly took the wind out of the sails of many a competitive model notwithstanding some limited counterblasts.

It is possible to argue the merits of certain allocations of shares, say, within education or to education, under second-best circumstances, but often more piecemeal strategies are required (for a discussion of the complexities see Ng 1977, McKee and West 1984).

One approach that coheres well with our broadly Popperian conditions for promoting the growth of knowledge is piecemeal welfare economics. In implementing this approach the strategy Sugden (1981, p. 111) suggests is for each (government) agency pursuing objectives other than the maximization of profit to work out independently

> its decisions in a framework of partial equilibrium, considering only those sectors of the economy that are most closely related to its own activities. This is the sense of the word 'piecemeal': no [agency] is asked to take an overall view of the economy or to make overall judgements about social welfare.

Needless to say, efficiency decentralized to the agency level collapses back into judgements of the *relative efficiency* of achieving more worthwhile agency goals per unit cost; to which we may add: subject to the constraint of optimizing the realization of the set of first-order agency goals including relative efficiency. Given the community-wide distribution of relevant knowledge required to make these sorts of judgements, especially in education, there is evidently no escaping, at least on the grounds of efficiency, the virtues of participation in educational decision-making.

Conclusion

Instead of summarizing, again, the main argument of this chapter, let us note a number of limitations on the conclusion. First, despite some of the advantages of more decentralization and less hierarchy in decision-making, there are natural limits to the amount of feedback that can usefully be processed, and limits, too, to the number of participants who should be involved. For this reason we have spoken of *sampling* various sources for feedback. Some version of the

Burnheim/Walker notion of statistical democracy seems helpful here (see Burnheim 1985, Walker 1987, 1988). The feedback-driven mechanism of error elimination and correction in decision-making does not really work as a perfectly competitive marketplace for the adjudication of all individual ideas. Moreover, even if it did, it would assume far too passive and idealistic an account of knowledge. For the growth of knowledge assumes some *power* over experimentation, a capacity to influence the organizational testing of hypotheses. A chief executive in a hierarchical organization enjoys this power by virtue of the organization's authority relations. However, another source of power can derive from being a *representative* of a large number of people (even when organized very loosely). For efficient decision-making well beyond the school level, representative participation seems more helpful.

These constraints set limits to the value of complete individual participation in all decisions and hence the amount of decentralization that is desirable in education. They do not undermine the efficiency of more valuable modes of participation in educational decision-making.

References

Argyris C. (1982). *Reasoning, Learning and Action*. (San Francisco: Jossey-Bass).
Argyris C. and Schön D. (1978). *Organizational Learning: A Theory of Action Perspective*. (Menlo Park: Addison-Wesley)
Beer S. (1966). *Decision and Control*. (New York: John Wiley).
Braybrooke D. and Lindblom C.E. (1963). *A Strategy of Decision*. (London: Collier-Macmillan).
Burnheim J. (1985). *Is Democracy Possible? The Alternative to Electoral Politics*. (Cambridge: Polity Press).
Chapman J.D. (1990). School based decision-making and management: implications for school personnel, in J.D. Chapman (ed.) *School-Based Decision-Making and Management*. (Basingstoke, Hampshire: Falmer Press).
Conway J.A. (1984). The myth, mystery and mastery of participative decision-making in education, *Educational Administration Quarterly*, 20(3), pp. 11–40.
Evers C.W. (1985). Hodgkinson on ethics and the philosophy of administration, *Educational Administration Quarterly*, 21(4), pp. 27–50.
Evers C.W. (1987). Ethics and ethical theory in educative leadership, in C.W. Evers (ed.). *Moral Theory for Educative Leadership*. (Melbourne: Ministry of Education).
Evers C.W. (1988). Policy analysis, values and complexity, *Journal of Education Policy*, 3(3), pp. 223–233.
Goodin R.E. and Wilenski P. (1984). Beyond efficiency: The logical underpinnings of administrative principles, *Public Administration Review*, 44 (November/December), pp. 512–517.
Lindblom C.E. (1979). Still muddling, not yet through, *Public Administration Review*, 39 (November/December), pp. 517–526.
Lipsey R.G. and Lancaster K. (1956). The general theory of second best, *Review of Economic Studies*, 24(1), pp. 11–32.
Locke E. and Schweiger O. (1979). Participation in decision-making: one more look, in L.L. Cummings and B.M. Staw (eds.), *Research in Organizational Behaviour*, Vol. I. (London: JAI Press).
McKee M. and West E.G. (1984). Do second-best considerations affect policy decisions, *Public Finance/Finances Publiques*, 39(2), pp. 246–260.
Mishan E.J. (1981). *Economic Efficiency and Social Welfare*. (London: Allen and Unwin).
Ng Y.-K. (1977). Towards a theory of third best, *Public Finance/Finances Publiques*, 32(1) pp. 1–15.
Popper K.R. (1963). *Conjectures and Refutations*. (London: Routledge and Kegan Paul).
Popper K.R. (1968). Epistemology without a knowing subject, in B. Van Rootselaar and J.F. Staal

(eds.) *Proceedings of the Third International Congress for Logic, Methodology and Philosophy of Science.* (Amsterdam: North-Holland).
Quine W.V. (1960). *Word and Object.* (Cambridge, Mass: MIT Press).
Schoderbek P.P., Kefalas A.G., and Schoderbek C.G. (1975). *Management Systems.* (Dallas: Business Publications).
Simon H.A. (1952). On the application of servomechanism theory in the study of production, *Econometrica*, **20,** pp. 247–268.
Simon H.A. (1976). *Administrative Behavior.* (London: Macmillan, third revised edition).
Strank R.H.D. (1983). *Management Principles and Practice.* (London: Gordon and Breach Science Publishers).
Sugden R. (1981). *The Political Economy of Public Choice.* (Oxford: Martin Robertson).
Walker J.C. (1987). Democracy and pragmatism in curriculum development, *Educational Philosophy and Theory*, **19**(2), pp. 1–10.
Walker J.C. (1988). Curriculum decision-making at the school level. Paper presented to Conference on School Decision-Making and Management, Woodend, 11–12 April, 1988.
Watanabe S. (1969). *Knowing and Guessing.* (New York: John Wiley).
Wonham W.M. (1984). Regulation, feedback and internal models, in O.G. Selfridge, E.L. Rissland, and M.A. Arbib (eds.) *Adaptive Control of Ill-Defined Systems.* (New York: Plenum Press).

6

Leadership and Learning: From Transformational Leadership to Organizational Learning

School education in Australia, as in most western democracies, is at present undergoing dramatic change. Restructuring in various shapes and forms has become the norm in almost every education system in the nation. In order to implement change and to continue to offer quality education in both substance and delivery, improved, better, stronger leadership is seen as central (Hallinger and Murphy 1992, Caldwell 1992, Murphy and Hallinger 1992, Beare, Caldwell, and Millikan 1989). And such leadership, although not considered the exclusive prerogative of the formal office-bearer, is nevertheless seen as mainly residing in the role of principal or vice principal.

We want to argue in this chapter that leadership, as currently debated and advocated in the education literature, is not helpful in meeting the challenges schools are facing, whether public or private. In particular, the notion of the transformational leader who is charged, amongst other things, with developing teachers' (and students') potential, to alter awareness, introduce vision and mission, and generally transform the organization and its members, is promising more than it can deliver. There are a number of reasons for why this conception falls short of its promise. In addition to its implicit great man theory of leadership who by dint of personality leads, as argued by Gronn (1994), there is the assumption that knowledge is concentrated at the top of the organizational hierarchy and that it flows downhill. This model is the organizational equivalent of putting all one's eggs in one basket and contains no discernible view of learning since there are no feedback mechanisms that allow learning from error.

In organization theory, however, it has long been recognized that risk-spreading is far more beneficial for the organization's growth and continuance over time. Hence, a more productive way of thinking about how an organization such as a school can meet challenges posed by an unpredictable future, and survive in a constantly changing environment, is to suggest a model of organizational learning and cognition that characterizes schools as learning organizations. Schools can be thought of as being made up of intricate nets of complex interrelationships that criss-cross formal positions of authority and power and carry knowledge and expertise in all directions, not just downwards, as suggested by TF leadership. In acknowledging that knowledge is located at every level of a school's organization, and by developing feedback loops to learn from mistakes, the liklihood of responding appropriately to difficult changes and challenges is much greater than is reliance on any one leader and his or her transformational skills. Such a model is also more congenial to the formal restructuring of many of the state school systems of education, which now operate with flatter structures rather than the traditional top-down bureaucracies that were characteristic of the past. In this model, it is the school that becomes its own transforming agent. Future challenges can be faced with a far greater expectation of success since we can draw on the much bigger pool of intellectual resources of the whole school community. Empowerment thus happens for the organization and not just for the individuals inhabiting it.

Leadership: Transactional and Transformational

While much has been claimed in terms of transactional and transformational forms of leadership, it is important that we know how much evidence and support there actually is for these forms in the primary leadership theory literature. Another way of putting this point is to ask how we can know the reality of leadership, a question rarely asked explicitly in the leadership literature (but see Hunt 1991). The answer to this question entails a view of what counts as knowledge, together with criteria of justification and evidence to support claims for leadership. It also tells us what methodological approaches are appropriate and defensible. Of central importance here is that claims made on behalf of, especially, transformational leadership, should not outrun the evidence provided to support them. And a most central part of such evidence is that TF theory must be able to provide an account of how transformational leaders can *learn* to become TF leaders. The question of how leadership skills, knowledge and behaviours are aquired is essential in order to provide a basis for training.

Our present concern with what is called leadership has risen to prominence especially since James McGregor Burns' 1978 landmark study of political leaders in which he developed the notions of TA and TF leadership (see also House, 1977). What he called *transforming* leadership is '... more potent [than transactional leadership]. The transforming leader recognizes and exploits an existing need or demand of a potential follower ... looks for potential motives in followers, seeks to satisfy higher needs, and engages the full person of the follower' (Burns 1978, p. 4). It is to develop and empower followers. Transactional leadership, in contrast, is characterized by an exchange of valued things. Unlike transformational leadership, it implies neither a binding nor elevating relationship of mutual engagement; rather, it emphasizes the satisfaction of basic needs and extrinsic rewards as main motivation for action, such as pay or status, for instance. TF and TA in Burns' conception are at opposite ends of the leadership spectrum, they are orthogonal, or mutually independent. Barely twenty years old, the conceptualization of leadership in terms of transactional (TA) and transformational leadership (TF) behaviours has in one form or other become the premise for much work subsequently done in general organizational as well as school settings especially in the 1990s (for discussions of Burns' work see Conger, Kanungo, and associates 1988, Bass 1990, Yukl 1989). Predating the TA/TF distinction is the notion of charismatic leadership, which originated in the work of the German social theorist Max Weber (1978; see also Gronn 1995, Hunt 1991, Lepsius 1986). Setting aside for present purposes permutations, variations and alternatives to Burns' as well as Weber's notion of charismatic leadership (Hunt 1991, p. 65 lists at least ten presently discussed approaches) it is the notion of transformational leadership as advocated in the work of Bass and Avolio (Bass 1985, Avolio and Bass 1988, Bass and Avolio 1990, 1993, 1994) that has gained much recent currency in organizational studies, and one which is currently drawn on for discussions on school leadership as for example in the work of Leithwood (1992, 1994), Leithwood and Steinbach (1995), Leithwood, Begley and Cousins (1994), and Leithwood and Jantzi (1990).

Building on the earlier work of Burns especially, Bass and associates were motivated to investigate those leadership processes that appear to go beyond those commonly identified with everyday organizational transactions. These processes were gathered under the umbrella term 'higher-order construct of transformational leadership', i.e. a range of individual factors that while distinct may yet be intercorrelated (Bass and Avolio 1993, p. 50). A factor analytic model that relies on operationalization of its concepts, measurement scales, and quantitative methodology, Bass et al.'s conception of leadership consists of seven basic factors. These include the categories of laissez-faire, transactional, and transformational leadership which in turn are categorized in terms of a higher-order factor of active and passive leadership. The latter is an expansion of the earlier (Bass 1985) model which, so the authors maintain, merely serves to encapsulate the whole range of leadership behaviours observed by followers, and is not a new two-factor model.

Transformational leadership is characterized by four factors, the so-called four 'Is' (Bass and Avolio 1993, pp. 51 ff.):

(1) *Charisma (idealized influence)*. Highly trusted and respected, followers want to identify with and emulate the leader. The flavour of this factor is best expressed in the corresponding sample item, 'Has my trust in his or her ability to overcome any obstacle'.
(2) *Inspirational motivation*. This factor comes close to the first depending on the degree of follower identification. The leader uses symbols and appeals to followers' emotions to reinforce awareness and understanding in the pursuit of shared goals. Sample item, 'Uses symbols and images to focus our efforts'.

(3) *Intellectual stimulation.* Followers are encouraged to question their old ways of doing things, their values and beliefs, including those of the leader and the organization, and to think of new ways to meet challenges. Sample item, 'Enables me to think about old problems in new ways'.

(4) *Individualized consideration.* Followers are treated according to their needs which may be raised to a higher level. They are helped to meet challenges and to become more effective in attaining goals. Learning opportunities are provided. Sample item, 'Coaches me if I need it'.

Transactional leadership, in contrast, is mainly defined by the following (see also Avolio and Bass 1988, pp. 30–33):

(5) *Contingent reward.* The leader rewards followers for attaining common goals and objectives. The interaction between leader and followers is one of positive reinforcement, based on an exchange of desired items. Sample item: 'Makes sure there is close agreement between what he or she expects me to do and what I can get from him or her for my effort'.

(6) *Management-by-exception.* This is almost the opposite of the above. The leader intervenes when mistakes are made or problems occur that need correcting. The intervention is characterized by negative feedback, punishment, or disciplinary action. Sample item: 'Takes action only when a mistake has occurred'.

(7) *Laissez-faire.* This is called the nonleadership factor, because leadership is absent. Decisions are delayed, not made, or happen by accident. There is no intervention of either a positive or a negative kind. Sample item: 'Doesn't tell me where he or she stands on issues'.

The optimal leader, Bass and Avolio argue on the basis of their research results, is someone who is at the active end of the leadership spectrum, and combines positive transactional behaviours with the four Is of transformational leadership. Good managers achieve good results; transformational leaders, however, in addition to producing higher levels of effectiveness, achieve results beyond expectation. The latter expresses a key concept of Bass *et al.*'s leadership model and is called the 'augmentation effect' (Bass 1985). Measuring transformational leadership behaviours, so they claim, allows for a higher level of precision in predicting the achievements, which go beyond those elicited by everyday transactional behaviours. In this sense, TF can be called a value-added model of leadership. And this includes that '. . . transformational leaders do not necessarily react to environmental circumstances — they create them . . . with their ability to concretize a vision, to excite others, to change the way problems are thought about, the transformational leader is able to get others to react over time in ways that earlier models neither anticipate nor elaborate upon' (Avolio and Bass 1988, pp. 36–37). This conception is also central to the work of Ken Leithwood and associates. Leithwood is best known for his advocacy of transformational school leadership which he has tested in many empirical studies. Leithwood et al. (1994, pp. 7–8) define TF as follows:

> The term 'transform' implies major changes in the form, nature, function and/or potential of some phenomenon; applied to leadership, it specifies general ends to be pursued although it is largely mute with respect to means. From this beginning, we consider the central purpose of transformational leadership to be the enhancement of the individual and collective problem-solving capacities of organizational members; such capacities are excercised in the identification of goals to be achieved and practices to be used in their achievement.

While the authors observe that, although conceptually, leadership can be spread throughout the school, their research into transformational school leadership was, in fact, based on principals or vice-principals (see also Leithwood 1994, pp. 498–499, Leithwood et al. 1994, Cousins 1994, p. 2). Thus claims about *school* leadership are claims about the *principal's or vice-principal's leadership.*

An important element of TF is that it has an emotional impact on followers. Bass et al. point out that in times of heightened uncertainty, followers are even more inclined to pay attention to the leader, '. . . charismatic leadership will have its greatest impact (cognitively and emotionally) in situations where threats to "steady state" are eminent' (Avolio and Bass 1988, p. 40). Given the turbulent political, economic, and social environment in which most schools have been operating and restructuring themselves, it is not surprising that TF appears to fit the bill. More specifically, the educational environment, according to Leithwood, no longer requires just transactional leadership, i.e. good management in how to accomplish set goals and objectives. This was appropriate in the context of the school effectiveness movement of the late 1970s and early 1980s, with its emphasis on classroom practices and curricula (see Hallinger 1992). With school restructuring, however, purposes and outcomes are unclear, as are the intellectual and other requirements to prepare schools for the twenty-first century and especially for site-based management and teacher empowerment. What is required now is 'commitment' so that 'frontline school staffs appreciate the purposes for change and . . . foster their commitment to developing, trying out, and refining new practices until those purposes are accomplished (or until they change)' (Leithwood 1994, p. 500, Rowan 1990).

Commitment leading to action is created by transformational leaders in a number of ways beginning with the articulation of a vision, or the right way of doing things that followers consider worthy of their efforts, thus increasing levels of effort. In Leithwood's conception, a vision, the leader's perception of what ought to be, needs to be useful and defensible, and 'provides relatively precise guides to action. ... Educational leaders use their vision of the healthy school much as a physician uses an understanding of the healthy well-functioning body' (Leithwood et al. 1994, p. 31). Second, transformational leaders are able to redefine the problematic or uncertain situation so that it appears that '... there is greater control over events than previously perceived by followers', and third, by suggesting alternative solutions to problems, the transformational leader deflects from followers' feelings of frustration and helplessness 'to feelings of there being a challenge' (Avolio and Bass 1988, p. 40). That challenge, in turn, leads to heightened effort on part of the followers. What is important in the charismatic relationship is that essential norms are shared and that there is a sense of working for the common good (also Leithwood 1992, p. 18, 1994, p. 501). A charismatic relationship has achieved its goals when 'subordinates have gone through a reexamination of their position in relation to the one espoused by the leader [and it is that] that motivates them to reconsider their needs and elevate their goals' (Avolio and Bass 1988, p. 42; also Leithwood et al. 1994, p. 8).

Although there are encouraging results regarding the effects of TA and TF on followers in a variety of organizational settings, according to Bass et al., a finding endorsed by Leithwood and co-workers, and despite their observation that TF is not rare at all, Avolio and Bass also note that

> very little is known regarding the process by which individuals become energized and under what circumstances a transforming leader will be most effective. In our opinion, this is one of the most important questions raised by Bass's model in that it not only focuses on how individuals are energized by a leader, but how the individual becomes the source of energy for others (Avolio and Bass 1988, p. 46).

These concerns are shared by Leithwood et al. (1994, p. 27). They note that the body of research evidence on school leader practices are both limited and uneven in quality. More importantly, there are no good studies of cause-and-effect relationships of leadership variables, and coherent theory explaining relationships among variables is absent. This leads them to conclude that 'There is, arguably, a greater need for research exploring these relationships than there is more descriptive research on effective practice ... Such research would help us understand how effective practice develops, a crucial matter about which current research has little to say' (Leithwood et al. 1994, p. 27). Indeed, and here he contradicts the findings reported by Bass et al., Leithwood points to one difficulty that crystallizes the problem in providing evidence for transformational leadership: 'Distinctions between management and leadership cannot be made in terms of overt behavior' (Leithwood 1994, p. 515). Accepting that this distinction is empirically problematic, Leithwood states that 'most of the overt practices of transformational leaders look quite managerial'. What makes for the difference is that 'Transformational effects depend on school leaders infusing day-to-day routines with meaning and purpose for themselves and their colleagues'. (Leithwood 1994, p. 515). We can really only tell the difference if we know the nature of the purposes and their effects which, of course, now depend on how people interpret what they see. It is at this opportune point that some of the difficulties with TF will now be considered in more detail.

Critical Issues in the Study of Leadership

Burns' (1978) early observation that leadership is one of the most observed and least understood phenomena on earth is still widely shared amongst contemporary leadership researchers. The major complaints are that there is little agreement on the concept of leadership and that there is practically no knowledge base for leadership studies because of their fragmented and often contradictory nature (Howe 1994). Furthermore, Howe (1994, p. 3278) notes that leadership studies are still largely tarred with the Theory Movement brush with its assumption of a logical empiricist, hypothetico-deductive structure of knowledge and its justification, accompanied by an overreliance on quantitative methodology. This, in his view, is a sorry state of affairs (Howe 1994, p. 3278). The use of questionnaires in particular, so prominently used by Bass, *inter alia*, is commented on critically by Hunt (1991, pp. 189, 209, 211), who argues that this methodology is used inappropriately since there is not yet any substantive knowledge of transformational leadership (also Yukl 1989, Smith and Peterson 1988). What the questionnaire measures actually represent is rather the describer's cognitive structure, that is, the respondents' implicitly held leadership theories (Gronn 1995, Lord, Foti, and Devander 1984). Thus, the attribution aspect casts serious doubt on anything real being measured in terms of the leader's behaviours. This is of particular significance in terms of the charismatic aspects of TF since this, by definition, required a strong emotional follower identification with the leader.

In their reply to some of these criticisms, Bass and Avolio (1993, pp. 54 ff.) have defended the use of questionnaires as appropriate 'at the early stages of developing a theoretical model' on the assumption that the researcher has to begin with some methodology somewhere. As for the attribution factor regarding charisma, the authors basically agree that it is both a behaviour and an attribution 'for it requires particular follower emotional reactions to the leader to be identified as such. We have no quarrel with this operational definition of charisma' (Bass and Avolio 1993, p. 58). Nevertheless, they support the idea of further research to disentangle leader behaviour from attribution, implicit theories, and follower perceptions.

It is quite clear from the above that Bass and Avolio believe that their basic analytic model is valid, and that it is the methodology that is not yet up to the task of measuring correctly. Let us now take a closer look at their leadership model,

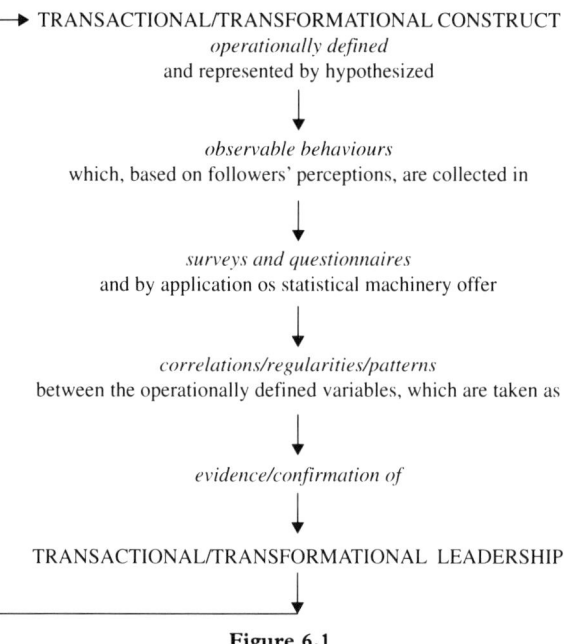

Figure 6.1

which can be represented as shown in Figure 6.1. Here we have an example of the kind of hypothetico-deductive structure Howe was referring to in his criticisms. As a central feature of *logical empiricism*, this account of scientific theory and practice is no longer accepted as valid in philosophy of science and epistemology. What are some of the reasons for this?

Roughly speaking, empirical evidence on this account is made up of singular observations, i.e. observation reports of behaviours, in the present case, which are hypothesized to be representative of either TA or TF leadership. These individual claims (observation reports) are at the bottom of the hierarchy; they form the *foundations* for claims to leadership at the top of the hierarchy. These claims are believed to be empirically testable, and the process of deduction and testing makes up the so-called hypothetico-deductive account (see Chapter 1; and Evers and Lakomski 1991 for a more detailed examination). In addition, an important feature of this empiricist account of scientific theory is the notion of *operational definition*. Since empiricist theory prescribes that its foundation is based on observation reports, every claim made — whether about observable or theoretical entities — has to be amenable to empirical definition. Now this appears to be straightforward with regard to such observables as sticks and stones, but seems problematic in relation to non-observable, theoretical entities, like leadership, for example, or any type of value such as good citizenship, equality or justice. The way out for logical empiricists was to operationalize them, that is, in the absence of the possibility of direct observation, to develop some measurement procedure which would capture the elusive entity. Operational definitions played a large part in the traditional scientific account of educational administration (the Theory Movement), and is still found in the work of Hoy and Miskel (1991).

Even though the conception of science in empiricist terms is no longer accepted as valid, it is important to recognize that Bass and Avolio and other leadership theorists were concerned with a problem of fundamental importance: namely, to attempt to explain how it is that leaders appear to be able to engineer social change successfully under conditions of uncertainty; or, to express it differently, *how to account for effective administrative practice*. A hypothetico-deductive account cannot accomplish this aim, and one of the reasons is its narrow conception of evidence, or *empirical adequacy*, which only admits observation reports. The problem with empirical adequacy is well illustrated in Leithwood's admission that TA behaviours cannot be told apart from TF behaviours. Leithwood correctly notes that behaviours need to be interpreted — they do not carry their meanings on their sleeves. However, since different observers/followers interpret behaviours differently, partially on the basis of whatever implicit theories they already have about, say, leadership (or anything else), there will subsequently be different and conflicting interpretations of what is to count as a TA or a TF behaviour. Since, as we saw earlier, logical empiricist theory relies essentially on observation reports as the final arbiters for adjudication of empirical claims, the only avenue open for further testing are further observations, possibly in different settings and circumstances. These, however, cannot provide any more substantive justification since the problem of identification arises anew in each new setting. No matter how many additional observations were to be carried out, as long as *empirical adequacy* remains the yardstick for adjudication many, conflicting, observation reports remain equally defensible. What this means is that in Leithwood's case, if there is no principled way of telling one leader behaviour from another, then any claim to have empirically identified transformational leadership effects is not justified. In the absence of justification, however, claims to leadership are nothing more than personal belief or opinion, which does not carry any empirical status, no matter how many empirical studies are conducted.

For Bass and Avolio's notion of charisma, the assumption of empirical adequacy is of equal concern. If it is conceded from the beginning that charisma may be an attribution as well, then this already undercuts the claim that it is an independent leader behaviour and thus a feature of leadership rather than followership. Moreover, if the authors are concerned to disentangle attribution, implicit theories, and follower perceptions, they will need more than observational evidence, which can be interpreted in many and conflicting ways; but since empirical adequacy is all the evidence admitted in form of specific observed follower reactions, there is no principled way to make the required distinctions. The epistemological resources, especially the conception of empirical adequacy, are insufficient to support the claims made on behalf of charismatic or transformational leadership.

In addition to aspiring to provide sound empirical evidence for leadership in the present example, empiricist science was hopeful of developing true lawlike and context-free generalizations about them that would allow for prediction, just like phenomena in the natural sciences. It would thus become eminently useful for the successful planning and improvement of social practices. Such hopes, however, are unwarranted because there are no foundations for knowledge, all knowledge is fallible, and there is always the possbility that we are wrong in our claims to know (for a more detailed discussion see Evers and Lakomski 1991, pp. 206–211). Prediction can thus never be large-scale for mainly four reasons. We have very limited understanding of social causes: (1) there are many interacting causes and (2) we do not know which cause may yield which outcome. This applies directly to the presumed effects of the actions of TF or any other leaders. Social life is extraordinarily complex and in constant flux (3), and there are theoretical limits to social prediction, as was shown by Karl Popper (4). These difficulties are recognized by Bass and Avolio and by Leithwood, albeit implicitly. Given these constraints, then, it is neither reasonable nor prudent to assume, as Bass and Avolio and Leithwood do, that the TF leader's vision and knowledge is a reliable base for correctly predicting the course of the organization's future.

Another implication of these difficulties is that the application of quantitative methodology to measure TF leadership is inappropiate. What is presumed by this kind of statistical-correlational methodology is that it can extract the *patterns* in the flux of everyday experience. But the very idea of TF leadership is that it does not follow any set pattern, that it is the most flexible and unpredictable form of leadership, which comes into its own precisely when there is turbulence and uncertainty (Evers and Lakomski 1995). Different situations and times elicit different responses, and TF at time x may no longer be TF at time y. It follows that TF leadership could not in principle generate the kinds of patterns that can be picked out by surveys or questionnaires. So the obvious question to ask is: if correlational or other statistical devices do not represent TF behaviours, what, then, do they represent? They represent aggregates of followers' 'confabulations' (Churchland 1983), or on-the-spot *plausibility judgements*.

Contrary to the common assumption that we have direct access to the contents of our own cognitive processes, which is the main feature of a particular theory of mind implicit in empiricist science, we know from social psychology and the brain sciences that this is not so (Nisbett and Wilson 1977). In such a theory of mind, it is presumed that just as knowledge of the external world is based on observation reports, so knowledge of our internal mental world is accessible by means of direct introspection. We have already seen that the first assumption is false. And so, indeed, is the second. It is no more than an extension of the fact that we always interpret external reality; our internal reality, the contents of our own minds, is similarly subject to the same condition of interpretation. This has to do with the simple fact that humans do not come ready-made with a vocabulary that describes both external and internal nature: we have to learn it from infancy onwards for both realms. This is how we acquire our theory of the world. It is through the acquisition of language that we are able to represent both outer and inner knowledge.

Consequently, given a questionnaire about assessing their principal's TA or TF behaviours, teachers, drawing on wider cultural perceptions of leadership, or perhaps on specific leadership theories they might have learnt, together with pre-given factors for each category, make up the answers, i.e. they invent them. In short, respondents do not have direct access to their mental processes, but they access prior learning of some kind. These are interpretations that are literally fabrications of our own minds that by themselves, may or may not refer to something real in the world. And this must not be misunderstood in terms of subjects' wilfully distorting replies, or wishing to mislead on purpose, although that may also play a role some of the time depending on the research conducted. Rather, confabulating is what we all routinely do, it is how we think as a matter of fact. The form in which this cognitive activity is represented is sentential (for the original definition of the sentential view see P. S. Churchland 1980, 1986). However, the manipulation of sentences does not exhaust, and is not identical with human cognitive activity or human reasoning and knowledge representation. Much of what we do goes on without ever being, or being able to be, represented in linguistic form; this is often described as practical or experiential knowledge, or simply knowing how to do something. Another way of putting this is to say that we are able to extract patterns from our environment that allow us to act in certain ways, like driving a car, without first doing a kind of sentence-crunching exercise that tells us in what sequence to carry out certain actions and behaviours. (Imagine a dancer.) We possess this capacity by dint of our massively interconnected brain, which in its circuitry allows for simultaneous, parallel distributed, readings of the environment. This is why we can recognize faces in a flash, and know how to react to certain administrative situations appropriately, for example, without being able to say why we did what we did. In short, quantitative methodology cannot measure TF leadership effects because it presumes that all cognitive activity is language-based activity whereas the kind of exceptional practice or problem-solving behaviours leaders (or anyone else) display precedes, or entirely eludes, linguistic representation. It is located in the brain's circuitry. Where does all this leave us with respect to leadership and organizational learning?

Organizational Learning

Several conclusions can now be drawn from the above discussion. Since there are no foundations for knowledge, and since whatever knowledge we do acquire is always in principle fallible, the best strategy for us to adopt is one that maximizes our ability to learn rather than putting all our energy into the prevention of error.

Leaders are fallible learners like the rest of us. Their vision may be based on faulty reasoning and incomplete information and they are not in a position to claim privileged knowledge by virtue of their formal position, although they may well be more knowledgable in some respects. Leaders, contrary to public opinion, are not born but made: they learn how to be leaders. It is this very feature that excercises the minds of leadership theorists, that is, a *causal* account of how it is that some people are apparently able to be such effective practitioners. Part of the answer lies in recognizing that human cognitive activity is not encapsulated in what can be said or represented in symbolic form.

An explanation for why this is so is offered by recent neuro-scientific models of human learning in form of network models of brain functioning (P. M. Churchland 1988, 1989, P. S. Churchland 1987, Sejnowski, Koch, and Churchland 1988; for an application of these ideas to education see Chapter 9). We know that people are good pattern recognizers, and that such pattern recognition is facilitated by the massively interconnected circuitry of the brain, and is independent of sentential processing, which it precedes. Given these very sketchy remarks about an enormously complex and as yet little understood enterprise, it should nevertheless have become clearer that there is little gain in maintaining a *hierarchical* view of knowledge distribution from leader to followers that puts great emphasis on the leader getting 'it right' by having the 'right' vision. If human learning and cognition are as indicated, then it follows that organizations such as schools ought to capitalize on the ways people naturally learn, and structure themselves accordingly (for further discussion see Chapter 5).

A much better hedge against error is, first, the recognition that error is always possible, incorporating the insights developed from Popper; second, to create structures inside the school, or other organization, that facilitate the double-checking of decision-making, and third, to create structures that allow learning from error to flow back into the organization so that it becomes a learning organization in its own right. In Argyris and Schön's (1984) terminology, organizations operate with single-loop, double-loop, and deutero-learning, or learning how to learn.

Organizational learning, although it relies on the learning of individuals, is not identical with the latter. In the study of organizations, which, according to Argyris and Schön (1984, p. 354), is a species of human learning in situations of interpersonal interaction, it is helpful to differentiate between two theories: the organization's *theory-of-action*, the espoused theory, and its *theory-in-use*, the 'lived' theory. The former indicates what the school, for example, is officially about, as espoused in its organization chart, policy statements and job descriptions. This implies the school's strategies for action, its norms, and its assumptions. This official theory is opposed to the often tacit theory-in-use, the way members actually go about doing their work which differs, often markedly, from what they *say* they do. In that sense it is the more important because real theory of the two. The theory-in-use can 'be inferred from observation of organizational behaviour — that is, from organizational decisions and actions' (Argyris and Schön 1984, p. 357). Note that the distinction between the two is the organizational equivalent to a distinction drawn earlier: being able to express actions and behaviours verbally and being an effective practitioner (such as a TF leader). Argyris and Schön's tacit theory-in-use captures the element of practice that is often circumscribed as 'knowing more than we can tell'.

The study of these practices over time give the best indication of organizational learning. 'Organization', in the authors' view, 'is on the one hand, made up of their subjective interpretations; on the other, there are the public maps of organization which indicate the organization's continuity over time. These maps are shared public referents and indicate things such as diagrams of work-flow, compensation charts, statements of procedure, etc., which are the media of organizational learning.' (Argyris and Schön 1984, p. 360). Sometimes, members perceive a mismatch between what they think the organization should be doing and what actually happens, and act to correct the mismatch between expectation and outcome (see Robinson 1995). This basic learning loop is an indication of one kind of organizational learning that serves to maintain the organization's stability:

> There is a single feedback loop which connects detected outcomes of action to organizational strategies and assumptions which are modified so as to keep organizational performance within the range set by organizational norms. The norms themselves — for product quality, sales, or task performance — remain unchanged (Argyris and Schön 1984, p. 361).

For this kind of individual learning to become organizational learning, members' discoveries, inventions or evaluations must, in turn, be incorporated into the organization's theory-in-use which will subsequently guide how people go about meeting the organization's goals. If this does not happen, then the organization cannot be said to have learnt. Single-loop learning best serves organizational effectivess, i.e. how to improve performance, existing goals and objectives. But organizations such as schools live in a constantly changing external environment that may force a revision of more than the accepted objectives. Organizational norms themselves may come under scrutiny. This can happen, for example, where declining enrolments in a single-sex school may raise the need to revise the school's fundamental philosophy and mission. This would need to happen via an appropriate, possibly conflict-laden, inquiry into the issue of single-sex versus co-education which, *inter alia*, would include ethical, social, economic, and philosophical considerations before the basic norm could be changed and new strategies implemented. This sort of learning designates double-loop learning which would result in new organizational maps and a changed theory-in-use. The distinction between single-loop and double-loop learning is best expressed as one that indicates relatively peripheral and relatively deep organizational learning, with the latter being close to 'deutero-' or 'learning how to learn', the third category. The latter signifies the kind of fundamental ongoing, cyclical evaluation of, and reflection on how an organization has learnt previously, both in terms of its effectiveness and change of norms.

On this account the school will be transformed to the extent that it manages to learn how to learn. And this can be accomplished best if the organizational design takes account of human fallible learning and cognition, and secondly, builds in feedback mechanisms to maximize the potential for error detection, and capacity to learn. If one takes an organizational learning point of view, then Heads of Department, for example, denoting a central structural feature of secondary schools, are important agents for organizational learning in terms of initiating both single- and double-loop learning.

But, finally, what of the transformational leader or principal? Is there a role left for such an individual? Principals, whether transactional or transformational, are in positions of formal power and can thus determine what is to happen in their schools. Here we need to remember that the holding of formal power does not equate with being epistemically privileged.

Conclusion

The conception of leadership, as currently debated in both its transactional and transformational guise, raises a number of difficulties. The particular concern of our argument was to point out that although leadership theorists such as Bass and Avolio direct their attention to a real and critically important social phenomenon, i.e. *the creation and execution of effective administrative practice under conditions of uncertainty*, their epistemological framework, including their concomitant theory of learning and cognition, is to narrow to explain the phenomenon under discussion. The claims made for both forms of leadership outrun the theoretical machinery available to support them. On a naturalist account of human learning as presently developed by the neurosciences, the kind of cognitive activity exemplified in being an effective practitioner/leader is a form of pattern recognition that is not language-based. This account also raises methodological difficulties for the study of leadership since quantitative-statistical methods are premised on the assumption of the validity of the sentential view. The naturalist view of learning and cognition, in acknowledging the fallibility of all knowledge, has implications for organizational design since it is more efficient to create feedback mechanisms for error correction rather than emphasize error prevention at the top. More participative structures that incorporate the knowledge of all may yet make the best sense in terms of facing the challenges posed by an inherently uncertain and unpredictable future.

References

Argyris C. and Schön D. A. (1984). Organizational learning, in D.S. Pugh (ed.) *Organization Theory*. (Harmondsworth: Penguin).

Avolio B.J. and Bass B.M. (1988). Transformation leadership, charisma, and beyond, in J.G. Hunt,

B.R. Baliga, H.P. Dachler and C.A. Schriesheim (eds.) *Emerging Leadership Vistas*. (Lexington, Mass.: Lexington).
Bass B.M. (1985). *Leadership and Performance Beyond Expectation*. (New York: Free Press).
Bass B.M. (1990). *Bass and Stodgill's Handbook of Leadership*. (New York: Free Press, third edition).
Bass B.M. and Avolio B.J. (1990). The implications of transactional and transformational leadership for individual, team, and organizational development, in R.W. Woodman and W.A. Passmore (eds.) *Research in Organizational Change and Development*, Vol. 4. (Greenwich: JAI Press).
Bass B.M. and Avolio B.J. (1993). Transformational leadership: a response to critics, in M.M. Chemers and R. Ayman (eds.) *Leadership Theory and Research: Perspectives and Directions*. (San Diego: Academic Press).
Bass B.M. and Avolio B.J. (eds.) (1994). *Improving Organizational Effectiveness through Transformational Leadership*. (Thousand Oaks: Sage).
Beare H., Caldwell B.J., and Millikan R.H. (1989). *Creating an Excellent School*. (London and New York: Routledge).
Burns J.M. (1978). *Leadership*. (New York: Harper and Row).
Caldwell B.J. (1992). The Principal as leader of the self-managing school in Australia, *Journal of Educational Administration*, **30**(3), pp. 6–20.
Churchland P.M. (1988). *Matter and Consciousness*. (Cambridge, Mass.: The MIT Press, revised edition).
Churchland P.M. (1989). *A Neurocomputational Perspective: The Nature of Mind and the Structure of Science*. (Cambridge, Mass.: The MIT Press).
Churchland P.S. (1980). Language, thought, and information processing, *Nous*, **14**, pp. 147–170.
Churchland P.S. (1983). Consciousness: transmutation of a concept, *Pacific Philosophical Quarterly*, **64**, pp. 80–95.
Churchland P.S. (1986). *Neurophilosophy: Toward a Unified Science of the Mind/Brain*. (Cambridge, Mass.: The MIT Press).
Conger J.A., Kanungo R.N., and associates (1988). *Charismatic Leadership*. (San Francisco, London: Jossey-Bass).
Evers C.W. and Lakomski G. (1991). *Knowing Educational Administration: Contemporary Methodological Controversies in Educational Administration Research*. (Oxford: Pergamon Press).
Evers C.W. and Lakomski G. (1995). Science in educational administration: a postpositivist conception, Invited Address, Division A (Administration) *American Educational Research Association Annual Meeting, San Francisco, 18 April 1995*.
Gronn P.C. (1994). The elixir of leadership, in F. Crowther, B.J. Caldwell, J.D. Chapman, G. Lakomski and D. Oglivie (eds.) *The Workplace in Education. 1994 Yearbook of the Australian Council of Educational Administration*. (Melbourne: Australian Council for Educational Administration).
Gronn P.C. (1995). Greatness re-visited: the current obsession with transformational leadership, *Leading and Managing*, **1**(1), pp. 14–28.
Hallinger P. (1992). The evolving role of American principals: from managerial to instructional to transformational leaders, *Journal of Educational Administration*, **30**(3), pp. 35–49.
Hallinger P. and Murphy J. (1992). *The Evolving Leadership Role of the Principal: International Perspectives*. Special Issue, *Journal of Educational Administration*, **30**(3), pp. 3–88.
House R.J. (1977). A 1976 theory of charismatic leadership, in J.G. Hunt and L.L. Larson (eds.) *Leadership: The Cutting Edge*. (Carbondale: Southern Illinois University Press).
Howe W. (1994). Leadership in educational administration, in T. Husen and T. N. Postlethwaite (eds.) *International Encyclopedia of Education: Research and Studies*. (Oxford: Pergamon Press second edition).
Hoy W.K. and Miskel C.G. (1991). *Educational Administration: Theory, Research and Practice*. (New York: Random House, fourth edition).
Hunt J.G. (1991). *Leadership — A New Synthesis*. (Newbury Park: Sage).
Leithwood K. (1992). The move toward transformational leadership, *Educational Leadership*, **49**(5), pp. 8–12.
Leithwood K. (1994). Leadership for school restructuring, *Educational Administration Quarterly*, **30**(4), pp. 498–518.
Leithwood K. and Jantzi D. (1990). Transformational leadership: how principals can help reform school cultures, *School Effectiveness and Improvement*, **1**(4), pp. 249–280.
Leithwood K., Begley P.T., and Cousins J.B. (1994). *Developing Expert Leadership for Future Schools*. (London, Washington, DC: The Falmer Press).

Leithwood K. and Steinbach R. (1995). *Expert Problem Solving — Evidence from School and District Leaders*. (Albany: State University of New York Press).

Lepsius M.R. (1986). Charismatic leadership: Max Weber's model and its application to the rule of Hitler, in C.F. Graumann, and S. Moscovici, (1986) (eds.) *Changing Conceptions of Leadership*. (New York, Berlin: Springer-Verlag).

Lord R.G., Foti F.J. and Devader C.L. (1984). A test of leadership categorization theory, internal structure, information processing, and leadership perceptions, *Organizational Behavior and Human Performance*, **34**, pp. 343-378.

Murphy J. and Hallinger P. (1992). The principalship in an era of transformation, *Journal of Educational Administration*, **30**(3), pp. 77-88.

Nisbett R. E. and Wilson T. D. (1977). Telling more than we can know: verbal reports on mental processes. *Psychological Review*, **84**(3), pp. 231-259.

Robinson V. (1995). Organizational learning as organizational problem-solving, *Leading and Managing*, **1**(1), pp. 63-79.

Rowan B. (1990). Commitment and control: alternative strategies for the organizational design of schools, in *Review of Research in Education*, **16**, pp. 353-392.

Sejnowski T. J., Koch Ch., and Churchland P. S. (1988). Computational neuroscience. *Science*, **241**, pp. 1299-1306.

Smith P.B. and Peterson M.F. (1988). *Leadership, Organizations and Culture*. (London: Sage).

Weber M. (1978). *Economy and Society: An Outline of Interpretive Sociology*. Vol. 1, G. Roth and C. Wittich (eds.) (Berkeley: University of California Press).

Yukl G.A. (1989). *Leadership in Organizations*. (Englewood Cliffs: Prentice-Hall, fourth edition).

7
Policy Analysis: Practical Reason or Empirical Science?

Methodological concerns have played a central role in public and educational policy analysis for as long as these disciplines have existed as discrete fields of inquiry and professional practices (Boyd 1988, Garson 1986, Mitchell 1984). Amongst the many problems still awaiting solutions none is more persistent than that of rationally selecting 'best policy' under conditions of imperfect knowledge and uncertainty. Traditionally, Laswellian policy science (Lerner and Lasswell 1951), or the *synoptic tradition*, sought to attack the problem by taking a global, historical view of the policy context, and to study the conditions of social change with systems analysis as its preferred theoretical framework and a narrow empiricism as its methodology (Garson 1986, p. 538). While equally mindful of the complexity of social life, defenders of the *anti-synoptic* tradition argued for an incremental, piecemeal approach to policy analysis (Braybrooke and Lindblom 1963), defended pluralism as its theoretical framework, and emphasized case and contextual studies as appropriate methodologies. The major concern of *anti-synoptic* writers was, and continues to be, however the alleged value-neutrality of Laswellian policy science, a concern that has subsequently led them to question the status of the field as a science, as well as the role of policy analysis in a democratic society.

One consequence of this development is that some analysts, motivated in part by reports of the negligible and even negative impact of policy research (e.g. Dunn, Mitroff and Deutsch) simply declared the science question dead: '*Resolved*: Policy analysis is not a science, is not scientific; indeed, scientific status is an inappropriate goal for policy analysis' (Landsbergen and Bozeman 1987, p. 625). Another consequence of the currently prevailing anti-science mood is the rise of alternative models of policy analysis that align themselves directly with postpositivist, interpretive social science (e.g. Callahan and Jennings 1983, Jennings 1983, 1987). Specifically, these *anti-synoptic* writers argue for the inclusion of political and ethical values, as well as the social context of policy analysis generally (e.g. Fischer and Forester 1987, Dunn 1983, Wildavsky 1979, MacRae 1976) and advocate that naturalistic approaches be adopted as the most appropriate to deal with these issues (e.g. Lincoln and Guba 1986). Combining many of the concerns of both the older *anti-synoptic* tradition as well as its contemporary interpretive expression is the policy analysis model suggested by William N. Dunn.

Dunn's transactional model of argument (Dunn 1981, 1982), based on Toulmin's practical logic (Toulmin 1958), is a recent and explicit attempt at providing a procedural model for rationally enabling policy choice. It represents a special case of the methodological paradigm that largely characterizes the newer approaches to policy analysis, that of practical reason with its emphasis on rhetoric, persuasion, and legal reasoning. Since policy makers and analysts are meaning constructing agents, so the argument goes, it is the epistemological feature of intentionality that provides the real grounds to understand social complexity and thus facilitates superior policy making. Unlike the 'statistical empiricism' of Laswellian policy science, the alternative methodological proposal of *practical reason* is believed to be antithetical to empirical scientific methodology and to lend itself neither to the formulation of laws nor to causal explanation. It is normative, not descriptive, more concerned with the adequacy and cogency of policy arguments than their truth.

But the claim of *practical reason* to represent human understanding in its essence is a matter of the soundness of the epistemology underwriting it. Following Quine's (1969) (see also Quine and Ullian 1978) 'Epistemology naturalized' argument, what makes for the soundness of an epistemology is that it be learnable; that is, its embedded theory of the mind must present an adequate account of human cognition and learning. An epistemology, then, is only as valid as its implied theory of learning, and there is no principled distinction to be made between the former and the latter. Epistemology is continuous with natural science (see Walker and Evers 1982, who introduced this argument into [philosophy of] education; Evers 1987, Walker and Evers 1988). In the present context, this means that the account of learning embedded in *practical reason* must be able to explain how we have come to know about such abstract objects as purposes and beliefs, which are taken to function in the explanation of behaviour.

It is the purpose of this chapter to argue that the epistemology underlying Dunn's procedural model, the classical *Justified True Belief* (JTB) account, is unsound by way of arguing that its implied view of learning and cognition, characterized by the 'folk psychological' (Stich 1983) propositional attitudes, is mistaken. Showing this requires demonstrating first that folk psychology is an empirical theory, and that it subsequently can be assessed in the manner of all theories (P. M. Churchland 1981, 1985, 1988a, 1989a, b, P. S. Churchland 1986). When evaluated as such, folk psychology's view of knowledge turns out to be sentential. But according to the neurosciences, whose business it is to explain brain functioning, there is no support for the assumption that the brain functions primarily as a linear 'sentence-cruncher', to use Patricia Churchland's colourful phrase (P. S. Churchland 1986). The consequence for Dunn's model is, then, that policy choice as an example of intelligent, rational action has little to do with overt linguistic behaviour, regardless of the form this takes. If so, it would seem advisable to hold the folk psychological categories of belief and desire at arms length and concentrate on how we in fact process information as a precondition for understanding policy choice as a specific example of cognitive action.

Policy Analysis As Reasoned Argument

In his introduction to the 'Symposium on Social Values and Public Policy' Dunn (1980–81) states that the policy sciences need to be based on a perspective that is neither value-neutral nor value-committed. The policy sciences must not be value-neutral since value-neutrality would deprive policy of political significance and take it back to the canons of positivism and the 'empirico-analytic sciences'. To be value-committed, on the other hand, is equally undesirable because this would deny the policy sciences of 'reasoned ethical discourse', which is Dunn's central concern. In his view, policy knowledge can only grow through open critical discourse which, in turn, presupposes rules or standards of assessment that enable the participants in policy-making to examine rival ethical claims without anyone thus dominating the outcome (Dunn 1980–81, p. 519). These rules or standards are spelt out in the transactional model of argument, which is Dunn's proposal for operationalising *practical reason* directed towards establishing the adequacy and cogency of claims to knowledge rather than their truth (Dunn 1981, 1982). Following Toulmin, there are no context-independent criteria for judging the merits of an argument, and this is as it should be, according to Dunn, given that policy analysis, as a practical science, is carried out in specific contexts.

Dunn offers his model as a reply and an alternative to the kind of social-scientific experimentation advocated by D. T. Campbell (1969) in his 'experimenting society'. Given that reforms are 'symbolically mediated and purposive social processes', Dunn believes they are more akin to arguments than to quasi-experimentation, or scientific experimentation generally, which directs questions to nature directly. The strength of the *transactional model*, as Dunn sees it, consists in broadening the range of standards thought suitable for challenging and assessing knowledge claims, including a number of specific tests that help determine their adequacy, cogency and relevance. These tests also have the function of 'plausible rival hypotheses' to the knowledge claims under examination. The model that best represents the way in which agents settle competing claims in social contexts is that of jurisprudential reasoning. Its standards include, amongst others, rules for making valid causal inferences. As a consequence, and unlike the replication of experiments, argumentation leads 'toward a pragmatic and dialectical conception of truth ... Knowledge is no longer based on deductive certainty or empirical correspondence, but on the relative adequacy of knowledge claims which are embedded in ongoing social processes' (Dunn 1982, p. 94).

Of central importance in the *transactional model* is the distinction between analytic and substantial arguments, which Dunn takes over from Toulmin (1958). Toulmin argues that formal logic is too narrow to be of any use in the practical assessment of arguments. Our claims to knowledge are sound when our supporting arguments are adequate. What is to count as an adequate argument is dependent upon the field from which it is taken: 'validity is an intra-field, not an inter-field notion' (Toulmin 1958, p. 255).

The merits of a procedural schema, based on the above described assumptions and proposed to guarantee rational policy choice if followed conscientiously, are said to be such that it allows (1) for the visual representation of the systematic structures of competing arguments; (2) for the critical analysis of the frames of reference or ideologies of those contesting knowledge claims; and (3) for the development of 'truth' and 'utility' tests by means of which knowledge claims can be tested. Finally, and most importantly, it allows for democratic knowledge creation via the critical exchange and challenge of knowledge claims in communicative action. The model is composed of six elements, which are data (D), claim (C), warrant (W), backing (B), rebuttal (R), and qualifier (Q). The first three elements parallel those of the classical syllogism, to be supplemented by the second set. Backing (B) denotes additional data, claims or arguments introduced in case the warrant is in doubt (Dunn 1982, p. 96). Warrants provide reasons for the acceptance of a claim (Dunn 1981, p. 42). Rebuttal (R) specifies conditions under which the adequacy or cogency of a knowledge claim can be challenged. Policy claims and rebuttals together 'form the substance of policy issues, that is, disagreements among different segments of the community about alternative courses of government action' (Dunn 1981, p. 42). Qualifier (Q) serves to indicate the degree of cogency or force of a claim.

Following the definition suggested by Weiss and Bucuvalas (1980), Dunn (1982) considers truth tests to be 'decision points concerning evidence; the grounds for accepting or rejecting truth claims include ... empirical as well as formal rational tests' (p. 100). Relevance or utility tests, on the other hand, 'are decision points concerning the delineation of an appropriate domain of inquiry or action', a definition Dunn expands to include 'the explicit or implicit purposes of knowledge claimants or their challengers' (p. 101). Truth tests that appraise the adequacy of a knowledge claim and challenge its causal assumptions are considered more problematic. Dunn suggests a classification of truth tests that takes account of alternative modes of explanation (von Wright), different knowledge-constitutive interests (Habermas), and competing standards for assessing ethical claims (MacRae). Specifically, Dunn (1982, p. 105) proposes that there should be different truth tests for different 'knowledge transactions', presumably based on the assumption that there are different kinds of knowledge to be transacted which have their corresponding purposes. These are empirico-analytic, concerned with logical consistency of laws, etc. and/or their correspondence to observed regularities; interpretive, directed toward human purposes, reasons, and motives; pragmatic denoting effective past action; authoritative, relating to the well-established status of those producing knowledge as well as its general acceptance, or the use of approved methods; lastly, critical in the Habermasian sense of liberating human actors from unexamined doctrines. According to Dunn, we learn what counts as an adequate argument under specific conditions in specific professions, knowledge is fallible and corrigible. This is quite in keeping with Toulmin's epistemological prescription to ignore scepticism and moderate our ambitions (Toulmin 1958, p. 248).

Applying the Procedure

It is easy to see why Toulmin's applied logic seems attractive to policy writers such as Dunn (and also Mason and Mitroff 1980–81) since here we appear to have a sound rationale for including the ordinary arguments of practitioners and those of the policy experts and scientists, thus satisfying the democratic ideal central to the *anti-synoptic* tradition. The model is all the more attractive since it is also combined with an assurance that the *transactional model* is not only context-sensitive, but also value-critical and intersubjectively valid. That is, it avoids the problems associated with straight-out partisan approaches without having to compromise on the right (democratic) values. Furthermore, we are also presented with a formal set of procedures by means of which we could settle any kind of claim or argument and thus guarantee policy choice. But can we?

Dunn acknowledges that stakeholders, subject to their different worldviews, ideologies, and frames of reference, are bound to challenge the presuppositions used to back up a warrant, that is, the reasons provided for a claim. He does not consider such challenges a disadvantage, but points out that in the empirico-analytic and hermeneutic sciences such questioning is not even possible, and it is this special feature that makes his model unique. While the potential for publicly challenging knowledge claims may be considered an advantage (accepting for the sake of the argument Dunn's claim regarding the alleged inability of both natural science and hermeneutics to challenge assumptions), whether or not it is of value depends on the model's ability to settle the challenge, to determine what is to count as the more adequate argument, given its own theoretical resources. Since policy analysis in Dunn's view aims *'to produce and transform policy-relevant information that may be utilized in political settings to resolve policy problems'* (Dunn 1981, p. 35) the model's claims stand or fall with the accuracy of its account of knowledge production and the practical performance it is said to guarantee.

Some problems are readily apparent. It is a commonplace to note that in the empirical world power, learning, and linguistic differences are distributed unevenly, disadvantaging some from the outset, and that gender, race, as well as age play a significant role in who gets to talk in social situations, and who is listened to, and who is included in the deliberation process in the first place. In addition, the implied assumption that all are committed to rational debate, that all share the same willingness to declare their ideologies or standpoints honestly, and — by implication — that none are thereby disadvantaged, is quite unrealistic.

For his model to work, Dunn must presume that power and inequality are suspended in order for 'good will' to prevail; that our standards and levels of learning and policy relevant knowledge are even; that we are all equally capable of putting our points of view; and that gender differences, race, and age are irrelevant; in other words, that we live in a perfectly rational world. Here Dunn's model resembles Habermas's ideal speech situation, which suffers from similar problems (Lakomski 1988). Central to the assumption of perfect rationality is the belief that human agents are in fact capable of 'knowing their own minds': their beliefs, thoughts, and intentions, which comprise self-knowledge, and secondly, that this self-knowledge is fundamentally linguistic. Only then is it reasonable to assume that rational deliberation could work at all. But before we consider the relation between beliefs and behaviour directly, let us first examine whether the decision procedure suggested by Dunn permits a policy choice to be made following its own prescriptions. Consider the following example.

According to recent research (Gill 1988), girls achieve better academically when they are in single-sex rather than co-educational classrooms (D). Therefore we might conclude that it is definitely better (Q) to educate girls in single-sex classrooms or even schools (C). Our warrant or reason for (C) is that the separation of boys and girls for purposes of academic instruction caused higher academic achievements for girls than had been the case when they were taught in co-educational settings. We could further back our warrant by saying that equality of opportunity is a fundamental principle that demands single-sex classrooms in the interests of academic achievement of females. We could further argue that whatever increases the academic achievements of *any* underprivileged group ought to be done, females being one such group. Recall Dunn's point that the empirical evidence is rarely in doubt, but that the underlying reasons and assumptions are challengeable. Suppose, then, that a group of parents opposes the claim that it is better to educate girls in single-sex classrooms on the assumption that short-term gains (higher academic achievements) do not outweigh longer-term social consequences such as presumed inadequate socialization. The parent group thus advocates co-education, rather than single-sex education, on the assumption that learning together is more important in terms of ultimately overcoming sexism than is segregation, even if such segregation yields better academic performance for the girls in the short term. We thus have two competing claims and must choose between competing policy prescriptions. How do we settle the conflict?

On Dunn's own account, the relevance and cogency of any person's or group's knowledge claim is always context-dependent. In the present example we have two different 'contexts' in terms of two different feminist theories. Following the logic of Dunn's model, what is relevant and cogent to a separatist feminist policy maker may not at all be relevant and cogent to a parent of a different feminist persuasion who believes that separatism is a mistaken strategy for overcoming sexism and its practices. On their own accounts, each theory is equally valid. But since Dunn is interested in producing and transforming *'policy-relevant information that may be utilized in political settings to resolve policy problems'* the question of which claim is the more 'adequate' must be settled. This is to be done by truth tests, by introducing 'different sets of assumptions and underlying presuppositions' (Dunn 1982, p. 103) that might challenge a claim's causal assumptions. But truth tests themselves have, as we saw earlier, their own relevant contexts. These are ultimately the different kinds of knowledge presumed to underlie the different purposes for 'knowledge transactions'. But if truth tests are thus tied to their own respective spheres of influence then, by definition, there are infinitely many, equally true arguments or theories. But since not all claims can be true, and if we cannot decide between true and false claims, we cannot decide the status of any claim. Knowledge is impossible and we end up with incommensurable theories. For if competing arguments, as well as relevance and cogency tests, are dependent on their contexts, and if there are specific truth tests for specific types of 'knowledge transaction' based on different kinds of knowledge then the standards of appraisal cannot possibly serve to arbitrate between competing claims. Both arguments and standards of appraisal are based on their respective knowledge foundations, and there is in principle no difference between the arguments to be assessed and that which is supposed to be doing the assessing. Not only can competing claims not be settled since each is restricted by its own epistemic sphere of influence (the principle of 'intra-field validity'), but claims within the same epistemic community face difficulties in terms of the support or evidence they themselves can marshall since whatever warrant or backing may be brought up is only as valid as the presumed conception of knowledge underwriting them, and *its* truth or validity cannot be presumed a priori. This means in the present example that no amount of new reasons or arguments offered, either for or against the above feminist claims, could in principle settle the issue of which educational policy to adopt. As a consequence, the *transactional model of argument* makes policy deliberation both relativist and arbitrary. Given the example presented here, efforts to overcome sexism and its practices become a matter of subjective preference, which can hardly rate as a satisfactory policy prescription. The inability to make a choice is the direct result of the model's epistemological assumptions, which are made more explicit in the following section.

The Regress of Reasons

As a procedure of justification, Dunn's model is faced with the threat of an infinite regress of reasons, characteristic of the JTB account of knowledge (see Armstrong 1981, Williams 1977, 1980, p. 5), which in Evers and Lakomski's (1991) definition is expressed as follows:

> Person x knows that p, where p is some particular claim to knowledge, if and only if
> (i) p is true

(ii) x believes that p, and
(iii) x is justified in believing that p for the reason q.

Displaying the justificatory structure of JTB shows how neatly Dunn's decision model parallels it. Of particular significance in the *transactional model*, as well as the JTB account, is condition (iii), because for q to be a justifying reason, it must itself be an item of knowledge, just as backings and warrants must be to support an argument. But if q is already an item of knowledge then JTB's definition is circular since it already contains an appeal to knowledge in the definiens. But more importantly, an infinite regress threatens since q can only be known by virtue of knowing something else. Faced with the problem of circularity as well as the infinite regress, the solution has been sought in the distinction between derived and immediate knowledge. JTB, then, is an account of derived knowledge, of claims known in virtue of being implied by further knowledge. The chain of implications is not infinite, however, as it stops at knowledge that is immediate or underived. It follows on this view that all our derived knowledge must rest on, or be derivable from, some foundation of immediate knowledge, traditionally the so-called sense-data, first-person sensory reports, and observation statements. Dunn's acceptance of a plurality of foundations is a variation on the theme that knowledge is believed to be in need of foundations, albeit different ones. This multi-foundationalist view has strong similarities with the paradigms view (see Walker and Evers 1988), although Dunn does not employ that concept.

In line with many interpretivist social scientists and philosophers who followed Kuhn's and Feyerabend's lead in raising decisive objections against logical empiricism, Dunn accepts that (social) reality is a matter of interpretation (with physical reality being a matter of direct observation), and that (social) theories are always underdetermined by data or evidence (Dunn 1982, p. 104). As a consequence, he also accepts that no theory can ever be justified evidentially, which leads to the conclusion that all theories are equally plausible or acceptable. There is thus no such thing as a true theory in the social sciences, and objectivity becomes unobtainable. But this conclusion only follows if evidence equals empirical evidence only. In addition to the fact that theories are always underdetermined by the available evidence, we also know that what is taken as empirical evidence is always theory-laden. So in comparing theories we cannot simply rely on empirical evidence as the neutral arbiter, as our 'touchstone' (Walker and Evers 1988). Since we are now in the business of actually comparing theories (observation being as theoretical as anything else), and since we know what makes for better or worse theories, the solution is that we use their traditional virtues, i.e. relative coherence, simplicity, and explanatory unity—Paul Churchland's (1985) 'superempirical virtues'—to argue for a global assessment of theories that includes empirical evidence as one criterion amongst several. On this account, then, the best theory is one that coheres best with our interpreted experience (see Evers and Lakomski 1991, pp. 37–44, BonJour 1985, Williams 1980). Additionally, Dunn's model relies on a theory of meaning in which terms are entirely determined by their embedding conceptual framework, an impossible position to hold. For if meaning were entirely determined by its conceptual role then we could never have learnt either of the feminist theories mentioned above. Nor could we have learnt about such abstract objects as intentions, thoughts, and reasons. Theories are complex networks of sentences which we do not learn all at once. We must have begun by learning simple terms or expressions first, but in order for us to understand these and their role in the theory, knowledge of the whole theory is already presupposed or we would not be able to assign them the function we do. Here we see that JTB promises more than it can deliver given the theoretical tools it has as its disposal. In presuming learning capacities — a theory of the mind — which postulate antecedents to learning that are themselves unlearnable, it comes out unknowable on its own account: it fails to be self-referential.

The Theoretical Character of Belief and Desire

In specifying conditions for claims to count as knowledge, an epistemology implicitly presumes a theory of the mind. What we can know depends crucially on the requisite perceptual and cognitive capacities that we developed as a species. And we have to learn from infancy onwards what these capacities are. We do not just 'have' observational judgments — they have to be learnt beginning with the prior learning of making complex perceptual discriminations, as Campbell (1974) and Popper (1981), for instance, have long argued. We also have to learn the requisite linguistic or propositional system within which we constitute our beliefs, desires, and thoughts. Since we first have to acquire that, it cannot itself be the medium of learning, leading to the conclusion that there has to be a type of learning prior to that involving the manipulation of sentences (P. M. Churchland 1989b, p. 155). And here our commonsense understanding of ourselves fails to provide an answer. In order to demonstrate where folk psychology falls short, it is necessary to make explicit what has been implied in the preceding discussion, namely, that it is an empirical theory, and one that — on the current evidence — might need to be replaced. Here the Churchlands' work provides the most telling arguments.

The propositional attitudes, so-called because they express a distinct attitude toward a specific proposition (P. M. Churchland 1988b, p. 63), i.e. *the thought that* [gum trees are lovely]; *the belief that* [humans act rationally]; *the fear that* [the Middle East crisis will turn into war] have long been taken to play a causal role in human behaviour and are central to folk psychology. Human conscious intelligence, according to folk psychology, is said to consist of the rational manipulation of these propositions by means of deductive inference. But rather than being the 'sort of super-causal *logical* relation' (P. M. Churchland 1988a, p. 213) they are alleged to be, the propositional attitudes display relationships typical of all theoretical explanations, as can be seen by comparing them to the logical pattern formed by physical laws. If the latter can be characterized by the 'numerical attitudes', both sets can be described as follows (P. M. Churchland 1988b, p. 64, 1981, pp. 70–71):

Propositional attitudes	Numerical attitudes
... believes that p	... has a length$_m$ of n
... suspects that p	... has a kinetic energy$_j$ of n

Either 'attitude' can be completed by putting a proposition in place of p or a number in place of n respectively. Only then do we have a determinate predicate. The logical relations holding between numbers and numerical attitudes also hold between propositions and propositional attitudes. Most importantly, where these relations hold universally, we are in a position to state laws — for the latter set of relations as much as for the former, numerical ones. In other words, we utilize the abstract relations that hold in the domain of certain abstract objects such as numbers, vectors, or propositions in order to help us state the empirical regularities between real states and objects, and that includes those between various kinds of mental states. Summing up, the full-blooded intentional idiom, contrary to popular opinion, possesses the same complex logical structure as the rest of our scientific theories.

An important consequence of folk psychology's empirical character is that it might actually be false. This possibility, however, seems quite counterintuitive because we appear to get by rather well by relying on its categories in terms of explaining and predicting everyday behaviour with a fair degree of success. Even if folk psychology were false, would it matter for our practices, including those that comprise 'policy analysis'? And how would we then explain our behaviour? These are difficult questions which cannot be answered at present. But to see that they are real, and that much depends on the answers, let us consider some of the fundamental problems for folk psychology that lend support to the claim that it might be a false empirical theory.

Problems for The Sentential Paradigm

In the following section I shall briefly note two major problems of the view that our cognitive activity consists in the manipulation of sentences. (For the original definition of the sentential view see P. S. Churchland 1980, 1986. See also I. Hacking's 1975 account of the reasons for 'lingualism's' favoured philosophical status). The first is what Stich (1983, p. 214) and P. S. Churchland (1986, pp. 388 ff.) call the 'infralinguistic catastrophe': much intelligent behaviour is displayed by organisms who do not have any overtly linguistic capacity, and that includes behaviour by human infants, adults suffering from left-hemisphere lesions, as well as deaf mutes (see P. S. Churchland 1983). To explain this, one could of course argue that this behaviour is not really 'cognitive' in the required sense. But this would beg the question since it is already assumed that intelligent behaviour is identical with linguistic behaviour. Secondly, while granting that the above is intelligent behaviour although it does not show itself linguistically, one prominent suggestion maintains that it is based on a 'thought language', Mentalese, in which organisms devoid of overt linguistic behaviour reason and solve problems (Fodor 1975). The assumption that organisms possess such a language of thought, which, according to Fodor, is a proper language with all relevant characteristics seems far-fetched, and its existence in principle impossible to ascertain since it is radically unlike the linguistic capacity we display as adults. Its categories can be reached *only* via those of our language, making the point of independently establishing them moot. (For a fuller discussion of Mentalese see P. S. Churchland 1980; see also Stich 1983, pp. 187–197.) But the more interesting and important issue here is that of how infants learn our language if Mentalese is assumed to be the template. If a child acquires, say, German by matching a German word with the appropriate word in Mentalese, the child can only learn those German words for which there are words in Mentalese. If matching is the process then one must conclude that the child does not in fact learn any new concepts at all in addition to those pre-existing in Mentalese. More absurd still is the further implication that all the new concepts gained through scientific discoveries must have already been present in Mentalese from the beginning, terms such as quarks, electrons, atoms, neurons (P. S. Churchland 1980) and what have you. The existence of a 'language of thought', then, would seem to stretch the bounds of credibility too far. Given these problems that have been noted albeit briefly, it is more reasonable to assume that cognitive activity does not equal 'sentence-crunching', an assumption that *inter alia*, enjoys the theoretical virtue of simplicity.

The second problem for the sentential view relates to that of accessing knowledge. For an organism to survive, it is essential that it have speedy access to knowledge relevant to respond to 'the four F's: feeding, fleeing, fighting, and reproducing' (P. S. Churchland 1987, p. 548). This goes as much for the cat's recognition of a mouse as food as it does for a driver's recognizing a red light as signalling 'stop'. Essential in either situation is that the relevant stimuli are recognized instantly. But how does the relevant information processing system know which bits of information to call up? What is its mechanism of sorting so that the correct response follows? The problem is far from innocent, notwithstanding the fact that we manage to respond by and large correctly and instantly to such stimuli in our everyday commerce with the world. Since this is so, calling up our mental store of sentences/beliefs in order to find those that may apply in a specific situation, and systematically eliminating those that do not by the relevant logical rules, would present a sure recipe for evolutionary disaster since such sorting would require a lot of time. Besides, the sentential/belief view presumes that we do have immediate access to our mental states, an assumption not born out by research in social and cognitive psychology (for relevant research see Nisbett and Wilson 1977). Borrowing N. R. Hanson's 'there is more to seeing than meets the eye-ball', we may now say that there is more to cognition than is expressed in sentences. But if the structures of knowledge are not sentences, then what are they? And if the propositional attitudes do not really explain what goes on inside our heads, and hence explain our behaviour, what kind of an explanation of behaviour could we possibly advance?

Neural Networks and Nonsentential Representation

The beginnings of an answer to both questions are suggested by computational neuroscience (for a brief overview see Sejnowski, Koch, and Churchland 1988) in what is termed 'Connectionism': the mind/brain's capacity for the parallel distributed processing of information, or PDP models of brain functioning (for an introduction see P. M. Churchland 1988b, ch. 7, P. S. Churchland 1986, ch. 10; a full account is Rumelhart and McClelland 1986). Secondly, given the PDP account, a most promising alternative to the folk psychological sentential explanation of behaviour is the *prototype activation model* developed by Paul Churchland (P. M. Churchland 1989a). In its simplest form this means that instead of calling up endless lists of sentences, i.e. rules or laws, the mind/brain 'calls up' a *prototype* of some cognitive situation (P. M. Churchland 1988a, pp. 217–218). Churchland acknowledges that the idea of prototypes is neither new nor uncontroversial. In the present context, it poses the particularly difficult problem of how it is represented in cognitive creatures such as ourselves if it is not represented linguistically. Here recent research into the functional properties of neural networks provides an answer. These are artificial networks that have been constructed to simulate essential features of the neuronal organization of the brain. What is so remarkable about them is that they have been extremely successful in learning assigned tasks such as differentiating mine from rock echoes, recognizing complex visual features, and transforming written text into speech, NETtalk, to mention just some (P. M. Churchland 1988b, pp. 156 ff.).

A simple network has three specific architectural features. It possesses (1) (neuron-like) processing units, (2) connections between these units, and (3) connection weights which are the differential strengths of connections between the processing units (P. S. Churchland 1987, pp. 550–553, P. M. Churchland 1988b, pp. 156–165). The means of communication between the processing units are signals such as (neuronal) firing rate that are numerical rather than symbolic. The bottom, input, layer of units are something like sensory units directly receptive to environmental input; then follows an intervening layer, the so-called 'hidden units', and finally the output layer. (In real brains there are between 5 and 50 intervening layers.) Each bottom unit emits an output through its own 'axon'. The strength of this output is a function of the unit's level of stimulation. Each axon branches out into a number of terminal branches and sends a copy of that output to each and every 'hidden unit'. The bottom units thus make a variety of synaptic connections with each of the intervening units. The strength each connection possesses is called its *weight*. What happens is that the set of activity levels (input vector) induced by stimulating the input units is transported upward to the hidden units, changing in the process by the influence of the output function of the bottom cells, the pre-existing pattern of synaptic weights, and the summing activity within each of the hidden units. A pattern of activations is thus produced at both the input level and another one at the hidden unit level. What kind of pattern results from this activity, for a given input, depends entirely on the configuration of synaptic weights which hit the hidden units (P. M. Churchland 1989a). The processes characterizing activity from bottom to hidden layer is repeated from hidden to top layer. The output vector, then, similarly is the result of the activation pattern generated by the hidden units. This description makes it clear that the most important characteristic of networks is their complete *interconnectivity*. It is possible to construct model networks with whatever number of input units, hidden units, and output units we wish. And we can also begin to appreciate the processing power embedded in the (modest) two-tier arrangement. Most important of all: the synaptic weights in the overall system can be modified in order to obtain the vector-to-vector transformation we want, networks can be trained to learn, as the example of Rosenberg and Sejnowski's (1987) NETtalk shows. The most remarkable thing to remember is that networks are not given any rules, laws, or generalizations prior to or during the learning process. The crucial factor in learning is the value of the synaptic connection weights which, in turn, are determined by a learning algorithm called 'back-propagation of error' or the *generalized delta rule* (P. M. Churchland 1988b, p. 159).

Basically, the strategy exploited by this algorithm is 'the calculated error between the *actual* values of the processing units in the output layer and the *desired* values, which are provided by a training signal.' (P. S. Churchland 1987, p. 551) The error signal of the output layer is fed back to the input layer and is used to adjust each weight in the network. This process is done countless times (via programming a conventional computer to act as 'teacher') by feeding the network with diverse examples of Fs, for instance, and by the network looking for a configuration of weights that will turn the neurons at the hidden level into a set of complex feature detectors to which the output units will respond in their turn. In being thus trained up the network slowly learns as the weights are adjusted after each back propagation of the previous error signal. Learning thus consists in minimizing the mean squared error over the training set of words, or Fs, or whatever else was fed into the system. In this manner, the network actually generates a set of internal representations for the relevant features to be recognized.

Policy Analysis As Naturalized Science

The most important results of studying the functions of artificial neural networks and PDP for our purposes are (1) that learning is not a matter of the linear processing of symbols and rules but is rather a global, or network, affair, and (2) that consequently, representations in the hidden units are not symbols either but are neuronal patterns of activation. Although neural network modelling is relatively young, leaving many questions still to be answered, it does suggest that human understanding and the grounds for action are not to be sought in sets of stored generalizations such as the propositional attitudes but are rather located in one or more prototypes as these were earlier described.

If natural language appears to be no more than 'a surface abstraction of much richer, more generalized information processes in the cortex, a convenient condensation fed to the tongue and hand for social purposes' (Hooker 1975, p. 217), what follows for policy analysis? The first result is that reliance on folk psychological categories as explanatory of policy behaviour, since mistaken, ought to be given up. We might need to continue using them since there is as yet no other medium of representation, but this communicative function must not be confused with its purported explanatory function. Indeed, what the preceding thumbnail sketch of brain functioning does show is just why commonsense or folk psychology is as successful as it is: it allows us to recognize problematic situations instantly *because* of the work done at the sublinguistic, neuronal level, i.e. given appropriate prototypes. Another consequence is that we can concede that a good policy analyst, as has been observed many times in the literature, may well be distinguished by his or her 'insight' or 'intuition' since such expressions can now be seen as place-holders for the complex computational processes of the brain. We know indeed more than we can tell, and we are beginning to explain why this is so.

Will knowledge of causal, neurological, detail shape and change the practice of policy analysis in view of the fact that natural language is the only medium we have for communication? Initially perhaps not. We appear to have to continue using an idiom not up to the task of representing our mental processes, or representing them falsely. But what we learn from the development of neural networks, PDP and the prototype activation model is that there is an alternative, empirical account of how cognitive creatures in fact acquire knowledge of their environment. And if this is the causal story then it follows that this will eventually include policy relevant knowledge as a possibly quite specific configuration of synaptic connection weights. We learn even more specifically that such knowledge generation is global given the network character of brain functioning, and that any policy decision made is the result not of rational deliberation or *practical reason* but of the successful relevant prototype being activated. And we may comment in passing that the distinction between analytic and substantive arguments, believed to be all important by Dunn (and Toulmin), simply evaporates. It is of no relevance in the neurological story of information processing.

How does the brain learn to recognize and subsume a specific problematic situation under a relevant policy prototype? Unlike an artificial network which is trained up by a 'teacher', for us our global environment steps in as the 'teacher'. In one sense, this amounts to saying that a policy analyst, for instance, needs to gather as many and as varied policy experiences as possible (leaving unspecified for the present what such an experience might be) since successful subsumption of policy relevant information is predicated upon already existing prototypes. 'Goodness' of policy choice would then be a function of how well trained a relevant neural network is, i.e. it is a function of the richness of existing prototypes.

This raises the question of what function such sentential structures as our best scientific theories have in relation to the practice of policy analysis in view of the fact that we appear to be learning without the aid of sentential structures. The short answer is that theories, even those we judge true by our coherentist standards, do not seem necessary in order to learn a complex practice. This is not to say that true theories are irrelevant to learning, only that since networks do not partition the inputs they receive into categories such as 'true' or 'false', or 'entailment', or 'implication', they do not possess any kind of privileged position. Since, on the other hand, true scientific theories do explain the way the world is, they must have some functional importance. But this is a big question in need of research. Tentative and speculative as these suggestions have been, and counter-intuitive as the elimination of folk psychology might appear, the way forward for policy analysis lies in the more detailed, empirical work of the neurosciences since they hold the key for the explanation of all cognitive behaviour including that of deciding which policy option is best.

References

Armstrong D. M. (1981). *Belief, Truth and Knowledge.* (Cambridge: Cambridge University Press).
BonJour L. (1985). *The Structure of Empirical Knowledge.* (Cambridge, Mass.: Harvard University Press).
Boyd W. L. (1988). Policy analysis, educational policy, and management: through a glass darkly?, in N. J. Boyan (ed.) *Handbook of Research on Educational Administration.* (New York and London: Longman).
Braybrooke D. and Lindblom C. E. (1963). *A Strategy for Decision.* (London: Collier-Macmillan).
Callahan D. and Jennings B. (1983). (eds.) *Ethics, the Social Sciences, and Policy Analysis.* (New York, London: Plenum).
Campbell D. T. (1969). Reforms as experiments, *American Psychologist,* **24**, pp. 409–429.
Campbell D. T. (1974). Evolutionary epistemology, in P. A. Schilpp (ed.) *The Philosophy of Karl Popper.* (La Salle, Ill.: Open Court).
Churchland P. M. (1981). Eliminative materialism and the propositional attitudes, *The Journal of Philosophy,* **LXXVIII**(2), pp. 67–90.
Churchland P. M. (1985). The ontological status of observables: in praise of the superempirical virtues, in P. M. Churchland and C. A. Hooker (eds.) *Images of Science.* (Chicago: University of Chicago Press).
Churchland P. M. (1988a). Folk psychology and the explanation of human behaviour, *The Aristotelian Society,* Supplementary Volume **LXII**, pp. 209–221.
Churchland P. M. (1988b). *Matter and Consciousness.* (Cambridge, Mass.: The MIT Press, revised edition).
Churchland P. M. (1989a). On the nature of explanation: a PDP approach, in P. M. Churchland *A Neurocomputational Perspective: The Nature of Mind and the Structure of Science.* (Cambridge, Mass.: The MIT Press).
Churchland P. M. (1989b). On the nature of theories: a neurocomputational perspective, in P. M. Churchland *A Neurocomputational Perspective: The Nature of Mind and the Structure of Science.* (Cambridge, Mass.: The MIT Press).
Churchland P. S. (1980). Language, thought, and information processing. *Nous,* **14**, pp. 147-170.

Churchland P. S. (1983). Consciousness: transmutation of a concept. *Pacific Philosophical Quarterly*, **64**, pp. 80–95.
Churchland P. S. (1987). Epistemology in the age of neuroscience. *The Journal of Philosophy*, **84**(10), pp. 544–553.
Churchland P. S. (1986). *Neurophilosophy: Toward a Unified Science of the Mind/Brain*. (Cambridge, Mass.: The MIT Press).
Dunn W. N. (ed.) (1980–81). Symposium on social values and public policy. *Policy Studies Journal*, **9**(4).
Dunn W. N. (1981). *Public Policy Analysis. An Introduction*. (Englewood Cliffs: Prentice-Hall).
Dunn W. N. (1982). Reforms as arguments, in E.R. House, S. Mathison, J.A. Pearsol, and H. Preskill (eds.) *Evaluation Studies Review Annual*, Vol. 7. (Beverly Hills: Sage).
Dunn W. N. (ed.) (1983). *Values, Ethics, and the Practice of Policy Analysis*. (Lexington: D.C. Heath).
Evers C. W. (1987). Naturalism and philosophy of education, *Educational Philosophy and Theory*, **19**(2), pp. 11–21.
Evers C. W. and Lakomski G. (1991). *Knowing Educational Administration: Contemporary Methodological Controversies in Educational Adminstration Research*. (Oxford: Pergamon).
Fischer F. and Forester J. (1987). *Confronting Values in Policy Analysis*. (Beverly Hills: Sage).
Fodor J. A. (1975). *The Language of Thought*. (New York: Crowell). (Paperback edition (1979) Cambridge, Mass.: Harvard University Press).
Garson G. D. (1986). From policy science to policy analysis: A quarter century of progress, in W.N. Dunn (ed.) *Symposium on Social Values and Public Policy*, *Policy Studies Journal*, **9**(4), pp. 535–544.
Gill J. (1988). *Which Way to School?: A Review of the Evidence on the Single Sex versus Coeducation Debate and an Annotated Bibliography of the Research*. (Canberra: Commonwealth Schools Commission).
Hacking I. (1975). *Why does Language Matter to Philosophy?* (Cambridge: Cambridge University Press).
Hooker C. A. (1975). Philosophy and metaphilosophy of science: Empiricism, Popperianism and realism. *Synthese*, **32**, pp. 177–231.
Jennings B. (1983). Interpretive social science and policy analysis, in D. Callahan and B. Jennings (eds.) *Ethics, the Social Sciences, and Policy Analysis*. (New York and London: Plenum).
Jennings B. (1987). Policy analysis: science, advocacy, or counsel, in S. Nagel (ed.) *Research in Public Policy Analysis and Management*. (Greenwich, Conn.: JAI Press).
Lakomski G. (1988). Critical theory, in J. P. Keeves (ed.) *Educational Research, Methodology, and Measurement. An International Handbook*. (Oxford: Pergamon Press).
Landsbergen D. and Bozeman B. (1987). Credibility logic and policy analysis, *Knowledge: Creation, Diffusion, Utilization*, **8**(4), pp. 625–649.
Lasswell H. D. (1951). The policy orientation, in D. Lerner and H. D. Lasswell (eds.) *The Policy Sciences*. (Stanford: Stanford University Press).
Lerner D. and Lasswell H. D. (1951). (eds.) *The Policy Sciences*. (Stanford: Stanford University Press).
Lincoln Y. S. and Guba E. E. (1986). Research, evaluation, and policy analysis: heuristics for disciplined inquiry, *Policy Studies Review*, **5**(3), pp. 546–565.
MacRae D. (1976). *The Social Function of Social Science*. (New Haven and London: Yale University Press).
Mason R. O. and Mitroff I. I. (1980–81). Policy analysis as argument, in W. N. Dunn (ed.) *Symposium on social values and public policy. Policy Studies Journal*, **9**(4), pp. 579–584.
Mitchell D. E. (1984). Educational policy analysis: the state of the art. *Educational Administration Quarterly*, **20**(3), pp. 129–160.
Nisbett R. E. and Wilson T. D. (1977). Telling more than we can know: verbal reports on mental processes, *Psychological Review*, **84**(3), pp. 231–259.
Popper K. R. (1981). *Objective Knowledge. An Evolutionary Approach*. (Oxford: The Clarendon Press, revised edition, reprinted with corrections and a new Appendix 2).
Quine W. V. O. (1969). Epistemology naturalized, in W. V. Quine *Ontological Relativity and Other Essays*. (New York: Columbia University Press).
Quine W. V. O. and Ullian J. S. (1978). *The Web of Belief*. (New York: Random House, second edition).
Rumelhart D. E. and McClelland J. L. (1986). *Parallel Distributed Processing: Explorations in the Microstructure of Cognition*. Vol. I. *Foundations*. (Cambridge, Mass.: The MIT Press).

Sejnowski T. J., Koch, Ch., and Churchland P. S. (1988). Computational neuroscience, *Science*, **241**, pp. 1299–1306.
Stich S. (1983). *From Folk Psychology to Cognitive Science: The Case Against Belief*. (Cambridge, Mass.: The MIT Press).
Toulmin S. (1958). *The Uses of Argument*. (Cambridge: Cambridge University Press).
Walker J. C. and Evers C. W. (1982). Epistemology and justifying the curriculum of educational studies, *British Journal of Educational Studies*, **30**(2), pp. 213–229.
Walker J. C. and Evers C. W. (1988). The epistemological unity of educational research, in J. P. Keeves (ed.) *Educational Research, Methodology, and Measurement*. (Oxford: Pergamon Press).
Wildavsky A. (1979). *Speaking Truth to Power: The Art and Craft of Policy Analysis*. (Boston: Little, Brown).
Weiss C. H. and Bucuvalas J. (1980). Truth tests and utility tests: Decision makers' 'frames of reference for social science research', *American Sociological Review*, **45**, pp. 302–312.
Williams M. (1977). *Groundless Belief*. (Blackwell: Oxford).
Williams M. (1980). Coherence justification and truth, *Review of Metaphysics*, **34**(2), pp. 243–272.

8

Administrative Decision-Making as Pattern Processing

There is a long tradition in educational administration that deems decision-making to be a fundamental feature of administrative work (Estler 1988, pp. 305–319). This tradition is, in turn, located within a broader cognitive perspective on organizational structures. After some scene setting, we propose to focus on two contrasting approaches to the representation of administrative knowledge. The first deals with symbolic representations and is fundamental to mainstream decision theory. It also has implications for administrator training and the interpretation of administrative behaviour. The second is new to administrative theory and arises out of recent work in cognitive science. This work seeks to explain human learning, reasoning, and knowledge on the basis of realistic models of brain processes. Using computer simulations of the information processing properties of neurons, connectionism or parallel distributed processing as it is sometimes called, offers an alternative, non-symbolic way of representing knowledge and accounting for learning and reasoning (Rumelhart and McClelland 1986a, b; Bechtel and Abrahamsen 1991). Following an outline of some important features of this alternative perspective on cognition, we develop a number of its consequences for reconceiving administrative knowledge, reasoning, and decision-making.

The Cognitive Perspective

Theories of knowledge and cognition have played a major part in theorising about organizations (see Chapter 11). For example, in the *Republic*, Plato envisaged a just society as being one which reflected, in its broad structure, the cognitive capacities of its citizens: a kind of meritocracy where people work at what they are most suited for. So, the story goes, philosophers, whose minds apprehend the Form of the Good, are clearly best suited to be a society's rulers. Versions of this doctrine, which posits an isomorphism between a hierarchy among individuals over cognitive performance, and some social hierarchy, have been around ever since.

Critical theorists who draw on the early work of Habermas have posited a link between accounts of knowledge, and organizational structure (Foster 1986). Thus scientific knowledge that is acquired through processes of manipulation and control of the environment is said to become the model for scientific theories of administration that imply structures for the manipulation and control of organizational members. Critique of these structures occurs through broadening our understanding of knowledge, particularly with regard to the equally legitimate contexts of communication and human freedom. Organizational structures that embody these wider cognitive interests are characteristically democratic and participatory, and conform to a particular conception of social justice.

In a number of books Christopher Hodgkinson (1978, 1983, 1991) has argued for organizational design to mirror a hierarchy he claims exists in moral knowledge. Since all decision-making is infused with values, an organization's most important decisions — those concerned with vision, broad direction or organizational mission, and general policy — should be made by people who apprehend the highest level of value, where matters of moral principle reside. As we move down the organizational hierarchy, less significant values come into play: the calculation of utility, the aggregation of majority wishes, or the satisfaction of emotion (Hodgkinson 1991, p. 157). In this way a hierarchy of values is isomorphic to an organization structured according to the kinds of morally relevant decisions that need to be made.

The organizational learning tradition associated with Chris Argyris's work is another influential approach in the cognitive tradition (Argyris 1982; Argyris and Schön 1978). Drawing on systems and control theory, the key idea is that organizations should be designed in such a way as to enhance the promotion of organizational learning. Since learning is assumed to occur, in part, through the process of feeding back information about the consequences of actions to policy, values, and decision structures, the demands of learning imply that characteristic feedback loops be incorporated into an organization's design. Strictly speaking, it is always individuals who learn, so the feedback loops are mainly those information flow paths that are hypothesised as significant for the development of an individual (or shared) cognitive perspective on organizational issues.

By far the most prominent position within the cognitive tradition, certainly within the last forty years, has been the decision-making perspective, especially that aspect of it that holds as an ideal the sort of quantitative modelling made possible by the mathematical apparatus of the decision sciences (McNamara and Chisolm 1988). In general terms, this approach recommends that structures be developed so as to enhance the efficiency and effectiveness of what is taken to be the central function of most organizations, namely, decision-making (Simon 1976, pp. 1–19, 172–197). We do not propose to examine here any of the usual decision-theoretic models, since these are rarely used in educational contexts (except for system-wide budgetary and planning purposes — for example Correa, 1975). Instead, we shall focus on a number of extensively used assumptions about reason and knowledge that figure not only as background to the decision-making perspective, but are part of the general cognitive tradition.

Representing Decision Knowledge

In decision theory it is useful to distinguish between normative and descriptive issues (Slovic 1990). Normative decision theory deals with the principles underlying optimal, or ideal, choice-making; with what decisions we *ought* to make given the information available. Descriptive theory, on the other hand, attempts to characterize what we actually do in choice-making, given the real cognitive and information constraints we operate under. Herbert Simon's work in administrative science has resulted in an elegant combination of both these aspects of decision theory into an account of choice-making as *bounded rationality* (Simon 1976; Newell and Simon 1972). Because of limits to both information and processing, the quest for optimality has costs. For example, the search for information can be expensive and time-consuming, and can reach a point where search costs exceed the cost of errors resulting from not having the information. Trading off normative demands against the limitations revealed by descriptive theory is the obvious practical outcome. It suggests that in practice, when uncertainties, complexities, and the costs of removing both are taken into account, decision-makers should settle for good rather than optimal decisions.

Allowing for this hedge against the limits of applied decision-making, we can still characterize the theory of instrumental reason that lies behind it roughly as follows: '... rational behaviour is simply behaviour consistently pursuing some well defined goals, and pursuing them according to some well defined set of preferences or priorities' (Harsanyi 1977, p. 42). Since means are used to achieve goals, some theory is presumed about the consequences of action, or what will follow from the means employed. Where definite outcomes are not known but alternatives admit of the assignment of probabilities, the decision is said to be made under conditions of risk. Where no assignment of probabilities is known, the decision is said to be made under conditions of uncertainty. Technically, decision theory is that part of the theory of rational choice that deals with decisions made under conditions of risk or uncertainty (Heap *et al.* 1992, pp. 349–50). In each case rational choices are said to maximize expected utility.

A less technical version of the same theory can be found embedded in commonsense, or folk-psychological accounts of rationality: rational people act *as if* they are employing what they believe to be the most likely means to achieve their most desired goals. However, note that for both commonsense, and versions with more arcane embellishments, rational decision-making is at bottom a matter of the appropriate selection and coordination of beliefs and desires — a result that appears to hold up across the cognitive tradition.

Over the last ten years 'belief–desire' theory, as it is often called, has come under serious questioning and debate in philosophy (Stich 1983, Churchland 1989, pp. 111–127). The debate has many strands, but the one we wish to focus on can be plausibly separated by considering how knowledge is assumed to be represented when it comes to training for administrative decision-making.

In educational administration it is customary to translate the informal desire component of decision into a priority-setting exercise defined over identifiable goals. To be effective, the formal constraints of ordinal ranking, consistency, and stability need to be met. To be applied in an organizational context involving communication and justification, the exercise requires some sharing of publicly accessible *representations* of goals, invariably the symbolic tokens of written language. These representations exist as mission statements, policy documents, school charters, and the like, serving as a locus for the development of training programmes.

Applied decision-making contexts place an even greater burden on the provision of symbolic representations of beliefs, since it is beliefs that reflect assumptions about the states of affairs necessary for bringing about desired goals. Symbolic theory formulations about funding, staffing, timetables, curricula, policy, learning, and organizations underwrite all sorts of decisions in educational organizations. Moreover, theories in these areas are typically seen as relevant, or useful, knowledge for administrator training purposes.

With such a strong emphasis on symbolic representations of knowledge, rationality in the decision-making tradition is mainly concerned with logical and quasi-logical relations among these objects. The clearest expression of this link between administration and symbol processing can be found, once again, in the work of Herbert Simon. For in addition to being a major influence in the development of the decision-making conception of administration, he has also produced work of fundamental importance in the dominant, symbolic, tradition of artificial intelligence (AI). The connection between decision-making and traditional AI is best captured by what Newell and Simon (1976, p. 111) call the Physical-Symbol System Hypothesis: 'A physical-symbol system has the necessary and sufficient means for general intelligent action'. Decision-making is a species of general intelligent action, and a physical-symbol system 'is a machine that produces through time an evolving collection of symbol structures' (Newell and Simon 1976, pp. 109–110). Since Newell and Simon follow the usual AI practice of specifying machines abstractly, by their symbol-processing properties rather than by hardware configurations, we end up with the result that intelligence is about dynamic relations among collections of symbols.

From about 1986 onwards, this version of AI has been seriously challenged by the arrival of a new, non-symbolic approach to cognition and knowledge representation. By way of introduction to this approach, it is instructive to reflect on a number of practical limitations to the traditional view.

Limits to the Symbolic Paradigm

Consider a simple example. You work with a computer all day, without difficulty, creating and manipulating data files at your keyboard. Drawing on your extensive experience, you attempt to *write down* for the benefit of a colleague, how to transfer part of a file from one disk to another. You get it wrong, yet in the normal course of work you make correct decisions about which keys to press (even in the absence of screen prompts) countless times. You *know* how to do it, you just cannot *say* how. Philosophers describe this as a distinction between knowing how and knowing that, and it manifests in all sorts of ways. For example, in most presented instances we can easily distinguish between something that is a chair and something that is not. However, it is a vastly more difficult exercise to construct a correct definition of a chair; that is, one which would apply to all our instances of classifying objects as chairs.

Or consider a much tougher problem. You are principal of a school where the teachers are running a classroom conditions campaign — a strict work to rule — because of funding cuts. But the campaign is beginning to affect adversely parent choices for new enrolments, a fact that will lead to even bigger funding cuts. What is the most reasonable decision to make? Unfortunately, there are no theorems of social science that permit the correct deduction of a sound strategy here. Indeed, there is no clear boundary around the class of premises relevant to such a deduction even if we had all the required symbolic formulations. Experience might suggest that you create a committee to advertise and promote the school, and make the organiser of the work-to-rule campaign a member — after all, this approach has worked before. But the resulting principle, perhaps expressed by the words 'conflict can be resolved by coopting opponents into the task of solving a common problem' looks a bit brittle. It admits of lots of counter-examples. The presence of counter-examples to a general principle does not, however, prevent the recognition and use of *relevantly similar* past experiences as guides to successful practice. It just means that we may not be able to find a symbolic formulation to express how we delineate the relevant from the irrelevant. But in fact non-symbolic similarity judgements form the basis of most human prediction, classification, and decision (Smith 1990).

Accounts of knowledge, decision, and training that fail to recognize this point end up with a chronic problem over how to understand the relationship between theory and practice. Options are closed in familiar ways. Knowledge is conflated with a model of knowledge representation, and theory with theory formulation. Together with the lack of any systematic account of how non-symbolic practical knowledge is to be represented, the dichotomy in administrator training is complete. The theory of representation trades in logical relations among, and semantic properties of, symbolic tokens. Here the training emphasis will be on meaning and the rules of symbol manipulation: on the *normative* standards of rationality. And because of the overwhelming linguistic/communicative demands of public rationality, lapses from norms still result in much descriptive decision theory being located on the same continuum as normative theory. Bounded rationality involves reasoning *as if* we are imperfectly manipulating symbolic representations in our heads.

The distinction between theory and practice arising out of symbolic views of knowledge is so pervasive in educational studies that it has been virtually institutionalized as a distinction between mind and brain. Mainstream education deals with *reason* and the development of mind. When familiar folk-psychological assumptions are not fruitful in explaining behaviour, especially serious difficulties in learning, an orthogonal explanatory model is invoked dealing with *causes* and the functioning of brains — the framework of Special Education. In our view, for both education and administration, the exception (or special) should be the rule.

Let us sum up what we see as the main limits to the symbolic paradigm of decision knowledge. First, a very large amount of decision relevant knowledge is non-symbolic and in fact may not admit of any formulation. (Think of decisions concerning judgements about quality.) Second, the symbolic paradigm has very few resources for accounting for this knowledge. Third, as a result, a deep puzzle invests the task of understanding the relationship between theory and practice. And finally, while our most developed theories of reason are in the symbolic tradition, the best recent descriptive theories of human learning treat learning as a *causal* process that involves neural information processing. The symbolic tradition therefore has great difficulty combining within the one coherent account of cognition these two divergently theorized elements.

Non-Symbolic Representations

In the early 1960s two research programmes competed in AI. The first, which came to dominate, was program-writing, or symbolic, AI. The second, associated with the work of Frank Rosenblatt (1962), aimed to develop an account of cognition based on the information-processing properties of networks of neurons. The objects of study, variously known as 'neural networks', 'connectionist networks', or 'perceptrons', are the mathematical and computer models of neurons constrained by varying degrees of biologically driven detail (Rumelhart and McClelland 1986a, b). Two-layer networks (see Figure 8.1) provide the simplest design options.

For this network, the input pattern S is a vector with four ($i = 4$) elements and the output pattern E is a vector with three ($j = 3$) elements.

Each input element is multiplied by a weight (w_{ij}) before it goes to some output node, which receives the weighted sum of all the inputs. Biologically, each element of S corresponds to a neuron's spiking frequency, nodes are neurons, and the weights between nodes are the strengths of the synaptic junctions between neurons. Teaching the network to associate a desired output pattern with some S is accomplished by adjusting the weights so that the output vector E produced by the network matches the target, or desired, output. These modest

110   EXPLORING EDUCATIONAL ADMINISTRATION

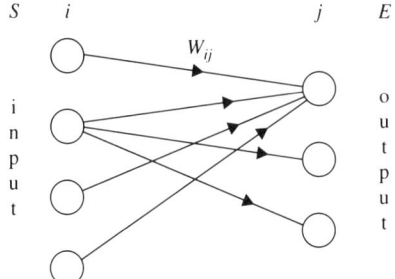

Figure 8.1 A two-layer pattern associator.

two-layer perceptrons thus have the capacity to learn, with the resulting knowledge residing in the value of the weights and their pattern of connectivity.

Interest in Rosenblatt's research programme evaporated fairly quickly when, in their 1969 book *Perceptrons*, Minsky and Papert were able to prove that perceptrons were severely limited in the kind of learning tasks they could perform. Admittedly, these limitations applied only to two-layer nets, but at the time no weight adjustment algorithm was known for nets of three or more layers. Indeed, it was not until the publication in *Nature* of 'Learning representations by back-propagating errors', by Rumelhart, Hinton, and Williams (1986) that such an adjustment procedure — the backpropagation algorithm (or delta rule) — became widely known. Nowadays, multilayer nets, with their powerful learning properties and their non-symbolic way of representing knowledge, are generating much interest in cognitive science. Although we know of no examples in administration, the following example drawn from another complex applied field, namely medicine, should illustrate the basis for this interest.

Consider the decision problem of correctly identifying acute myocardial infarction in patients who present to a hospital with severe chest pain. Baxt (1990) reports the study of 356 such patients: 236 who turned out not to have infarction and 120 who did. Could a multilayer neural network be trained to extract patterns in patient data more efficiently to yield better diagnostic decision-making than that provided by well-trained physicians? To test this possibility, Baxt constructed a four-layer net as shown in Figure 8.2. The 20 input nodes of the first layer correspond to 20 different input elements composed of items on a patient's record: e.g. age, sex, shortness of breath, hypertension, etc. Some items were coded 1 or 0, depending on their absence or presence (or whether they admitted of binary coding); others were given continuous numbers between 0.0 and 1.0. The correct, or target, output was set at 0 or 1, for absence or presence of infarction. Random weights were initially assigned between the layers, and the network was trained up on half the sample (60 with and 118 without infarction). Training consisted of feeding in the 178 20-element vectors in random order, over and over again, and adjusting all the weights according to the backpropagation algorithm whenever there was a difference between the net's

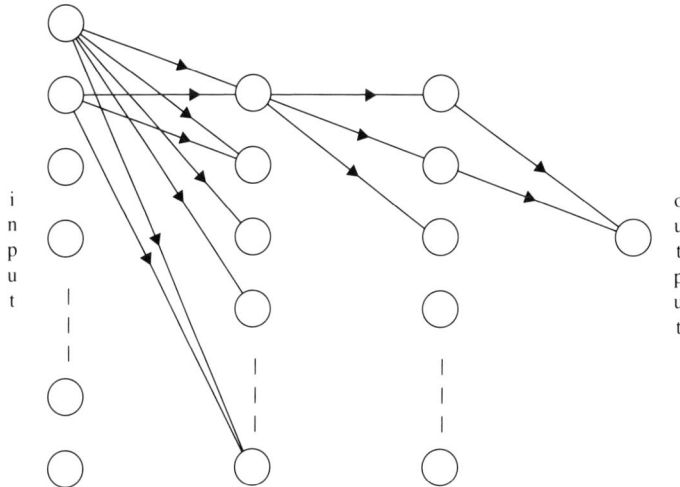

Figure 8.2 A 20 × 10 × 10 × 1 four-layer neural network. Not all connections are shown.

output from a given vector, and the correct target of 0 or 1. When the net's performance on the training schedule eventually converged to the target output values, the trained net was then tested on the other half of the data set for diagnostic decision-making. As Baxt (1990, p. 480) reports:

> The network correctly identified 92% of the patients with acute myocardial infarction and 96% of patients without infarction …This is substantially better than the performance reported for either physicians or any other analytical approach.

This practical, diagnostic, knowledge is not located at any *particular* place in the network; rather, it is *distributed* across all the weights in their geometrical configuration. The resulting pattern of weights thus forms an internal representation of the network's knowledge of the diagnostic distinction between patients with and without coronary occlusion.

Unlike traditional AI, which simulates cognition through the manipulation of semantically rich symbolic representational structures, connectionist models are able to extract significant regularities present in data sets by their capacity to learn, without relying on a prior symbolic coding of knowledge. They end up behaving *as if* they are following a rule; but in fact no rule has been coded. Networks have been constructed and trained up to perform tasks like giving the past tense of English verbs (Rumelhart and McClelland 1986c, pp. 216–271), distinguishing valid from invalid argument forms in the propositional calculus (Bechtel and Abrahamsen 1991, pp. 163–174), pronouncing written English text (Sejnowski and Rosenberg 1986, pp. 663–672), and distinguishing rocks from mines on the basis of sonar evidence (Churchland 1988, pp. 157–62).

A corresponding analysis of learning to make sound administrative decisions from experience might run, in general terms, as follows. Problem scenes are classified according to some breakdown into relevant salient features applicable to a range of problems. These features are the elements of an input vector, more like a visual pattern or picture, which is then processed to produce a decision output of expected consequences. A mismatch between expectation and experience drives revisions to processing. We think again, perhaps recognizing some elements as more important than we first allowed, perhaps expanding the vector to cover features we think we overlooked. We make more decisions, learning over time through pressure of persistent feedback from experience. In the end, we may exhibit great skill, effortlessly seeing through complex administrative scenes to some underlying pattern which we immediately associate with a (we hope) good decision. In this way, a great deal of ordinary, everyday, decision-making comes to resemble pattern recognition and processing rather than the logical manipulation of theory formulations, with an administrator's practical knowledge residing in the relevant weights and geometry of the brain. To the extent that neural networks accurately reflect biologically realistic properties, they have much to tell us about non-symbolic learning from experience, the representation of knowledge of practice, and the real cognitive processes of decision-making.

Synthesizing the Symbolic and the Non-Symbolic

If the symbolic tradition of knowledge representation, reasoning, and decision-making encounters difficulties in meshing with plausible causal accounts of learning and the role of non-symbolic knowledge in the human cognitive economy, the connectionist approach has an opposite problem. It fails to give a plausible account of the sheer cognitive utility of the symbolic. (Indeed, it has often been argued that language is essential for thought.) For example, administrators often use theory formulations (for example, budget categories, semantically significant symbolic descriptors, etc.). to *identify* the salient features of the problem scenes presumed in network learning.

More damagingly, there is an important limitation to decision-making based on network experiential learning. Inferences, or generalizations, from the training subset of experience are only valid insofar as the statistical structure of the training sample matches the statistical structure of the world of application. However, many of the most important administrative decisions in education are either (1) designed to change the course of future experience, or (2) are of a non-routine, atypical nature, where the training sample is least generalizable. For these cases, where professional rather than mere algorithmic judgement is called for, administrators are dealing with hypothetical, or often *counterfactual*, circumstances. It is precisely here, where symbolic representations can be seen to go proxy for experience, that we can appreciate the utility of inferential reasoning as normative-ruled-based symbolic manipulation. Without the niceties, experience trains for dealing with similar experiences, whereas theory formulations, together with the (symbolic) apparatus for validly manipulating formulations, permits a knowledgeable evaluation of options and choices beyond the reach of training data. The challenge, as we see it, is to combine both sources of knowledge within a neurophysiological theory of human learning; that is, to combine the representational power of symbols with the learning power of nets by giving a neural network account of symbol processing.

This is about where the literature on connectionist AI is at the moment. As Geoffrey Hinton (1991, p. 3) remarks in the Preface to his collection *Connectionist Symbol Processing* (a Special Issue of the journal *Artificial Intelligence*), '... the problem is to devise effective ways of representing complex structures in connectionist networks without sacrificing the ability to learn the representations'.

Although progress has been rapid, there is much work still to be done on the development of connectionist models if they are to provide useful insights into the range of human cognitive phenomena, or even just decision-making aspects. Nevertheless, the new AI does imply some significant breaks from earlier cognitive alternatives. For example, the assumption that symbolically oriented normative decision theory differs only in degree from descriptive theory, looks wrong in view of the absence of any apparent structural isomorphism between the causal detail of neural processing, revealed by our best descriptive theories, and any symbolic representations being processed. That is, the language of thought does not appear to mirror the language of logic, or natural languages. So, despite its normative value as an account of rationality, it is unlikely that normative theory (or even the familiar belief–desire theory of folk psychology) can provide deep insights into the cognitive dynamics of decision-making.

Also, for purposes of understanding the relationship between theory and practice in administrative training, it would be more reasonable to regard symbolic theorizing as a species of practice — namely the practice of operating on, say, graphemes or pictorial displays of symbolic tokens. Symbol-based reasoning could then be reconstrued as pattern completion, with the design and presentation of representationally rich symbols (seen as informative ways of compressing knowledge of experience) becoming a much more significant factor in training programmes. A picture may not only be worth a thousand words; it may *be* a thousand words.

On these, and many other matters, the new AI can be expected to provide fresh insights. As it does, a systematic reconceptualisation of all that is familiar about administrative decision-making will be inescapable (see Chapters 9 and 11).

References

Argyris C. (1982). *Reasoning, Learning and Action*. (San Francisco: Jossey-Bass).
Argyris C. and Schön D. (1978). *Organizational Learning: A Theory of Action Perspective*. (Menlo Park: Addison-Wesley).
Baxt W. G. (1990). Decision-making: The diagnosis of acute coronary occlusion, *Neural Computation*, 2(4), pp. 480–89.

Bechtel W. and Abrahamsen A. (1991). *Connectionism and the Mind*. (Oxford: Blackwell).
Churchland P.M. (1988). *Matter and Consciousness*. Cambridge, Mass.: MIT Press, revised edition).
Churchland P.M. (1989). *A Neurocomputational Perspective*. (Cambridge, Mass.: MIT Press).
Correa H. (1975) (ed.). *Analytical Models in Educational Planning and Administration*. (New York: David McKay).
Estler S. (1988). Decision making, in N.J. Boyan (ed.) *Handbook of Research on Educational Administration*. (New York: Longman).
Foster W. (1986). *Paradigms and Promises*. (New York: Prometheus Press).
Harsanyi J.C. (1977). Morality and the theory of rational behaviour, in A. Sen and B. Williams (eds.) (1982). *Utilitarianism and Beyond*. (Cambridge: Cambridge University Press).
Heap S.H. et al. (1992). *The Theory of Choice*. (Oxford: Blackwell).
Hinton G.E. (1991) (ed.). *Connectionist Symbol Processing*. (Cambridge, Mass.: MIT Press).
Hodgkinson C. (1978). *Towards a Philosophy of Administration*. (Oxford: Blackwell).
Hodgkinson C. (1983). *The Philosophy of Leadership*. (Oxford: Blackwell).
Hodgkinson C. (1991). *Educational Leadership: The Moral Art*. (New York: SUNY Press).
Minsky M. and Papert S. (1969). *Perceptrons*. (Cambridge, Mass.: MIT Press).
McNamara J.F. and Chisolm G.B. (1988). The technical tools of decision making, in N.J. Boyan (ed.) *Handbook of Research on Educational Administration*. (New York: Longman).
Newell A. and Simon H.A. (1972). *Human Problem Solving*. (Englewood Cliffs: Prentice Hall).
Newell A. and Simon H.A. (1976). Computer science as empirical enquiry: symbols and search, in M.A. Boden (ed.) (1990). *The Philosophy of Artificial Intelligence*. (Oxford: Oxford University Press).
Rosenblatt F. (1962). *Principles of Neurodynamics*. (New York: Spartan Books).
Rumelhart D.E., Hinton G.E., and Williams R.J. (1986). Learning representations by back-propagating errors, *Nature*, **323**, pp. 533–536.
Rumelhart D.E. and McClelland J.L. (1986a) (eds.) *Parallel Distributed Processing*. Vol. I. (Cambridge, Mass.: MIT Press).
Rumelhart D.E. and McClelland J.L. (1986b) (eds.) *Parallel Distributed Processing*. Vol. II. (Cambridge, Mass.: MIT Press).
Rumelhart D.E. and McClelland J.L. (1986c). On learning the past tenses of English verbs, in D.E. Rumelhart and J.L. McClelland (eds.) *Parallel Distributed Processing*. Vol. II. (Cambridge, Mass.: MIT Press).
Sejnowski T.J. and Rosenberg C.R. (1986). NETtalk: A parallel network that learns to read aloud, in J.A. Anderson and E. Rosenfeld (1988) (eds.). *Neurocomputing: Foundations of Research*. (Cambridge, Mass.: MIT Press).
Simon H.A. (1976). *Administrative Behavior*. (New York: Free Press, third edition).
Slovic P. (1990). Choice, in D.N. Osherson and E.E. Smith (eds.) *Thinking*. (Cambridge, Mass.: M.I.T. Press).
Smith E.E. (1990). Categorisation, in D.M. Osherson and E.E. Smith (eds.) *Thinking*. (Cambridge, Mass.: MIT Press).
Stich S. (1983). *From Folk Psychology to Cognitive Science*. (Cambridge, Mass.: MIT Press).

9
Educating the Brain

Educational theory has been much influenced by sentential models of knowledge representation. The classical 'justified true belief' account of knowledge is an account of knowing that p, where p is some proposition expressed by a sentence. So if education is thought to involve the acquisition of knowledge then on the classical view this will mean coming to believe true sententially expressed claims for good reasons. Major alternatives still leave intact sentential representations; for example, within the foundational tradition, falsificationist approaches have their testable hypotheses and observation reports. Acquiring knowledge is a matter of acquiring our best theories — represented as networks of variously associated sentences. And for non-foundational, coherentist epistemologies the elements usually reckoned as subject to coherence criteria of theory choice are sentences.

If education is thought to involve the development of reason then sentential models of knowledge can also exert an influence through theories of reason, notably theories for which reason is a matter of valid use of language within particular kinds of discourse, for which good reasoning is a matter of manipulating a symbol system according to its *rules*. On this analysis, rational behaviour is *rule-governed* behaviour, the rules being either explicit or implicit.

This model of knowledge representation, when taken as theoretically basic, generates a number of difficulties. For example, how do we mesh theories of knowledge in which talk of logical relations among propositions (understood to be expressible sententially) is currency, with causal theories of learning? Or if 'knowing that' is distinct from 'knowing how', how is the latter to be represented? Or, perhaps more generally, how is theory, construed sententially, to be related to practice?

In what follows, we shall cite some examples of these problems and worry over one main issue; namely, how to understand the way symbols, rules and propositions are related to behaviour and human practice. We shall then suggest that language does not reflect what is basic about the nature and structure of human knowledge. Work in cognitive neurobiology, using brain behaviour as a constraint on theories of learning and information processing, has led to the development of powerful non-sentential accounts of knowledge representation. After sketching some of the work in this area, notably some features of parallel distributed processing accounts of cognition, we shall try to give some very brief indication of their consequences for the examples we began with and for the linking of theory and practice in general.

Education, Knowledge and Reason

In *Conditions of Knowledge*, Israel Scheffler (1965) offers an account of the relationship between epistemology and education. As a first approximation he connects learning and teaching, on the one hand, with knowledge and belief, on the other, as follows:

> Learning that Q involves coming to believe that Q. Under certain further conditions (truth of 'Q' and, for the strong sense of *knowing*, proper backing of 'Q'), it also involves coming to know that Q. Teaching that Q involves trying to bring about learning that (and belief that) Q. under characteristic restrictions of manner, and, furthermore, knowing that Q . . .) (Scheffler 1965, p. 13)

The account of both knowledge and belief assumed here is propositional, with 'Q' a placeholder for some sentence. Moreover, the distinction between knowledge and belief is the familiar classical one: knowledge is belief that is justified and true. When we add in the characteristic restrictions on manner of teaching, restrictions that are designed to ensure the relevance of *reasons* and so promote what Scheffler (1965, p. 13) calls 'the free critical judgement of the student's mind', we have a basis for imposing epistemological constraints on pedagogy. Thus, for classical foundationalism, the regress of reasons in reverse might function as the progress of learning.

Scheffler offers improvements to the classical story. He also thinks it is *incomplete* in that it omits *procedural* knowledge, or 'knowing how'. Since truth is a property of sentences, and perhaps justification likewise, though for strings of sentences, 'knowing how', which is a matter of skilled performance, will be distinct from sententially expressed propositional knowledge. However, this leads to a discounting of 'knowing how' as the relevance of reasons is a constraint on teaching; the giving of reasons is a matter of expressing propositions. Scheffler's remarks on the place of reasons in skills imply a compromise:

> Where skills are involved, reasons pertain to the basis for following one procedure rather than another, choosing one rather than another step or strategy, or building up such-and-such a facility rather than another for the sake of achieving a given broad competence (Scheffler 1965, p. 107).

This suggests that in the case of skills reasons are *external* to procedure. The currency of reason appears to be the sentential descriptions of practice, with good reasons being determined, for example, by sound arguments connecting descriptions of means and ends. Needless to say, skilled practitioners can offer entirely wrong reasons for their skilled performance.

Paul Hirst is another, more robust, defender of proposition-like knowledge, at least in his earlier writings. His forms of knowledge thesis is *defined* on propositions and he views liberal education as involving an initiation into all the different kinds of true propositions (Hirst 1974, pp. 30–33). He thinks that '. . . Intelligible thought necessarily involves symbols of some sort and most of it involves the symbols of our common languages' (Hirst 1974, p. 83).

Our common languages are not uniform in their structure, however. Modes of discourse, or language-games, can be distinguished by their distinct rules for forming expressions. It is these differences among language-games that mark out the forms of knowledge. Teaching a mode of discourse does not require one to teach its rules explicitly (Hirst 1974, p. 83). But having mastered a language game one could be said to know its rules implicitly, or perhaps to have tacit knowledge of them.

In his earlier work, these rules would be propositional. In his 1983 paper 'Educational theory' (Hirst 1983), where he wants to defend a distinction between theoretical knowledge and practical knowledge, he denies that all rules of practice can in principle be expressed sententially:

> Practical knowledge consists of organized abilities to discern, judge and perform that are so rooted in understanding, beliefs, values and attitudes that any abstracted propositional statements of those elements or of rules and principles of practice must be inadequate and partial expressions of what is involved (Hirst 1983, pp. 11–12).

However, when pressed on the connection between non-propositional knowledge and reason, he reverts to a position similar to Scheffler's on skills. It is practical *principles* that are reasonable or not, and these are abstracted from practice. (On this score, Scheffler and Hirst are both more 'intellectualist' than Ryle.) The most important educational consequence of this epistemological stance is that having separated out the practical and the theoretical, there is a problem in figuring out how to put them together again for educational theory, which is a practical theory. For if theory is sentential then on this analysis educational theory will always be inadequate as a guide to educational practice.

A final example of the important role of sentential models of knowledge and reason in education can be found in the theory of reason put forward at one time by Richard Peters. Since sentential models abstract nicely from any details of how human brains might actually work in their acquisition of knowledge or their reasoning activity, Peters (1974, p. 120) wastes no time on the neurosciences: 'I propose to ignore also physiological findings about the development of the brain, even though these deal, of course, with necessary conditions for the development of rationality'. They are ignored because he thinks that physiological models of brain functioning seem 'conceptually corrupt from the outset' in crediting nervous systems with 'plans, strategies, information' (Peters 1974, p. 120). Instead, we can best '... understand rational behaviour and belief as informed by general rules. It is behaviour and belief for reasons — and for reasons of a certain kind' (Peters 1974, p. 121).

Telling a physiological story, even in arbitrarily great detail, merely gives us a causal account of human *movement*, of what *happens* to people. To explain human actions — signing a cheque, bidding at an auction, or solving a problem — we need to invoke the rules that underlie and make intelligible the associated bodily activity. Indeed, it is only when we cannot posit appropriate rules, when rationality appears to lapse, that appeals to causation are methodologically sound.

Since Peters thinks that rule-following cannot be explained naturalistically, he freely acknowledges one problem for his theory. How do entities like propositions, which are expressed by sentences functioning as reasons, bring about the bodily movements that are part and parcel of responding to reasons? He thinks that 'in this respect wants and wishes are equally mysterious in their mode of operation' (Peters 1974, p. 116). (Presumably there is some token–token equivalence here but reasons do not, of course, match types of causes.)

To sum up, we have mentioned three puzzles for sentential models of knowledge in education: how are reasons related to skills, how can theoretical knowledge be connected with practical knowledge to give practical theory, and how can reasons bring about specific behaviours?

Brain Processes

Just as the performance characteristics of motor vehicles can be rendered less mysterious by an informed look under the bonnet, so there are grounds for supposing that the study of brains is not entirely irrelevant to giving an informed account of human knowledge and reason. But because language has a mortgage on commonsense, the place to look is at the edges of reason, the anomalies that defy familiar belief–desire folk-theoretic accounts. In having to deal with the phenomena of dyslexia, mirror writing, or indeed a range of select cognitive oddities that may boggle our sheltered imaginations, educators have little choice but to include brain functioning as an explanatorily relevant factor (see Gaddes 1985). Yet when we get down to the level of brain information-processing, there is little that readily corresponds to folk theorem views of knowledge, or the sentential dynamics of reasoning. Since this level also underwrites what presents as *normal* cognitive functioning, parsimony counsels a closer look.

In taking a closer look at the detail of brain functioning it might be thought that acknowledging its relevance begs the question in favour of reductionism; that a general defence of reductionist strategy has gone missing. Our view, however, is that this concern puts matters around the wrong way. As in the case of such successful reductions and/or eliminations in science as occurred with alchemy, phlogiston, and vitalism, the merits of a purported reduction depend very much on the merits of the new reducing theory. But since theoretical merit will in turn depend partly on the plausibility of such detail, the closer look that follows is not incidental, but rather integral, to a reductionist strategy.

You are listening to an academic paper. The speaker's larynx activates the air in a succession of compressions and rarefactions of various frequencies and amplitudes which impinge on your ear. The number of cycles per second, or hertz (Hz) that your ear is capable of detecting covers the range 20–15,000 Hz, with middle C at 523 Hz. Frequency determines pitch. The amplitude of sound waves determines loudness, and is measured in decibels (a logarithmic scale). Sound waves enter the external ear canal and strike the tympanic membrane, or ear drum, causing it to vibrate. This mechanical energy is then transmitted to the inner ear by a set of three small bones called ossicles — one connected to the tympanic membrane, one to the oval window of the cochlea, and one in between. This middle ear arrangement permits the efficient transfer of mechanical energy from air to the fluid-filled cochlea, for without it, incoming sound would be mostly reflected back (Kandel and Schwartz 1985, ch. 31).

It is in the inner ear that this mechanical energy is converted to electrical energy for transmission through the nervous system to the brain. The nervous system may be regarded as consisting of two parts: a central nervous system (CNS) comprising the brain and spinal cord (and also the retina), and the peripheral nervous system (PNS). Cells called 'neurons' are the basic elements of the nervous system and are usually classified into three kinds: motor neurons, sensory neurons, and interneurons (all the rest). The single most important feature of neurons is their capacity to receive, integrate, and transmit signals, which they do by virtue of a specialised anatomy. See Figure 9.1.

The branching structure of dendrites receives signals, which are combined at, or on their way to, the cell body, or soma. If of sufficient strength, the signals will cause the soma to fire, sending an action potential, or spike, down the axon to its presynaptic bulbs. These bulbs form a synapse (usually) with the dendrites of other neurons. In the 1940s Donald Hebb hypothesized that if the firing of one

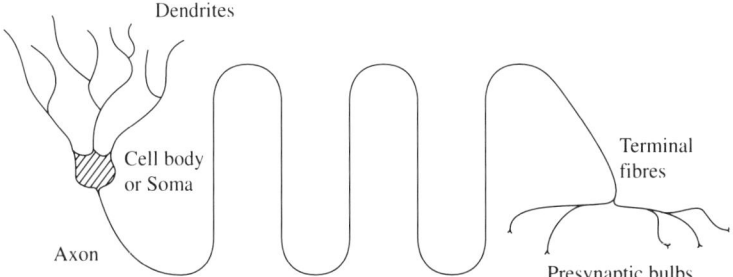

Figure 9.1 (*Source*: Churchland P. M. (1988). *Matter and Consciousness*. Cambridge, Mass.: MIT Press, revised edition, p. 131.)

neuron repeatedly contributes to the firing of another then this will be reflected in properties of the synapse, the site, he thought, of long-term memory. We now know that the bulbs contain neurotransmitters, which are released to cross a gap (of about 20 ångströms) to the receptor sites of dendrites (or sometimes soma). Recent work by Kandel and associates on the learning behaviour of the sea hare, Aplysia, shows that after conditioning, the openings of presynaptic bulb vacuoles that contain neurotransmitter become enlarged, thereby facilitating more rapid release of their contents. Learning produces changes in neuron anatomy at the molecular level. Depending on the nature of the synaptic junction, the release of neurotransmitter will either decrease (inhibit) or increase (excite) the probability of the next cell firing (see Kandel and Schwartz 1985, ch. 62).

On this brief sketch, the only signal variable is spiking frequency, the number of times a neuron fires per second. In the case of hearing, if loudness is coded as spiking frequency, how is the frequency of sound coded? The answer is that the cochlea contains an array of hair cells — sensory neurons that depolarize under mechanical stimulation — responsive to different sound frequencies. The highest sound frequencies cause spiking of neurons closest to the oval window, the lowest, spiking at the end of the cochlea. To maintain this information, the metrical displacement of receptors must be preserved, at least topologically, by some kind of topographical map throughout transmission. As Churchland (1986, p. 125) remarks,

> This preservation of neighbourhood relations is called a cochleotopic or tonotopic map, and the orderly mapping of neurons with sound frequencies is preserved at each synaptic station on the way to the cortex.

Such topographical maps appear to exist for all sensory modalities and are repeated throughout the cortex:

> How many cochleotopic maps are on the human temporal cortex is not yet known, but on the basis of other primate studies it would be surprising if there were fewer than six on each hemisphere (Churchland 1986, p. 127).

Since these maps may synapse with other sensory maps and other processing arrays in the brain, there is scope for matching and association of what is heard with past stored and present incoming information from a range of sources.

Parallel Distributed Processing

How do we learn from what we hear (or see or taste) and how is that knowledge represented? Information in the central nervous system appears to be defined by a combination of geometrical structure among neurons and spiking frequency control (which is a function of synaptic connection strengths). In the last ten years, but especially in the last three, a great deal of work has been done constructing models of neural networks with these properties, and attempting to simulate cognitive phenomena. The two key ongoing research questions are:

1. What range of cognitive phenomena can be performed by computer models of neural networks?
2. To what extent can these models be designed to reflect real biological constraints?

In what follows, we shall give some details of three such models.

The first is due to Paul Churchland (1988, pp. 156–165).Consider the problem of trying to interpret a sonar signal in order to distinguish between a rock and a mine. The sonar emits a signal, which echoes off underwater objects in the distance. Since mines and rocks are not identical, there may be some detectable difference to be found if echoes are correctly analysed. This is the sort of task that can typically be done well by a network. See Figure 9.2.

The input units on the left-hand side receive echo signal power at various frequencies. Since geometry is important, we may represent input order $(i_1 = 0.14, \ldots, i_{13} = 0.04)$ as an input vector I. Each input node is connected with each hidden unit, and the hidden units are in turn all connected to two output units, one for mine, one for rock. The connection strengths between nodes determine the amount of signal that is transmitted. The network decides there is a mine if the output vector is $\langle 1,0 \rangle$, and a rock if the output vector is $\langle 0,1 \rangle$. The learning task is to train up the network to make these decisions.

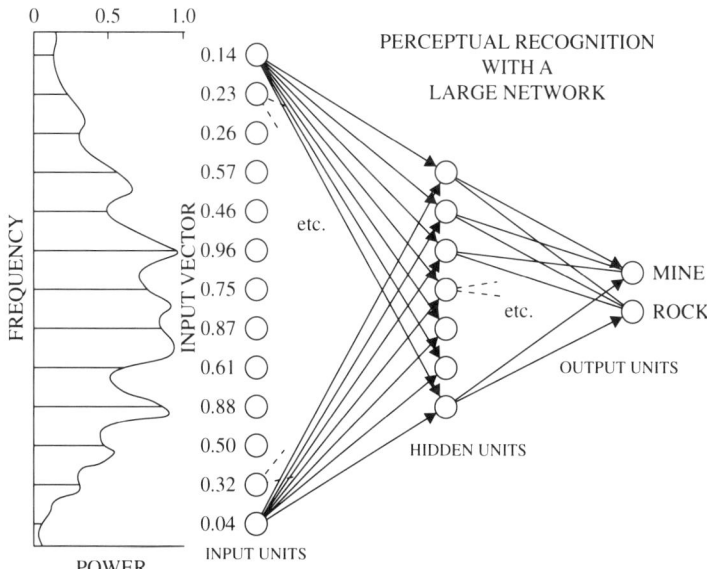

Figure 9.2 (*Source*: Churchland P. M. (1988). *Matter and Consciousness*. Cambridge, Mass.: MIT Press, revised edition, p. 159.)

We begin with a sample of 50 rock echoes and 50 mine echoes and feed them, in mixed order, into the network. Before we start, we set all of the weightings at some arbitrary number between -1 and 1. On its first trial, the network might produce some incorrect output vector, say $\langle 0.51, 0.49 \rangle$. We then adjust the weights using a mathematical formula (for example, what we describe later as the delta rule) so that the network's error is reduced. After a large number of trials the network settles down to give highly reliable output. The network, thus trained, can now be used to interpret new signals that were not part of its training schedule, and it will perform these new tasks again with very high reliability. It has learned to distinguish mines from rocks.

There is, however, no one point at which this knowledge is located. The network's knowledge of the distinction, and the identifying characteristics of mines and rocks, is distributed across the entire network. There is no coded rule, as such, that it uses for drawing the distinction, although *we* may use rules to describe how the system operates. Indeed, we may try to formulate necessary and sufficient conditions for some object presenting as, say, a mine. But then this would be a linguistic representation that the network itself makes no use of. Moreover, no one connection can be regarded as essential. If modelled synaptic junctions, or weights, are gradually removed at random, the network, like the human brain, degrades gracefully, diminishing slowly in function.

Precisely how are such networks trained? Consider a simpler, one layer system with four input signals $\langle a,b,c,d \rangle$ synapsing on three purkinje cells to produce outputs $\langle x,y,z \rangle$ in a physical array simplified, but typical, of the organization of the cerebellum cortex for all creatures. See Figure 9.3.

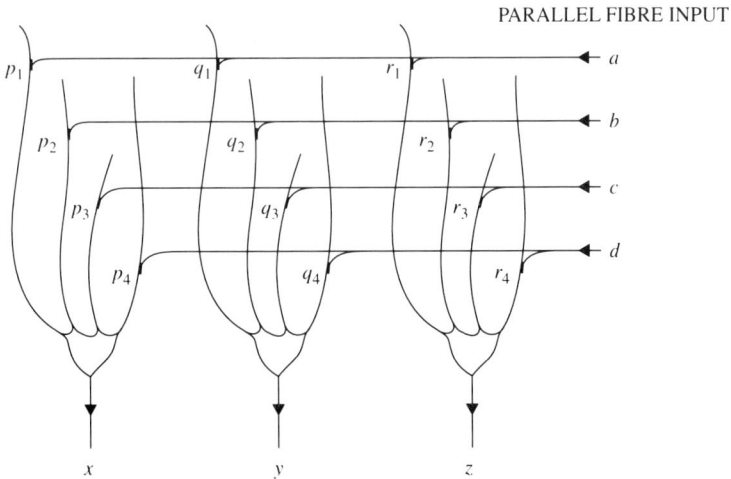

Figure 9.3 (*Source*: Churchland P. M. (1988). *Matter and Consciousness*. Cambridge, Mass.: MIT Press, revised edition, p. 152.)

The various p_i, q_i and r_i are the synaptic weights that affect the spiking frequency of the signals from the axons to the dendritic pathways of the three cells. We can therefore calculate the output signals as follows (although note that this calculation ignores background levels of activation and firing threshold levels, major sources of non-linearity):

$$x = p_1 a + p_2 b + p_3 c + p_4 d$$
$$y = q_1 a + q_2 b + q_3 c + q_4 d$$
$$z = r_1 a + r_2 b + r_3 c + r_4 d$$

But to those familiar with matrices, this is just

$$\begin{bmatrix} x \\ y \\ z \end{bmatrix} = \begin{bmatrix} p_1 & p_2 & p_3 & p_4 \\ q_1 & q_2 & q_3 & q_4 \\ r_1 & r_2 & r_3 & r_4 \end{bmatrix} \begin{bmatrix} a \\ b \\ c \\ d \end{bmatrix}$$

In the case of Churchland's sonar interpreter, we have a three-layer network comprising an input layer, a hidden layer, and an output layer. There are also two sets of weights: one set, say M, between the inputs and hidden nodes, and another, say N, between the hidden nodes and outputs. (If this were a simple linear system we would have

Output = N. M. Input

with the matrix product able to be represented by a single new matrix, for calculation purposes). Roughly speaking, the delta rule takes the difference between the actual output, say $\langle 0.51, 0.49 \rangle$ and the correct output $\langle 1, 0 \rangle$, multiplies each difference by the signal strength at each preceding node, and adds that product to the previous connection strength between each node. In repeating this procedure again and again on the elements forming the matrix representation, the network can gradually be trained (see Rumelhart and McClelland 1986a, ch. 11; Bechtel 1989; Nadel *et al.* 1989; Pfeifer *et al.* 1989; Pinker and Mehler 1988).

A second, more complicated, example of network learning was demonstrated when Rumelhart and McClelland (1986b, ch. 18) trained a large network to generate the past tense of English verbs. For some linguists, first-language learning is assumed to be guided by implicit rules innate to the learner. These rules are thought to make rapid learning of language possible given its sheer complexity and the modesty of learning samples and learner cognitive skills. However, no rules of use were coded into the Rumelhart and McClelland network, which simply extracted its own regularities over successive trials.

Although their full account is long, complex, and controversial, some basic ideas can be briefly extracted and described. A large 460-binary-unit vector called a Wickelfeature representation was devised to code phonological representations of the root form of verbs. An identically structured vector was also used

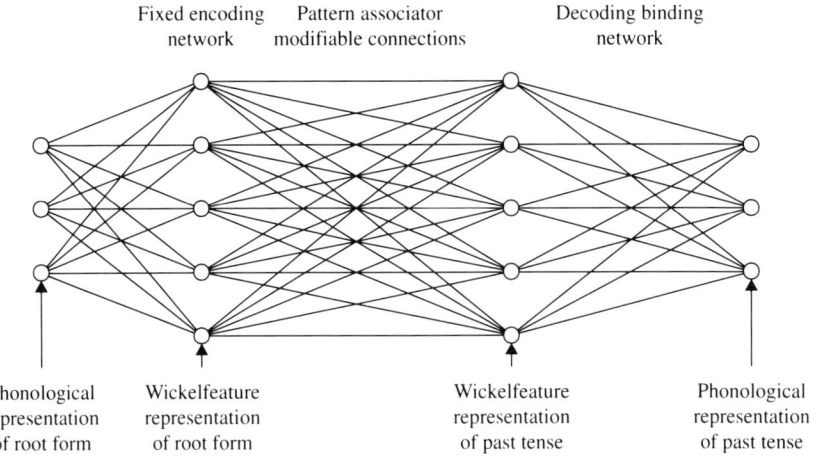

Figure 9.4   (*Source*: Rumelhart D. E. and McClelland J.L. (1986). *Parallel Distributed Processing: Explorations in the Microstructure of Cognition*. Cambridge, Mass.: MIT Press, vol. 2, p. 222.)

for past-tense representations, which could then be decoded to a corresponding phonological representation. The learning procedure then consisted in adjusting the connection strengths between the input and output Wickelfeature vectors according to whether the output was the correct past tense of the coded input verb. Figure 9.4 captures the structure of the network.

Various training simulations were run on particular subsets of 10, 410, and 86, of a total of 506 verbs, with the network achieving almost error-free parsing on even the largest subset after about 200 trials — that is, all the verbs in the training set fed through the network 200 times. Its learning development also appeared to mirror the features of children's learning of past tenses; for example, after a small number of trials it mistakenly regularized irregular pairs to produce results such as run/runned and go/goed.

For all its impressiveness, this network is many orders of magnitude less impressive than the powerful neural apparatus children bring to language learning. Resulting mature language use may be describable in terms of rules, but evidently knowledge of these rules, either implicit or explicit, is not essential for language learning or competent linguistic performance.

The Rumelhart and McClelland model operates under more realistic neurophysiological constraints than the linear sonar interpreter. Of these constraints, simulated neural threshold behaviour seems to be the most important. If we add in threshold retuning due to habituation and sensitization, we can model sudden global network changes such as perceptual gestalt switching. Stephen Grossberg (1982) has developed networks which oscillate over ambiguous figures. Parallel distributed processing also suggests the possibility of a neural perspective on what folk psychology would call rational belief change. Success in philosophy of science at specifying rational criteria for theory change has been modest. We think it is worth exploring the possibility that a network representation of the dynamics of rational theory change can amount to a nonsentential account of the application of coherence conditions of theory choice.

This brings us to the last example of network performance we want to mention. Can networks be trained to reason? Paul Churchland (1989) thinks that they can certainly be trained to recognize valid inferences: 'It is, after all, a matter of recognising instances of prototypical *patterns*. And nets are wonderfully skilled at doing that'. Churchland has in mind William Bechtel's work. Bechtel has constructed and trained a neural net to discriminate a class of valid from invalid inferences within the propositional calculus, to a level of skill roughly equal to his beginning undergraduate logic students (see Bechtel and Abrahamsen 1991, pp. 163–175). Bechtel's network consists of 14 input units, 2 layers of hidden units and 3 output units. See Figure 9.5.

Simple numerical inputs were defined for atomic sentences A,B,C,D and

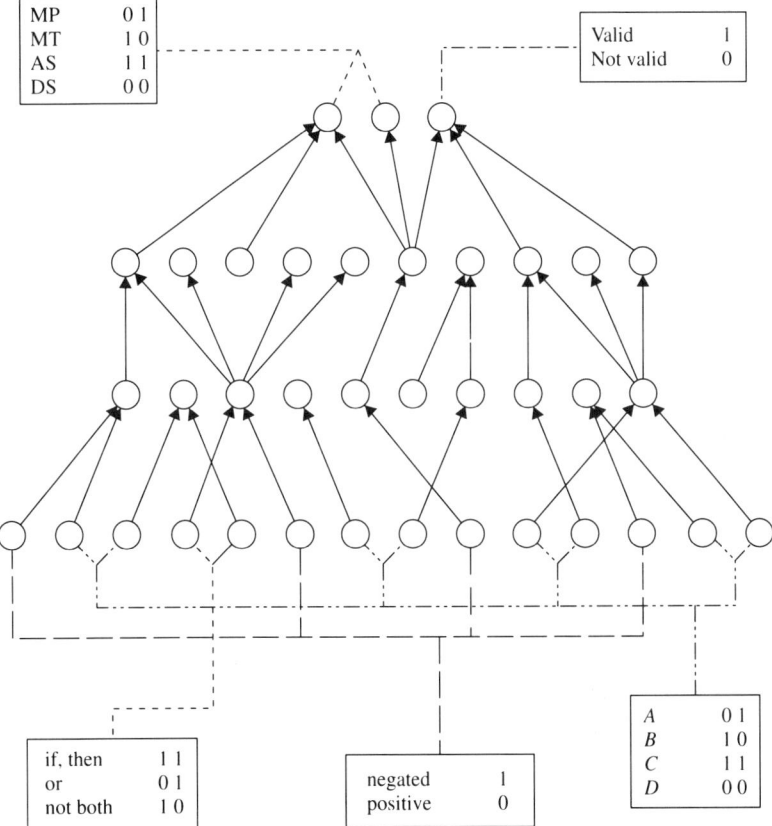

Figure 9.5 (*Source*: Bechtel W. and Abrahamsen A. A. (1991). *Connectionism and the Mind*. Oxford: Basil Blackwell, p. 169.)

connectives *if... then*, *or*, and *not both*, by assigning to each a binary pair, while negation was indicated by a single value, 0 or 1. The argument

If A, then C, C, therefore A
can thus be coded as the pattern
00111011011001

using the following interpretation for each position in the 14-place input vector:

position	1:	0	=	no negation
	2–3:	0,1	=	A
	4–5:	1,1	=	if... then
	6:	0	=	no negation
	7–8:	1,1	=	C
	9:	0	=	no negation
	10–11:	1,1	=	C
	12:	0	=	no negation
	13–14:	0,1	=	A

The correct 3-position output vector for this pattern is 010, where the first two numbers code one of 4 possible argument forms and the last codes for validity as follows: modus ponens 01, modus tollens 10, alternative syllogism 11, and disjunctive syllogism 00; valid 1, invalid 0. So the above argument is modus ponens, invalid.

How was the network trained and how did it perform? From the atomic sentences and the connectives

> ... 576 possible problems were constructed. These were divided into three sets of 192, one of which was used for initial training. The training consisted of 3000 epochs, each consisting of the 192 trials, with the patterns being presented to the network in a random sequence. After training the second set of 192 patterns was presented to the network as a test. An answer was judged to be correct if the value on all output units was on the correct side of 0.5. The network was correct on 139 patterns, or 76% of the time. (Chance would be 12.5%.) The network had obviously generalised, although not perfectly. After testing with the second set of patterns, the network was then trained for another 5000 epochs on these patterns. It was then tested on the third set of 192 patterns. This time the network was correct on 161 patterns or 84% of the time (Bechtel and Abrahamsen 1991, p. xx).

Although the training schedule is quite long, this is an impressive performance for such a small network. Options for increasing its *rate* of learning would include alternative coding strategies at the input phase, increasing the number of nodes in the hidden layers and also the number of hidden layers, and improving the weight adjustment feedback algorithm. Improvements would also be necessary to expand the number and size of arguments or patterns in the propositional calculus that the net could deal with.

Despite their limitations, all three networks discussed enjoy one great advantage: the theory of knowledge representation coheres in a theoretically fruitful way with attempts to understand the causal structure of learning. And this provides some hints for dealing with our earlier difficulties in educational theory.

Knowledge and Unity

From a network point of view, there is no physical difference in the representation of 'knowing that' and 'knowing how'. All the knowledge a neural network possesses is located in its geometry and connection weights. A person's knowledge of bike riding, one may assume, is coded more efficaciously by the adjustment of network weights from a range of spiking frequencies issuing in the sensory organs associated with feedback-induced balance and motor coordination than from firings along the auditory nerve, however eloquent and inspiring. But there is nothing fundamental about this difference; it merely reflects the causal proximity of relevant events in the brain. (Unusual 'neural wiring' in people can result in very unusual learning skills: for example, learning to play a piano where training has been almost entirely a matter of *listening* to music.)

Hirst's puzzle over the nature of practical knowledge arises mainly because of his views on knowing that. In the absence of any systematic body of sententially expressed theory, or law-like generalizations concerning complex tasks like teaching or administration, what is the status of knowledge acquired by practice? In the case of first-language acquisition it was once fashionable to posit a rule-based language acquisition device. But this move seems implausible with teaching or administration. However, as we have seen, neural nets are quite capable of learning from experience without the benefit of rules. It is, of course, nice to have sentential representations of good generalizations, important facts, and the like, since a well-turned train of auditory spiking vectors can cause systematic alterations to neural connection strengths and organization.

There is a double bonus here, in that if nets find it easy to learn associations among *these* inputs then resulting patterns of causal mapping may enable them to go proxy for full-blooded experience. We can indulge in armchair theorizing. As we assume that the vast bulk of human learning proceeds without the benefit of sententially formulated true theory, then as we say, it's nice to have, rather than necessary, for the learning of complex practices. In treating theoretical knowledge as having the same kind of non-sentential network representation as practical knowledge, a difference based on linguistic formulation does not seem as fundamental or as problematic as Hirst's analysis implies.

Concerning Peters's separation of reasons and causes, we can now offer a line of response. Donald Davidson's theory of rational cause is an improvement over dualism, but it is anomalous monism precisely because of the global or holistic nature of reason: reason is Quinean and isotropic. Now network representations are nothing if they are not global. An understanding of their dynamics holds the promise of a monistic account of the link between the macrophenomena of reason and the developing microstructures of cognition — without the anomalies. At least there is a systematic research programme that seems capable of delivering results within the framework of a naturalistic theory of education (see Evers 1987).

Part of the great utility of natural language is that it does effortlessly abstract from causal detail. This is hardly surprising given the span of time over which it has developed. But in allowing natural language to underwrite educational theory, a certain shallowness of analysis comes to invest the treatment of ordinary language educational terms like 'knowledge', 'learning', and 'reason'. It is one of the strengths of the new emerging network models of cognition that the detail is addressed and found relevant.

References

Bechtel W. (1989). Formal logic and pattern completion, unpublished paper.
Bechtel W. and Abrahamsen A. (1991). *Connectionism and the Mind*. (Oxford: Blackwell).
Churchland P. M. (1988). *Matter and Consciousness*. (Cambridge, Mass.: MIT Press, revised edition).
Churchland P. M. (1989). In correspondence with C. W. Evers.
Churchland P. S. (1986). *Neurophilosophy: Toward a Unified Science of the Mind/Brain*. (Cambridge, Mass.: MIT Press).
Evers C.W. (1987). Naturalism and philosophy of education, *Educational Philosophy and Theory*, **19**(2), pp. 11–21;
Gaddes W. H. (1985). *Learning Disabilities and Brain Function: Neuropsychological Approach*. (London: Springer-Verlag).
Grossberg S. (1982). *Studies of Mind and Brain*. (Boston: D. Reidel).
Hirst P. H. (1974). *Knowledge and the Curriculum*. (London: Routledge and Kegan Paul).
Hirst P. H. (1983). Educational theory, in P. H. Hirst (ed.). *Educational Theory and its Foundation Disciplines*. (London: Routledge and Kegan Paul).
Kandel E. R. and Schwartz J. H. (1985) (eds.). *Principles of Neural Science*. (New York: Elsevier).
Nadel L. *et al.* (1989) (eds.). *Neural Connections, Mental Computation*. (Cambridge, Mass.: MIT Press).
Peters R. S. (1974). *Psychology and Ethical Development*. (London: Routledge and Kegan Paul).
Pfeifer R. *et al.* (1989) (eds.). *Connectionism in Perspective*. (Amsterdam: Elsevier).
Pinker S. and Mehler I. (1988) (eds.). *Connections and Symbols*. (Cambridge, Mass.: MIT Press).
Rumelhart D. E. and McClelland J. L. (1986a). *Parallel Distributed Processing: Explorations in the Microstructure of Cognition*, Vol. I. (Cambridge, Mass.: MIT Press).
Rumelhart D. E. and McClelland J. L. (1986b). *Parallel Distributed Processing: Explorations in the Microstructure of Cognition*, Vol. II. (Cambridge, Mass.: MIT Press).
Scheffler I. (1965). *Conditions of Knowledge*. (Chicago: Scott, Foresman).

10
On Theory in Educational Administration: Beyond Greenfield's Subjectivism

Debate over the nature of theory in educational administration has often been between the somewhat narrow and austere models bequeathed by logical empiricism, and those arising out of the subjectivist approach associated with the work of Thomas Greenfield. In what follows, we review some historical and methodological features of this discussion. As a way forward, we propose a view of theory formulation as data compression, and link it to an account of learning and cognition as pattern processing. The result, we suggest, is a perspective on administrative theory that both coheres with natural science and all its explanatory resources, but is able to make use of natural language formulations of subjective understandings and all the practical utility they enjoy.

Overview

It is useful to divide the development of theory in educational administration over the last forty years into two roughly equal periods. The first, which runs from the early 1950s to the early 1970s, spans the interval of intellectual dominance of traditional behavioural science conceptions of educational administration. The second, which covers the time up to the present, is characterized by theoretical diversity and divergent accounts of the nature of administration, its methods, and its purposes. Although traditional views still predominate in the field's literature and administrator training programmes, systematic alternatives are now readily at hand.

Distinguishing the boundaries of intellectual movements is always difficult. Influences can be diffuse and apply with varying force at different times and places. Nevertheless, Thomas Greenfield's address, given twenty years ago, to the International Intervisitation Programme (IIP) at Bristol may be regarded as a major catalyst for the transition that took place: it was both a clear and systematic attack on the key assumptions of traditional administrative science, and a suggestive and influential pointer for future research (see Gronn 1983).

Greenfield's address, and subsequent work, had such extensive implications because two matters he singled out for special attention were among the most central in behavioural science. The first was epistemological, and concerned the *objectivity* of knowledge and procedures for justifying knowledge. The second was ontological, and concerned the reality of objects central to the domain of administrative theory, namely organisations. His position on both matters in the IIP paper owes much to Weber. For Weber (1947, p. 104) '... subjective understanding is the specific characteristic of sociological knowledge'. Greenfield's (1975, p. 9) corresponding comment is that 'it is impossible for the cultural sciences to penetrate behind social perception to reach objective social reality'. And again, for Weber (1947, p. 102) '... there is no such thing as a collective personality which "acts" ', to which Greenfield (1975, p. 9) adds 'only individuals acting on their interpretations of reality'.

However, this kind of analysis would have had little persuasive effect on a view of social explanation that focussed on the science of measurable behaviours if Greenfield had not been able successfully to imply an extension of these methods of sociology to traditional science itself. Two philosophical developments transpired to make this feasible. Since we have presented and discussed these developments in some detail in Chapters 1 and 2 especially, it is sufficient for present purposes to remind the reader of their main thrust. The first development consisted of a number of powerful arguments directed at logical empiricism as the epistemological framework that characterized both traditional science and the conception of science adopted by the Theory Movement. Of particular note was the problematic nature of the relationship between theory and its supporting evidence. Since all our observations are 'theory-laden', that is, expressed in some theoretical vocabulary, there is (a) no 'theory-free' way of seeing the world, and (b) we are, subsequently, in the position of being able to reject an observation rather than the theory under test when observation contradicts theory. Also, the same set of observations can be used to support many different theories, since there are always more possible ways of construing data than there are data points. This raises the problem of which of these many alternative construals (theories) is being conformed by the data. Given these consequences, there seem to be no adequate grounds for arbitration between competing theories, making theory choice either arbitrary or impossible. The second development arose from Kuhn's account of the growth of knowledge, the paradigms view. The important result here is that his conception of justification is paradigm-specific, meaning that paradigms contain their own criteria of what is to count as knowledge. Since there is no additional epistemological explanation of how we are to arbitrate between the knowledge claims and justifications of competing paradigms, it seems plausible to argue that paradigm shifts can be accounted for in the realm of sociological knowledge (for extended discussion of the paradigms view see Evers and Lakomski 1991, ch. 10).

Greenfield demonstrates familiarity with all these arguments, which are deployed, throughout his work early and late, to make a systematic and detailed case against scientific objectivity (see particularly Greenfield 1979, 1980). In summary, his defense of subjectivity in educational administration includes two components: a separation of social science from natural science, with Weber's subjective method of understanding clearly applicable to the former; and the deployment of postpositivist philosophical arguments against logical empiricism and its objectivist version of knowledge and justification.

Theory in Educational Administration

Thomas Greenfield and the writers of the Theory Movement accepted the point that epistemology shapes both the content and structure of administrative theories. Daniel Griffiths's important early book, *Administrative Theory* (1959), illustrates the consequences of such shaping particularly well for logical empiricism. And Greenfield's Bristol paper contains a well-known summary of different epistemological influences on social theory, in the form of a table, under the heading: 'Alternative bases for interpreting social reality' (Greenfield 1975, p. 7). Later on, we propose to reproduce and extend Greenfield's original table, inserting a third basis of interpretation, additional to the 'natural system' and 'human invention' perspectives; namely, a coherentist view. However, for now, we want to explore some ways in which the language of theory formulation affects views of what theories can be about.

Logic and Theory Formulations

One of the fruits of natural science is a developing knowledge of the invariant features of nature. These are patterns that seem to hold up regardless of physical context, and are captured by the great universal generalizations of scientific theory, for example the conservation laws, the second law of thermodynamics, the law of gravitational attraction, and so on. Theory formulations that express these generalizations make much use of mathematics. However, in the early part of this century, mathematics itself was subject to rigorous scrutiny by Bertrand Russell and Alfred North Whitehead, who sought to show that it could be derived from some relatively modest logical apparatus. While their attempt was not successful owing, in part, to the incompleteness of mathematics (proved by Gödel in 1930), their system of logic exerted a powerful influence (first through the logical positivists of the Vienna Circle) on philosophical accounts of scientific theories. In particular, the elegant predicate calculus, with its simple syntax and clear semantics, proved a natural vehicle for expressing and studying theories (for an influential example see Hempel 1958). The fact that this symbolism has only two ways of expressing quantification, namely 'all...' and 'at least one...', poses few problems for those branches of science, such as physics, that trade in relatively context-free invariants. But when the universal quantifier 'all...' is needed to cover generalizations that are true (when they are) only under very restricted conditions, then these formulations become very cumbersome indeed when all the contextual circumstances are spelt out. Consider, for example, what additional qualifications are required to put an 'all' in front of: Teachers with principals who maintain emotional detachment will be more loyal than teachers with principals who are excitable (Hoy and Miskel 1987, p. 5). It would therefore be useful to have a more flexible model of theory formulation for areas of inquiry where patterns are highly context-dependent.

The logic used by Feigl, Hempel, and other logical empiricist philosophers of science, has a further consequence for social science. It is commonly held that recourse to inner mental episodes was excluded in favour of behavioural phenomena because of the empiricist demand for observable evidence. True, but this is only part of the story. A more complete explanation is that standard quantificational logic is not neutral with respect to, for example, Weber's method of understanding. The reason is somewhat technical but worth pursuing briefly. The point is that this logic (and the whole of mathematics) is *extensional*. In an extensional system, any referring expression that is part of a true sentence can be replaced by any other and the sentence will still be true, provided only that the new expression refers to the same object as the old. Thus, consider the following sentence:

(1) Monash Secondary College is a good school.

Suppose (1) is true, and suppose it is also the case that

(2) Monash Secondary College = The largest school in Clayton.

Then because (1) is an extensional context, the following can be deduced by substituting into (1) the co-referring expression 'The largest school in Clayton':

(3) The largest school in Clayton is a good school.

Now consider a typical non-extensional, or intensional, context:

(4) Jones believes that Monash Secondary College is a good school.

Despite the fact that (2) is true, we cannot substitute the co-referring expression into (4) to deduce:

(5) Jones believes that the largest school in Clayton is a good school.

So although (2) and (4) may be true, it does not follow that (5) is true. The reason is that Jones may not know (2), may not know that Monash Secondary College is the largest school in Clayton.

Theory builders with an eye for semantically clear formulations will therefore want to exclude referentially opaque constructions like (4) and (5) from their theories. However, it is not just belief constructions that go. Also excluded are all the other inner mental attitudes (or *propositional attitudes*) that Jones may possess: desiring, wanting, knowing, feeling, thinking, understanding, and their ilk, for they all generate non-extensional, or referentially opaque, sentences. Small wonder that Weber's method of understanding fell on hard times; it could not even be formulated within the logical apparatus of positivist theory building.

Understanding Theory in a World of Causes

One response to the use of extensional logic in social science theory formulations is to deny that the advantages of clarity outweigh the loss of antecedently intelligible mentalistic discourse. After all, what could be more utterly familiar from early childhood on upwards than talk of knowing, thinking, wanting, and understanding? Indeed the very teaching of logic and its austere standards is done with intensional discourse in an intensional framework of promoting knowledge and understanding. And while we are at it, why should inner phenomena like reasons and desires be observable? Surely they can be known inferentially from behaviour in the same way that scientific unobservables such as electrons or quarks are known inferentially from the behaviour of observable nature.

Let us suppose, for a moment, that there are no epistemological or semantical problems with the language in which the method of understanding is conducted. Following Greenfield's approach, let us suppose further that we can inquire into the nature of organisations by interpreting the actions and words of individuals, by 'trying to understand what people within them think of as right and proper to do' (Greenfield 1975, p. 11) and that we are entirely successful. That is, we are able to develop accounts of people's inner motivations, beliefs, and desires that they would recognize as appropriate reflections of the meaning of what they do and why they do it.

Despite having a complete set 'of meanings which people use to make sense of their world and action within it' (Greenfield 1975, p. 7), there remains a deep puzzle as to how these subjective meanings connect up with the causal push and pull of the world's inventory of material objects which, of course, includes flesh-and-blood people. Greenfield does not address this problem, either in his Bristol paper or in any subsequent work, which is surprising because Weber worried about it in the context surrounding the passages Greenfield usually quoted (see Weber 1947, pp. 87–115, and also 10–29 where Parsons, in his Editor's Introduction, canvasses the problem). But without some way of knitting together subjective intensional explanations of human meanings, reasons, and intentions with scientific extensional explanations of the (caused) patterns in nature, it is hard to see how an understanding of human subjectivity could ever explain anything that ever *happens* in the real material world. A strong methodological dualism here threatens to detach Greenfield's method of understanding from the business end of explanation altogether.

Theory as Compression Algorithms

So far we have canvassed two approaches to theory building in educational administration. The first, associated with the logical formalism of logical empiricism, appeared clear and elegant with regard to, say, physics, but quickly became cumbersome and awkward when applied to the highly contextualized phenomena of the social world. The second, which posited a fundamental distinction between the natural sciences and the social sciences and made use of explanation in terms of understanding inner motives and reasons, appeared to become irrelevant when the prospect of a fundamental dualism was pressed. In what follows, we suggest a more naturalistic way through this impasse by invoking some ideas from information theory.

Natural science is at its most impressive when it can produce simple formulations that capture lawlike regularities in nature. However, the impressiveness of science on this score should not blind us to the fact that there are many, more modest, regularities in the world essential for social life. Human communication is sustained by whole networks of regularities concerning the sounds and graphemes that make up spoken and written language. Additional networks maintain the stability of beliefs, desires, and expectations. Indeed, if human behaviour were random, social life would be impossible. The patterns that make organization possible are not (relative to the data) of a piece with the lawlike generalizations of physics. They admit of numerous exceptions. Greenfield is right when he protests that administrative science has not produced a single significant social law. But neither do we have randomness. Instead there is a continuum of patterning, with universal context free regularity at one end and randomness at the other. The logical apparatus favoured in the heyday of administrative science was most valuable only at an endpoint.

Following a suggestion of Daniel Dennett's (1991), based on the work of Gregory Chaitin (1975), we can less wastefully specify a place for this middle-level degree of order in theorizing. Consider first Chaitin's definition of randomness: 'A series of numbers is random if the smallest algorithm capable of specifying it to a computer has about the same number of bits of information as the series itself' (Chaitin 1975, p. 48). The basic idea is that the series is random if it cannot be *compressed* into an expression containing fewer bits. Conversely, data are *patterned* if they can be compressed. Thus the series of numbers

010101010101010101010101010101010101010101010101010101010101

is patterned because it can be compressed into

THIRTY REPETITIONS OF 01

The size of the series is 62 characters (counting a blank at either end); but the size of the compression algorithm is only 26 characters (assume a conversion of characters to binary digits). Similarly, if the observational data of science are expressed as a series of binary digits, then true sentences expressing lawlike generalisations will make for massive data compression.

Compression algorithms are also quite common in the social arena. In fact, our whole commonsense belief–desire explanatory framework compresses data. Explaining that Jones went to the fridge to get a beer because she was thirsty provides a masterful compression of all the present physical detail into just that which might be relevant to explain (and predict) Jones's behaviour. A child misbehaving in class might be understood by others, using belief–desire theory, to be seeking *desired* attention by behaving in a way *believed* to attract it. Explaining such behaviour in terms of the child's posited subjective coordination of inner beliefs and desires will economically account for, or compress into a short formula, a great deal of otherwise complex, disparate, and irregular activity.

Not all the data will fit this pattern. We now know better than to apply the theory to, say, the child who struggles to avoid producing mirror writing. The usual strategy, however, is to keep the compression algorithm going for data that fit the pattern, and treat all the exceptions as 'special' and subject to a different theory, or model, of explanation. Hence we have the common distinction between Special Education with its grouping of educationally relevant cognitive phenomena into taxonomies suggested by neuroscience and the study of causes, and mainstream education, where talk of reasons prevails.

The great utility of many of the belief–desire, or folk-theoretic, or inner subjective accounts of complex social and organizational life that are given is that they are empirically adequate for many cases, much of the time. Methodologically, this amounts to treating the complex data of social life as exhibiting a simple pattern against a background of 'noise', with the exceptions being the noise. If we want to account for more of the noise, the total pattern usually becomes more complex (though it achieves greater compression because of the larger amount of data now represented). Consider a familiar example from physics. For most purposes, the dynamical behaviour of particles can be described in a simple way by Newton's laws of motion, with the exceptions — objects whose motion is near the speed of light — appearing as noise. To include (and therefore reduce or eliminate) the noise, a more complex pattern needs to be described — one that allows for relativistic effects. In practice we use both theories: Newton's for low velocities, where ease of calculation does not compromise empirical adequacy, and Einstein's for high-velocity particles. The example is instructive in other ways. Thus we can say that at low velocities the world behaves *as if it were Newtonian even though it is not* (it is really relativistic). We treat Newtonian mechanics *instrumentally*, and relativistic mechanics *realistically*.

We want to make a similar point in relation to social science. Human beings use the intentional stance, or belief–desire theory, for a fair number of everyday matters because in these cases the social world behaves *as if* it were an aggregation of subjectively coordinated beliefs and desires. However, lurking in the background is a more intricate, but comprehensive, pattern in the causal fabric of human brain, and other physical processes (just as neural functioning is as much responsible for normal writing as it is for mirror writing).

Unfortunately, there are two good reasons why the analogy with the physics example breaks down. First, there is no known compression algorithm on the causal data of microcognition that is comparable to relativity. For purposes of prediction and explanation in social science we rarely have any other choice but to rely on the locally effective data compression of subjectively oriented folk-theoretic theory formulations. Second, many of the acts actually performed in administration are accomplished by language, or more precisely, by speech acts. That is, the very constitution of some administrative acts is dependent on the mentalistic framework embedded in the language by which they are effected: for example, requesting that someone offer reasons for a proposal.

Our suggestion therefore is to be fairly pragmatic about administrative theory. After all, it is the empirically adequate expressions of patterning that are most useful in accomplishing organisation. Since actions need to be taken with some sense of an outcome more probable than chance, any way of discerning a regularity in the passing show of experience will be better than none at all. A more relaxed approach to the structuring of developing administrative theories than that entailed by extensional quantificational standards of logical empiricism is therefore reasonable. Yet the way to improve theory lies in amending it to accommodate anomalies, exceptions, or the unexplained, to extend its reach with the goal of achieving comprehensiveness. This is why, in the physics example, the world is regarded as *really* relativistic, rather than Newtonian. Since it is highly unlikely that improvements to administrative theories in the direction of greater comprehensiveness will *exclude* science, it is important to constrain the interpretation of proposed theory formulations with the requirement that they be compatible with, or cohere with, or be on the developing explanatory agenda of, natural science. The resulting double standard means praising the simple falsehoods for their great local utility while being willing to recognize truth on the side of the big, but more complex, picture.

Theory into Practice

One weakness of Greenfield's approach to theory, we claimed, is that it leaves problematical how a person's inner subjective motives and reasons are related to their behaviour. Unless we can fill out the detail there will be a difficulty in explaining how people are guided by theory or, more practically, how we are to construct administrator training programmes. Some of Greenfield's suggestions — for example, spending time in a monastery, or as a patient in a mental hospital — imply a lacuna here (Greenfield 1980, p. 112). We want to conclude by indicating some ways in which viewing theory formulations as compression algorithms might, in the context of a more suitable, naturalistic, theory of learning, throw some light on relating theory to practice.

Interestingly, puzzles over relations between theory and practice are common to both subjectivist and traditional science approaches to administration because they share a common approach to the form of theory formulations; namely, a *symbolic*, or language-like, approach. This perspective is then applied to inner processes of thought, reflection, and decision, which then appear as processes involving the rule-based manipulation of structures with symbolic properties. However, while this is a congenial way of representing theoretical knowledge, it is barely adequate as a model for knowledge of practice. Certainly, for perhaps the vast bulk of successful practice, practitioners would be hard pressed to give symbolic formulations of the practical knowledge they possess, and it is doubtful if the required formulations could be given. In fact, until recently, there has been no adequate (non-symbolic) model for representing knowledge of practice.

But now consider the new emerging neural network models of representation that are gaining currency in cognitive science (Rumelhart and McClelland 1986; Churchland 1989; Clark 1993). The new approach, known as connectionism (or parallel distributed processing) attempts to construct models of cognition based on what is known about information processing in the brain: that is, how thought might actually be instantiated in neural wetware. The picture we get from this approach is not one of the brain as a sequential combinatorial processor of symbol-like tokens, but rather the brain as a pattern processor, with learning being like pattern extraction or discovery. There are many models, but a common one can be described with the help of Figure 10.1. The columns of circles, called layers, denote neurons, and the arrows connecting them, called weights, denote the strength of the synaptic junctions between the neurons. Incoming signals are expressed in the form of a column vector, with each element matching a receptor neuron in the input layer. Each element is then multiplied by the various weights, and added by a summing function, to give the transformed pattern, or signal (representing neuron spiking frequencies) at the hidden layer. A similar transformation takes place en route to the output layer. These transformations can be used to drive network learning if the system is augmented in two ways. First, the network's output is matched against what should be the correct output (or decision) for the corresponding input. And second, the difference between output and expected output is used to change all the weights so as to reduce that difference. If there is a pattern in the flow of input data, and the network is appropriately designed, the steady pressure of feedback-induced weight adjustment will cause the net to learn the pattern, representing it internally in the resulting configuration of weights and their values. For each input, the net can then 'make a decision' in the form of producing the required matching output response.

This type of learning by backpropagation has been used to train nets to learn

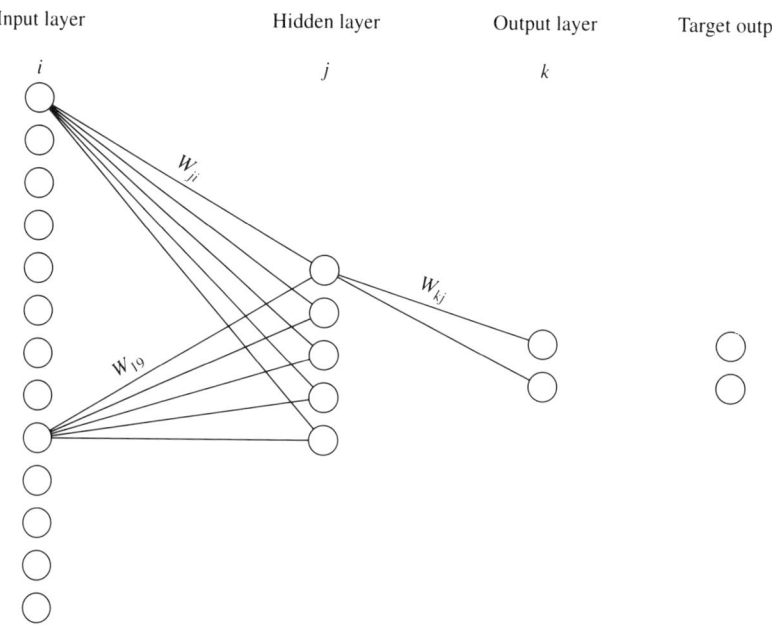

Figure 10.1 A three-layer feedforward net with 13 input nodes, 5 hidden nodes, and 2 output nodes. The target output nodes are not part of the net's design. Weights W_{ji} and W_{kj} are adjusted by backpropagation of error to minimize their contribution to the difference between the output layer and target output. Not all connections are shown.

the past tenses of English verbs, to learn to pronounce written text, to identify valid from invalid argument forms in logic, to distinguish rocks from mines, for speech and writing recognition, and for making medical diagnoses from clusters of symptoms, to give just a few examples (see Bechtel and Abrahamsen 1991). To the extent that these computer models of neural networks are biologically realistic, we might conjecture some corresponding attributes of human learning from experience. Thus, administrative problem scenes might be classified according to some breakdown into relevant salient features applicable to a range of problems. These features (which would be discerned from much prior processing) would, in turn, be the elements of an input vector, more like a visual pattern or picture, to be processed to produce a decision, or output response representing expected consequences. In the event of a mismatch between expectation and experience, we think again, making adjustments to our assessment of features and the importance we place on them, or our view of what will happen if we act in certain ways. We make more decisions attending, perhaps unconsciously, to results and making further adjustments. Over time we may come to exhibit a high level of skilled performance concerning the most complex of administrative scenes (Chapter 8 contains a more detailed discussion).

The beauty of this learning is that it does not require any prior formulated

Table 10.1: Theories of social reality

Dimensions of comparison	A natural system	Human invention	Naturalistic coherentism
Philosophical basis	Realism: the world exists and is knowable as it really is. Organisations are real entities with a life of their own.	Subjectivism: the world exists but different people construe it in very different ways. Organizations are invented social reality.	Naturalism: the natural world exists because that is the best, most coherent, explanation of phenomena. Organizations are real patterns of human association.
The role of science	Discovering the universal laws of society and human conduct within it.	Discovering how different people interpret the world in which they live.	Discovering empirically adequate compression algorithms for capturing comprehensively the regularities in social life.
Basic units of social reality	The collectivity: society or organizations.	Individuals acting singly or together	Individuals responding to other individuals or groups, and the natural world.
Method of understanding	Identifying conditions or relationships that permit the collectivity to exist. Conceiving what these conditions and relationships are.	Interpretation of the subjective meanings that individuals place upon their action. Discovering the subjective rules for such action.	Exploring networks of expressions that fit the behaviour of individuals and account for their own theories of behaviour.
Theory	A rational edifice built by scientists to explain human behaviour.	Sets of meanings that people use to make sense of their world and action within it.	The most coherent network of claims about the world.
Research	Experimental or quasi-experimental validation of theory.	The search for meaningful relationships and the discovery of their consequences for action.	Search for non-random features of the world.
Methodology	Abstraction of reality, especially through mathematical models and quantitative analysis.	The representation of reality for purposes of comparison. Analysis of language and meaning.	Full range of techniques that will establish the existence of patterns.
Society	Ordered. Governed by a uniform set of values and made possible only by those values.	Conflicted. Governed by the values of people with access to power.	Patterned. Governed by networks of causes describable in various ways.
Organizations	Goal-oriented. Independent of people. Instruments of order in society serving both society and the individual.	Dependent upon people and their goals. Instruments of power that some people control and can use to attain ends which seem good to them.	Dependent upon individual and resultant collective goals, with stability ensured by memory and records.

Table 10.1: *Continued*

Organizational pathologies	Organizations get out of kilter with social values and individual needs.	Given diverse human ends, there is always conflict among people acting to pursue them.	Dissonance among individuals and between organization and society.
Prescription for curing organizational ills	Change the structure of the organization to meet social values and individual needs.	Find out what values are embodied in organizational action and whose they are. Change the people or change their values if you can.	Promote structures for enhancing organizational learning.

compression algorithms expressing the pattern linking inputs and correct outputs. Rather, the network extracts what pattern there is and compresses it into the learned weight configuration. Herein lies a person's practical expertise, though all of us would have difficulty translating our own internal configurations of weights into some external linguistic representation of that knowledge. Notice, however, that the practical expertise so gained has an important limitation: it fails where the statistical structure of experience does not match the statistical structure of the world of future application. But, of course, the most interesting administrative problems are quite often counterfactual; that is, they call for judgments on what is to be done if we alter current practices. Advanced levels of professional practice typically call for judgments on the non-routine.

We can improve matters by presenting to the eye powerful symbolically expressed summaries of the more extensive experience of others. After all, symbolic theory formulations are also able to be processed by suitably trained networks — those that can read. In this way, reading can, with profit, go proxy for some first-hand experiences. However, really massive gains over experience come from manipulating theory formulations so that they apply to matters beyond experience, enabling us to think hypothetically and act accordingly. The imagination so augmented can explore policy and decision option spaces that have not been lived through, perhaps not with precision, owing to the friability of current administrative theory and its context dependence, but with a modest prospect of beating chance. The result is a less demanding, but more plausible, role for theory as a guide to practice.

The implications of a naturalistic account of theories wedded to a naturalistic account of cognition (all flowing from a naturalistic coherentist view of knowledge and its justification) are yet to be worked out in detail in educational administration. However, Table 10.1, adapted from Greenfield's IIP paper, provides some suggestive contrasts with the symbolic/cognitive approach, in both its human invention form and its systems scientific form, and an agenda for further work.

References

Bechtel W. and Abrahamsen A. (1991). *Connectionism and the Mind* (Oxford: Blackwell).
Chaitin G. (1975). Randomness and mathematical proof, *Scientific American* **232**(5), pp. 47–52.
Churchland P.M. (1989). *A Neurocomputational Perspective.* (Cambridge, Mass.: MIT Press).
Clark A. (1993). *Associative Engines.* (Cambridge, Mass.: MIT Press).
Dennett D. (1991). Real patterns, *Journal of Philosophy*, **88**(1), pp. 27–51.
Evers C.W. and Lakomski G. (1991). *Knowing Educational Administration: Contemporary Methodological Controversies in Educational Administration Research.* (Pergamon Press: Oxford).
Greenfield T.B. (1975). Theory about organisations: A new perspective and its implications for schools, in Hughes M.G. (ed.) *Administering Education: International Challenge.* (London: Athlone Press, cited as reprinted in Greenfield and Ribbins 1993).
Greenfield T.B. (1979). Ideas versus data: How can the data speak for themselves?, in Immegart G.L. and Boyd W.L. (eds.) *Problem-Finding in Educational Administration.* (Lexington: Lexington Books).
Greenfield T.B. (1980). The man who comes back through the door in the wall: Discovering truth, discovering self, discovering organisations', *Educational Administration Quarterly*, **16**(3), pp. 26–59. (Cited as reprinted in Greenfield and Ribbins 1993).
Griffiths D.E. (1959). *Administrative Theory.* (New York: Appleton-Century-Crofts).
Gronn P. (1984). *Rethinking Educational Administration: T.B. Grenfield and His Critics.* (Geelong: Deakin University Press).
Hempel C. (1958). The theoretician's dilemma: a study in the logic of theory construction, in Feigl H., Scriven M. and Maxwell G. (eds.) *Minnesota Studies in the Philosophy of Science*, Vol.II. (Minneapolis: University of Minnesota Press).
Hoy W.K. and Miskel C.G. (1987). *Educational Administration.* (New York: Random House, third edition).
Rumelhart D.E. and McClelland J.L. (1986) (eds.) *Parallel Distributed Processing.* Vols I and II. (Cambridge, Mass.: MIT Press).
Weber M. (1947). *The Theory of Social and Economic Organisation.* (New York: Free Press, translated and edited by T. Parsons).

11

Cognition, Values and Organizational Structure: Hodgkinson in Perspective

In a number of earlier pieces (Evers 1985, 1993; Evers and Lakomski 1991, ch. 5, 1993) we have reviewed Christopher Hodgkinson's principal works (Hodgkinson 1978, 1983, 1991) and debated some of his most important and influential theses, particularly his values taxonomy as a decision guide for moral leadership. Hodgkinson (1986, 1993) has responded with characteristic vigour, wit, and ingenuity. This chapter seeks to continue the debate with a focus on Hodgkinson's most recent attempt to relate his theory of the place of values in administrative decision-making with his theory of leadership and its consequences for organizational design. We explore the idea that these theories are related centrally by an isomorphism that exists between the structure of value as expressed in Hodgkinson's values taxonomy, and the stratified leadership/followership structure he posits for organizational design. We propose to examine the nature of this isomorphism and its intellectual foundations in Hodgkinson's thought against the background of a number of alternative isomorphisms. The first is that proposed by Plato between the structure of knowledge and the structure of society. The second, on which we spend most time, is that which Simon (1976) thought should obtain between the structure of decision-making and preferred organizational design. The third, involving only a brief excursion, is that proposed by critical theory between kinds of knowledge and democratic patterns of governance. Last, and of equal brevity, comes our own proposed isomorphism: a relationship between a naturalistic theory of reason (which sees reason as constrained by the demands of our account of the growth of knowledge) and organizations structured by the requirement of promoting organizational learning.

Cognition and Organizational Structure

Theories of human cognition have long played an important part in understanding organizations, in justifying particular organizational configurations, and in defining the nature of leadership. Plato's *Republic* (1970, p. 488) contains perhaps the first comprehensive presentation of these themes. In a revealing parable against democracy he asks us to imagine a ship where all the sailors, though deficient in nautical skill...

> ... are quarrelling together about the pilotage — each of them thinking he has a right to steer the vessel, although up to that moment he has never studied the art, and cannot name his instructor, or the time he served his apprenticeship.

The point Plato goes on to make is that without adequate knowledge, the crew is not able to select a skilled person even if such a person were present. Control of the selection process should be in the hands of those with appropriate knowledge. However, in a properly run ship, it is the captain who has this knowledge, who is able to make these choices, who is at the apex of the correct relationship between knowledge and power.

Plato's theory of knowledge, in the *Republic*, assumes a broad distinction between knowledge of examples and knowledge of principle. More technically, this amounts to a distinction between knowledge based on the perception of many instances, and knowledge of the form, essence, or idea, underlying each instance. True knowledge is of the forms, or essences, and is acquired by the exercise of the intellect, suitably trained. Leadership of the state is therefore the province of those whose intellects permit an education sufficient to engender apprehension of the Form of the Good, the form by which a particular state is adjudicated to be morally just. And these intellects are possessed by philosophers, who are therefore the natural rulers of the just state.

In developing this isomorphism between cognitive capacity and leadership capacity, Plato is aided by an objectivist theory of knowledge. Thus when it comes to the question of whether something is good because a philosopher-king says so or whether the philosopher-king says so because it is good, the theory of forms implies the latter. A just society, or organization for that matter, thus models in its structure, the cognitive capacities of its members to meet its knowledge demands.

Despite its appealing anti-democratic direction, Hodgkinson has qualified reservations about this Platonic argument:

> It is futile to wish with Plato for the advent of the philosopher kings, but it is neither unreasonable nor impractical to seek programs which will move leaders in the Platonic direction (Hodgkinson 1991, p.134).

One ground for Hodgkinson's caution here is that his theory contains no analogue of Plato's theory of forms. For Hodgkinson, organizational decisions are made against the background of a continuum of legitimate interests, or valid points of view, rather than against one abstract ideal imposed by the leader on all. After all, philosopher-kings can make mistakes, or be unworthy; so '...it would be safer to assume some bias in the unworthy direction and some use for philosophy as a therapeutic against pathology' (Hodgkinson 1978, p. 153).

A further reservation may be found in an important difference between the Platonic theory of reason, and Hodgkinson's. For Plato, cognition is largely a matter of reason, with the whole range of forms — ethical, mathematical, scientific, aesthetic — being apprehended by the intellect. But Hodgkinson maintains a distinction in ways of knowing, at least between the ethical and the scientific, between facts and values. To rework Plato's parable, on Hodgkinson's ship, the technical, scientific matters of navigation *can* be left to a trained crew; it is the captain's duty to determine the purposes of the voyage and the destination. Generalizing to organizations, it is the tasks of 'mobilizing, managing, and monitoring' that management performs (Hodgkinson 1991, p. 64). These are the technical and rational tasks of finding and acting on optimal means for achieving given ends. Administration and leadership are about purposes and destinations, about 'philosophy, planning and politics' (Hodgkinson 1991, p. 64).

In adequately achieving these tasks, two different epistemologies are required; one for knowledge of means, and one for knowledge of ends. Hodgkinson thus stands somewhere between Plato, whose rationalist view of knowledge models even empirical science on the ideal of mathematics, and Simon, whose sternly empiricist view of knowledge leaves no room for ethics, save as the expression of affective preference. In accepting the traditional fact/value distinction, Hodgkinson embraces an empiricist view of science, not unlike Simon's. Hence his separation of management as means-oriented from administration as ends-oriented. However, while empiricism provides a broad set of procedures for adjudicating among competing claims about means, Hodgkinson wants to ensure that a helpful and practical adjudication of values can occur in the absence of some version of Plato's strong objectivity thesis.

His first two books, *Towards a Philosophy of Administration* (1978) and *The Philosophy of Leadership* (1983) develop and defend a values taxonomy that may also function as a decision procedure. Hence his classification of values into three types corresponding to principled matters of will (Type I), judgments based on reason (further subdivided into Type IIa and Type IIb) and expressions of affect (Type III) forms a hierarchy — $I > IIa > IIb > III$ — reflecting the moral superiority of different types of values (Hodgkinson 1978, p. 116). His most recent book, *Educational Leadership: The Moral Art* (1991) places important restrictions on the scope of any such procedure by arguing that administration is an art. This identification serves two important purposes in Hodgkinson's thought. The first is more familiar, against Simon, and differentiates administration from science along the fact/value dimension. The second signals the non-algorithmic nature of administration, complicating any straightforward application of his values hierarchy (Evers 1993).

We use the term 'non-algorithmic' to capture an implied epistemological balance between the extremes of arbitrary subjectivism and Platonic objectivism over ethics. Maintaining the viability of this balance requires Hodgkinson to see cognition as wider than mere reason. It also embraces the will, and the affective domain. Part of his argument for this wider perspective involves showing the inadequacies of Simon's theory of reason, given the demands of administrative decision-making.

For Simon, the fundamental problem of administration was the persistence of 'principles of administration' that did not allow for rational decision-making. Simon likened these principles to proverbs that occur in pairs:

> For almost every principle one can find an equally plausible and acceptable contradictory principle. Although the two principles of the pair will lead to exactly opposite organizational recommendations, there is nothing in the theory to indicate which is the proper one to apply (Simon 1976, p. 20).

Although he does not deny the importance of action carried out as the result of applying principles, Simon draws attention to the neglect of choice prior to action, in other words, the *processes of decision*. More specifically, a 'general theory of administration must include principles of organization that will ensure correct decision-making...' (Simon 1976, p. 1). *Rational decision-making* is thus the centrepiece of Simon's theory of administration which, as instrumental reasoning, finds expression in his *means–end schema*.

Following classical positivist doctrine, Simon differentiates sharply between facts and values, factual propositions being validated by agreement with observation, and value judgements being validated by human fiat. The business of determining *questions of policy* is one to be decided by reference to a system of intermediate values and the appraisal of their relative weights. This involves both ethical and factual considerations. *Questions of administration*, on the other hand, are to be decided by comparing possible lines of action in terms of this value system; and this is to be done by considering *all* relevant facts of the matter. Rationality in administrative behaviour consists in making a correct administrative decision, that is, a decision that selects appropriate means to reach designated ends. An administrator acts rationally insofar as s/he selects such effective means. Administrative decision-making thus consists of alternatives chosen that are appropriate means for reaching desired ends where these ends, in turn, are often merely instrumental to more final objectives. Hence Simon talks about a series or hierarchy of ends. Rationality in his system means constructing such means–ends chains.

The theory of organizational structure that is implied in satisfying the constraints of bounded rationality in decision-making is clear in its broad outline: whatever its other purposes, an organization's material configuration should reflect what is required for effective decision-making. That is, a flow chart linking elements of the abstract decision process should be isomorphic to the organization's functional design (Simon 1976, pp. 240–241). A basic difference between Simon's and Hodgkinson's theory of organizational structure can therefore be derived from the distinction between instrumental reason and the features of Hodgkinson's hierarchy of values.

Instrumental Reason and the Complexity of Goals

Since all human behaviour is motivated in some sense, and since administration is merely a subset, it too is motivated, or fundamentally value-laden (Hodgkinson 1991, p.114). Values, in turn, according to Hodgkinson (1991 p. 84), are

> ... concepts of the desirable, and because administration is essentially motivational and valuational, it is intrinsically philosophical. It is squarely in the realm of morality ... all leaders are philosophers and educators ... because they are engaged, commonly or otherwise, in the education of desire — in persuasion about ends and the conversation of means into ends.

Facts in Hodgkinson's conception of leadership are, as he says, 'all that is the case', borrowing a Wittgensteinian term. Unlike values, which are always in conflict, facts can *never* be in conflict. They 'pertain to the objective world of nature and science. They are by definition true' (Hodgkinson 1991, p.89). Since values are always in conflict, and since we all bring our own divergent values to bear on the facts, Hodgkinson believes that the task of leadership is to find some 'working resolution of value conflict'. In his value theory, and especially the *hierarchy of values*, he attempts to do just that. Without discussing these in detail (see Evers and Lakomski 1991, ch. 5), let us note here that his hierarchy of values raises the fundamental question of how to choose the right values and how to justify this choice. Suffice it to note that Hodgkinson, like Simon, sharply differentiates between facts and values according values, as *concepts of the desirable*, exclusivity in terms of defining administrative leadership or action. His position is thus also characterized as accepting the fact/value dichotomy, ironically characteristic of classical positivism.

Simon's hierarchy or series of ends is described as being always incomplete and sometimes inconsistent. This is so since the ends to be attained are often incorrectly stated because alternative ends are not considered. Rational decision-making, according to Simon, always requires the comparison of alternative means in terms of the respective ends to which they will lead, a central feature of instrumental, or means–ends reasoning. This means that 'efficiency', by which Simon understands the attainment of maximum values with limited means, must be a guiding criterion in administrative decisions. A further limitation concerning the means–end schema to which he draws attention is that a complete separation of means from ends is not possible since alternative means are not 'valuationally neutral'. In addition, the means–ends terminology tends to gloss over the time element in decision-making insofar as only one end at a time can be considered as compared with the many ends over a period of time. As Simon points out, rational choice poses two problems. The first is that if a particular end is to be realized *now* what alternative ends must thus be given up? And secondly, how does this limit the ends that may be realized at other times? Having listed these difficulties, Simon nevertheless believes that 'A theory of decision in terms of alternative behaviour possibilities and their consequences meets all these objections' (Simon 1976, p. 66).

Individuals at any point in time are confronted with a large number of alternative behaviours:

> Decision, or choice ... is the process by which one of these alternatives for each moment's behavior is selected to be carried out. The series of such decisions which determines behavior over some stretch of time may be called a *strategy* ... The task of rational decision is to select that one of the strategies which is followed by the preferred set of consequences. It should be emphasised that *all* the consequences that follow from the chosen strategy are relevant to the evaluation of its correctness, not simply those consequences that were anticipated (Simon 1976, p. 67).

Decision-making thus involves three steps: (1) all alternative strategies must be listed; (2) all the consequences that follow upon each of these strategies have to be determined; and (3) these sets of consequences have to be evaluated in comparative fashion. Simon acknowledges that it is obviously impossible for the individual to know *all* the alternatives or *all* their consequences, and thus notes the discrepancy between actual behavior and his model of objective rationality (Simon 1976, p. 67). It is here that knowledge has an important function. It is to 'determine which consequences follow upon which of the alternative strategies' (Simon 1976, p. 68). This entails selection from the whole class of possible consequences a more limited subclass, or even (ideally) a single set of consequences correlated with each strategy. Since individuals, however, do not know the alternative behaviour consequences of their behaviours, there being no 'reverse causality'; they can merely form 'expectations' of future consequences which are in turn based on known empirical relationships.

It can now be seen that the means–end distinction in Simon's conception is not quite the same as the distinction between facts and values. The connection between the two sets is as follows:

> A means–end chain is a series of anticipations that connect a value with the situation realising it, and these situations, in turn, with the behaviours that produce them. Any element in this chain may be either 'means' or 'end' depending on whether its connection with the value end of the chain, or its connection with the behavior end of the chain, is in question. The means-character of an element in a means–end chain will predominate if the element is toward the behavior end of the chain; the end-character will predominate if the element is descriptive of the consequences of behavior. If this be so, terms that are descriptive of the consequences of a behavior may

be taken as indicia of the values adhering to that behavior (Simon 1976, p. 74).

Defined in this way, the means–end relationship does not allow for a sharp split between facts and values although it remains true that for Simon they remain analytically quite separate. He has it both ways because the justification of provisional or intermediate ends, although instrumental, eventually backs all the way up to some final end or ends beyond the scope of his science of administration.

Summing up, 'rationality is concerned with the selection of preferred behavior alternatives in terms of some system of values whereby the consequences of behavior can be evaluated' (Simon 1976, p. 75). 'Knowledge about the consequences of behavior was thus identified as a primary influence on choice'. A second influence consists in the preferences of the behaving individual for one set of consequences rather than another. 'The problem of choice is one of describing consequences, evaluating them, and connecting them with behavior alternatives' (Simon 1976, p. 77).

Against Instrumental Reason

Instrumental reason has certainly had its critics in educational administration. The best known are critical theorists, who have maintained that capitalism owes its stability against internal contradictions at least in part to the common but ideological belief that human cognition consists of instrumental reasoning, modelled after the fashion of science construed positivistically. For this pattern of reasoning, while analytically well equipped to evaluate competing means, fails adequately to problematize ends, and so promotes an uncritical acceptance of a social system's dominant goals or ends. Following Habermas (1972), critical theorists customarily distinguish two additional forms of reason and knowledge, namely, that associated with human communication, or hermeneutics, and that associated with human liberation, or emancipatory knowledge, and argue that these are essential for a more complete understanding of social life and its material possibilities (For a systematic presentation of this position see Foster 1986.)

The organizational structure isomorphic to this account of cognition is different again to that of Plato, Simon, or Hodgkinson. Rather, it places heavy emphasis on participatory democratic structures, with the hallmark of leadership being the capacity to empower all organizational participants.

Corresponding to this account of cognition is an epistemology based on the methods of transcendental deduction. Essentially, this is the method of analysing the *presuppositions* of thought in a particular domain (or in all domains). Habermas's account of knowledge reflects ultimately his view of what are the knowledge conditions for human flourishing. His defense of a democratic ethic is based on an analysis of the presuppositions of human communication.

Hodgkinson's critique of the limits of instrumental reason employs a quite different methodology. His key premise, argued for in a variety of ways is, as we have seen, the value-ladenness of administration. In our view, many of his arguments against Simon employ the technique of showing that certain of Simon's fundamental distinctions outrun their available posited evidence. For example, he argues that the theoretical decision to choose instrumental reason is either made on instrumentally reasonable grounds or it is not. If it is not then this 'does of course imply at least another level of rationality' (Hodgkinson 1978, p. 58), and hence instrumental reason is not a sufficient account of rational choice. If the choice is made instrumentally then where does the regress stop for justifying the chain of ends invoked for its use? As Hodgkinson (1978, p. 58) remarks, 'some system of values' is required. The alternative simply makes the choice random.

Another argument looks at the distinction between role and character. Too strong a focus on the abstract features of reason distracts from the obvious point that decisions are made by humans with beliefs, wants, desires, and personalities. An organizational role in practice is always *an interpretation* of what is to be done, filtered through a person's character, including of course moral aspects of character (Hodgkinson 1991, p. 93). As there is no way of factoring out the moral from a person's character-laden interpretation of their place in organizational life, administrative roles are inescapably value-laden. According to Hodgkinson (1991, p. 93), the only scope for minimizing this values influence is by a retreat into managerialism, but he is skeptical of the required disengagement of personality being achieved.

However, were such a move successful, it would succeed only by limiting choice in decision-making to that of a 'mere conduit'. Decisions concerning goals and purposes still have to be made somewhere, and for Hodgkinson this will occur at the level of administration, which is above management. Ultimately then, it is the open nature of organizational choice that demands a broader account of cognition than instrumental reason. Partitioning off the managerial function to the instrumental merely serves to highlight Hodgkinson's point about the embeddedness of values in administration.

These arguments indicate how a broadened account of cognition maps onto a preferred structure for organization. We can now provide some elaboration to show why an isomorphism holds between his values taxonomy and his theory of organizational structure. It is useful to begin with the place of instrumental reason in organizational decision-making, for here we have, at its most sophisticated, the detailed calculation of consequences at the executive-managerial level, where alternative means are modelled for their efficacy in producing intermediate ends. Where these ends are taken to be concepts of the desired, this decision procedure corresponds in general terms with consequentialist ethical reasoning — the sort of reasoning associated with the justification of Hodgkinson's Type IIa values and characteristic of humanism, pragmatism, and (one variety of) utilitarianism (Hodgkinson 1991, p. 97). The implementation of these decisions occurs in the context of group processes governed by knowledge, culture, and organizational morality that issues at the level of the collective. For organizational stability, the justification of these values is made in terms of aggregating viewpoints towards some consensus. This kind of moral reasoning is associated with Type IIb values. At the lowest level of organization, the level of *individual* rank and file, moral motivation is governed by affect, the self-justifying Type III values. Needless to say, it is at the very top of the organizational hierarchy, the level of administration, where the various concepts of good are validated by what is right: justification in terms of ideals and matters of principle — the domain of Type I values. For Hodgkinson, these are more than matters of reason — they belong to the realm of will. Hodgkinson's view of human cognition therefore includes the elements of emotion, reason, and will. These elements map onto his hierarchy of values which in turn maps onto his proposal for organizational structure.

Comments and Critique

In our earlier papers we have expressed reservations about much of Hodgkinson's theory, particularly the separation of fact and value that underlies both his taxonomy and his argument against the possibility of a science of administration. We think that his own best arguments for the value-ladenness of administration extend naturally to imply the value-ladenness of facts. We have also queried the effectiveness of his taxonomy as a device for ethical decision-making. In this discussion we shall take up his recent 'moral art' solution to the problem of effectiveness, of striking a helpful and practical balance between idealist objectivity and non-cognitivist subjectivity in ethics.

The problem appears acute, for on the one hand his sharp fact/value distinction (based on a belief in the naturalistic fallacy) denies obvious empirical checks on the *rightness* of decisions, while on the other he is not availing himself of the usual non-naturalistic moral epistemologies. Administrators do not discern the 'form of the good' by some special faculty. Nor does he suppose they engage in the transcendental deduction of moral imperatives after the fashion of Kant. Yet his theory requires that leaders be able to make superior moral judgments. How, exactly, are they able to do so?

It seems to us that Hodgkinson's strategy is to challenge the basic methodological assumption behind this question. The demand for a familiar moral epistemology amounts to the demand for an applicable justification procedure. But this appears to beg the question against the justification of non-algorithmic judgements. Sound judgement in literature, music, and art, for example, can be acknowledged ahead of any agreement on an appropriate epistemology. In fact one can argue that being able to produce these judgements should be a test of the adequacy of any proposed epistemology. So why not extend the same epistemic latitude to moral leadership? It is simply a mistake to equate the absence of a specifiable justification procedure with the presence of an arbitrary subjectivity.

The main strength of the 'moral art' solution, therefore, is that it appears legitimately to combine a measure of moral objectivity with methodological agnosticism over the details.

We have a reply to this line of argument, although in fairness to Hodgkinson, when fully worked out it presumes the very naturalism in ethics that he wants to dispute. But we can note some preliminary limitations. For a start, the argument is provisional and can be threatened by the supply of detail in the form of a comprehensive moral theory that specifies how leaders should act. A number of the theories that figure in Hodgkinson's taxonomy aim to do precisely that. Now he has objections to all of these but that does not exhaust the domain of options. To shore up the provisional nature of his defense, he needs to argue for a *principled agnosticism* in the justification of moral claims.

More seriously, researchers in the utilitarian tradition, for example, not only see their theory as a comprehensive account of moral justification; they also see it as *excluding* other moral theories (Singer 1979). And likewise for the Kantian tradition (Rawls 1972). But this combination of comprehensiveness and incompatibility counts against the possibility of constructing a hierarchy of values based on contested systematic patterns of justification. Where comprehensiveness is taken as an epistemological virtue, being only partly right is not sufficient for an ethical theory. The incompatible theories that appear in Hodgkinson's taxonomy all prohibit their being ranked, especially by a morally significant relation of 'better than'.

One response, suggested by Hodgkinson's (1991, pp. 145–152) discussion of conflict resolution, is to make judgements about the structure and operation of the whole values taxonomy part of the moral art, since it is leaders who have to make the applications. However, note that as the domain of the non-algorithmic becomes wider and more theoretically contentious, the analogy with literature, music, and art begins to break down. For the analogy was reasonable precisely because some consensus over expert judgement obtained. Once that consensus is breached, the strategy of methodological agnosticism in justification loses its force.

Moving to some of the general features of our own approach, we see naturalism as providing useful guidance on ethics. People have to learn right from wrong, good from bad, just from unjust, in the same way that they learn everything else, for, like Hodgkinson, we are skeptical of the existence of some special ontological realm of the moral and of some special faculty for discerning it. A theory of ethics therefore needs to cohere with our best scientific accounts of human learning. Although we see learning as best explained in terms of the functioning of the central nervous system in its interaction with the environment, no account of the full neural detail is required to appreciate the relevance of *empirical* checks on what is learned. For us, objectivity emerges out of our most coherent response to interpreted experience. Through a natural process of conjecture, refutation, and coherent adjustment, we build up and improve our global theory for responding to a spectrum of issues, empirical, ethical, aesthetic, and so on. The conditions for improving ethical theory and ethical judgment are therefore the conditions for improving knowledge in general. These are the social arrangements that permit the nurturing and testing of knowledge claims, and promoting their revision if found wanting.

Since, on our view, all knowledge is fallible and a candidate for improvement, the crucial organizational corollary of this epistemology is the need to maintain structures that promote organizational learning (Argyris and Schön 1978). But these structures also instantiate a substantive ethics that places a premium on, for example, tolerance, respect for persons, the value of knowledge, freedom of speech and association, and equity of access to receiving and contributing to knowledge. On our view, moral leadership does not consist in the burden of providing a vision that must be right. Such a demand would outrun the capacity of any reasonable moral epistemology. Rather, it consists in articulating the best (but fallible) moral perspective and maintaining the organizational relations for correcting and improving that perspective in the face of changing circumstances and further improvements in knowledge. The isomorphism that thus follows from our position is a relationship between our fallibilist epistemology and theory for promoting the growth of knowledge, and the organizational structures that promote the communication and testing of conjectures and the coherent processing of adjustments that result from the feedback of refutations, or the discovery of error. This model of promoting soundness in decision-making, moral or otherwise, is best exemplified in the organizational learning tradition, and more recently (although the ethical dimension is underdeveloped) the total quality management movement.

There are many points at which Hodgkinson would dissent from our naturalist Deweyan ethic of promoting the growth of knowledge, and its consequences for structuring relations between cognition, values, and organization. However, the clarity of his writing, the boldness of his conjectures, and the enthusiasm he displays in the pursuit of further understanding, provide a touchstone with which we can readily agree.

References

Argyris C. and Schön D. (1978). *Organizational Learning: A Theory of Action Perspective*. (San Francisco: Jossey-Bass).

Evers C.W. (1985). Hodgkinson on ethics and the philosophy of administration, *Educational Administration Quarterly*, 21(4), pp. 27–50.

Evers C.W. (1993). Hodgkinson on moral leadership, *Educational Management and Administration*, 21(4) pp. 259–262.

Evers C.W. and Lakomski G. (1991). *Knowing Educational Administration: Contemporary Methodology Controversies in Educational Adminstration Research*. (Oxford: Pergamon Press).

Foster W. (1986). *Paradigms and Promises*. (Buffalo: Prometheus Press).

Habermas J. (1972). *Knowledge and Human Interests*. (London: Heinemann).

Hodgkinson C. (1978). *Towards a Philosophy of Administration*. (Oxford: Blackwell).

Hodgkinson C. (1983). *The Philosophy of Leadership*. (Oxford: Blackwell).

Hodgkinson C. (1986). Beyond pragmatism and positivism, *Educational Administration Quarterly*, 22(2), pp. 5–21.

Hodgkinson C. (1991). *Educational Leadership: The Moral Art*. (Albany: State University of New York Press).
Plato (1970). *Republic*. (New York: Macmillan, translated by J.L. Davies and D.J. Vaughan).
Rawls J. (1972). *A Theory of Justice*. (Cambridge, Mass: Harvard University Press).
Simon H.A. (1976). *Administrative Behavior*. (New York: Free Press, third edition).
Singer P. (1979). *Practical Ethics*. (Cambridge: Cambridge University Press).

12

What Price Democracy? An Examination of Arrow's Impossibility Theorem in Educational Decision-Making

The current rhetoric in education emphasizes decentralization of decision-making and increasing participatory decision-making at the school-level. Although we believe that a case for school-based decision-making in democratic fashion can be made, such a case must incorporate, and have an answer to, some fundamental problems of actual decision procedures by means of which lofty ideals of participative democracy would have to be put into practice. With few exceptions, actual voting, or other decision procedures have either been left unspecified, or where mentioned, have not been examined in terms of their inherent technical properties.

In this chapter, we want to raise some dilemmas that result from inherent, technical features of egalitarian decision procedures. These have been brought to light by the economist Kenneth J. Arrow in form of his so-called 'Impossibility Theorem'. The result of this theorem states that if we do want egalitarian decision-procedures then we have to give up the idea that the decisions made are rational or decisive. If we want to maintain a degree of rationality, we may adopt a consensus procedure — at the cost of being able to make decisive choices. If, on the other hand, we want to have decisive choices, we ought to opt for dictatorship. According to Arrow, such generally shared objectives as equality of power, decisiveness, and rationality are in irreconcilable conflict, and compromises are unavoidable.

Since Arrow's work is not yet discussed in the educational literature, and because his arguments clearly apply to decision-making in education, the major part of this chapter describes the Impossibility Theorem and the various attempts at eliminating it, or at relaxing the constraints he suggested for the axiom. In the final part we show how problematic the selection of a principal in fact is, when examined in the context of various possible, egalitarian, decision-procedures.

Arrow's Theorem

Taking his departure from what appeared to be 'just a mathematical curiosity' (Plott 1976, p. 511) in majority rule, the economist Kenneth J. Arrow (1951, 1963) undertook what must surely count as one of the most significant investigations into the conditions of rational voting in a capitalist democracy. The question that guided what has come to be known as 'Axiomatic Social Choice Theory' seemed harmless enough. Arrow wanted to find out (1951, 1963, p. 2) 'if it is formally possible to construct a procedure for passing from a set of known individual tastes to a pattern of social decision-making, the procedure in question being required to satisfy certain natural conditions'.

His quest was decidedly normative in that he wanted to find a set of ethical norms, or axioms, which could be imposed on the social choice process, and to find that process which would satisfy the axioms (Mueller 1976, p. 185). Although Arrow posed the problem in terms of finding a suitable constitution by way of voting procedures, his study raises fundamental problems for egalitarian social choice processes in general that incorporate the preferences of individuals in order to reach the best choice.

The result of Arrow's inquiry was, in Plott's terms (1976, p. 512) 'broad, sweeping, and negative', and gained him the Nobel Prize in economics in 1972. Simply stated, what does not exist in Arrow's view, and that of other political and social choice theorists, is a procedure for voting, or making formal choices, that is simultaneously rational, decisive, and egalitarian. In fact, he proved that these criteria of an ideal system are incompatible, and that the only procedure that does satisfy all of them is that of a dictatorship. The general impact of Arrow's results were that, all of a sudden, he had opened

> a gigantic cavern into which fall almost all of our ideas about social actions. Almost anything we say and/or anyone has ever said about what society wants or should get is threatened with internal inconsistency. It is as though people have been talking for years about a thing that cannot, *in principle*, exist (Plott 1976, p. 512).

To understand the impact of what Arrow has proved, let us begin by considering some anomalies of voting, and that of majority rule first.

Majority rule is a time-honoured, simple, and egalitarian way of aggregating individual preferences by ranking them in pairs of candidates or alternatives. It runs into problems as soon as more than two alternatives are involved, giving rise to what, following Condorcet, is called the 'paradox of voting'. Imagine a three-member committee consisting of Tom, Dick and Jane who have to rank three candidates x, y, and z. Using notational form, we have

    Tom $xPyPz$ (x is preferred to y, y is preferred to z)
    Dick $yPzPx$
    Jane $zPxPy$

Which candidate should be chosen by the committee? Of course, the candidate everyone wants, the one preferred by the majority. But the committee preference looks like this: $xPyPzPx$. In other words, we encounter a cycle. X defeats y, y defeats z, and z defeats x, all by two votes to one (e.g. Blair and Pollak 1983, p. 76). This outcome is called the 'majority rule cycle' or the 'majority preference cycle' (Plott 1976, p. 514). Efforts that attempt to avoid it by adding more members to the number of candidates to be ranked have proven to be unsuccessful. Detailed discussions of this problem can be found in Sen (1970), Mueller (1979), and Plott (1976) in particular. The existence of a cycle means that majority rule while being egalitarian, does not permit a best choice to be made, it is indecisive. Unfortunately, the problem does not disappear when turning to another, commonly used, procedure, the Borda count, or point system. Since it is well enough known, it shall suffice here to pinpoint the difficulty. Following Plott (1976), consider a four-option, seven-person group that, by allocating the relevant numbers to the candidates, gets a clear winner and a ranking. Now suppose the least preferred candidate gets dropped off the list, necessitating an adjustment of scores from 1–3, rather than the previous 1–4. It now emerges that the group preference is exactly the *inverse* of that reached in the first counting. This is a very odd yet valid result according to the procedure. Which procedure and/or candidate is to be chosen? Again, a rational choice is not possible.

The problems inherent in majority rule and the Borda count led Arrow to ask whether such inconsistent social preferences arise in *all* voting systems or just in the ones described. Since it is practically impossible to list all possible ones, and the relevant configurations of individual preferences and their orderings, Arrow suggested instead some general principles, or axioms, that all procedures ought to meet and that would find general agreement. For ease of understanding, we follow a combination of Mueller's (1979) and Plott's (1976) reformulation of the axioms rather than Arrow's more technical account. The first principle is that of

(1) *Unanimity*, or the *Pareto Principle*. If an individual preference is unopposed by any contrary preference of any other individual, this preference is preserved in the social ordering. Another way of putting it is to say that if everyone preferred option or candidate x over y, then society, or a committee, should neither be indifferent between them, nor perversely prefer y over x. This axiom has been misunderstood in the sense that it is seen to rule out conflict. This it does not do since, according to Plott (1976) 'if' has to be read as 'when'.

(2) *Nondictatorship*. No individual should be in a position such that whenever s/he expresses a preference between any two alternatives and all others express the opposite preference, his or her preference is always preserved in the social ordering.

(3) *Preference Transitivity*. If society prefers option x to option y, and it prefers option y to option z, then option x is preferred to option z. In notational form: $(xPyPz)$ (xPz).

(4) *Indifference Transitivity.* Insert the term 'indifferent' in place of 'prefers' in principle 3. In notational form, $(xIyIz)$ (xIz).
(5) *Universal or Unrestricted Domain or Scope.* This principle requires that a social choice be capable of aggregating every possible configuration of voters' preferences. It must be sufficiently general to be able to deal with all possible conflicts and controversies between opinions.
(6) *Independence of 'Irrelevant' (Arrow) or 'Infeasible' (Plott) Alternatives.* The social choice between any two alternatives must depend on the rankings of individuals over only these two alternatives, and not on their rankings over other alternatives (Arrow's example 1951, 1963, p. 26).

Now Arrow's Theorem states that there is no constitution that satisfies all of these principles. He proved that the only ones who do are *dictatorial* (Arrow 1951, 1963, p. 63). There is no doubt in the literature that the theorem is true. For simplified accounts see Sen (1970, 1982), Mueller (1979, p. 187), and Plott (1976, p. 524). The obvious question to be raised is, of course, what does all this mean, and how reasonable are the axioms anyway that apparently exert such enormous constraints?

To begin with, the intuition that underlies Arrow's proof is well captured by Mueller (1979, p. 188):

> The unrestricted domain assumption allows any possible constellation of ordinal preferences. When a unanimously preferred alternative does not emerge, some method for choosing among the Pareto preferred alternatives must be found. The independence assumption restricts attention to the ordinal preferences of individuals for any two issues, when deciding those issues. But, as we have seen in ... majority rule, it is all too easy to construct rules which yield choices between two alternatives, but produce a cycle when three successive pairwise choices are made. The transitivity postulate forces a choice among the three, however. The social choice process is not to be left indecisive (Arrow, 1963, p. 120). But with the information at hand, individual ordinary rankings of issue pairs, there is no method for making such a choice that is not imposed or dictatorial.

The literature is replete with attempts to disprove this impossibility result and escape its 'unpalatable implications' (Blair and Pollak 1983, p. 78), and 'Arrow-dodgers' (Sen 1982, p. 165) have attempted, in particular, to relax Arrow's axioms in order to bypass, or eliminate, the result. The general agreement is, however, that while the discussion continues, the outlook is bleak. Just how significant and devastating Arrow's result is can be seen when one considers that his constraints are quite weak in that they constitute no more than plausible necessary conditions only. In fact, they appear to be weaker than appropriate for 'reasonable notions of distributional equity' (Mueller 1979, p. 188). The first principle, *unanimity*, as well as (2) *nondictatorship*, need not be discussed in the present context if we want to continue with such notions as individualism and democracy. The postulate of *transitivity*, axioms (4) and (5), demands that social choice produces a *consistent* social ordering. Arrow's reasons for this requirement are (1) 'that some social choice be made from any environment', (Arrow 1951, 1963, p. 118), and (2) transitivity guarantees 'the independence of the final choice from the path to it' (Arrow 1951, 1963, p. 120). As several writers have argued, however, these two requirements are really quite different ones, and are not necessary for transitivity. The first one is reasonably easy to defend since deadlocks are undesirable in a democracy (see Mueller 1979, p. 190). In order to make choices, however, Mueller for example argues that it is not necessary to assume a social preference ordering based on individual preference rankings. One only needs a social choice *function* in order to select the best alternative from a set (see also Sen 1970, pp. 47–55). Full transitivity is not necessary. It is sufficient to rely on 'quasi-transitivity' or 'acyclicity'. '"Quasi-transitivity" requires transitivity of the preference relation, but not of indifference; acyclicity allows x_1 to be only 'at least as good as 'x_n' even though $x_1 P x_2, x_2 P x_3 \ldots, x_{n-1} P x_n$.' (Mueller 1979, p. 190). The penalty for accepting quasi-transitivity, however, is that it produces an oligarchy that can force its preference on all other members. Adopting acyclicity, on the other hand, means that veto power is given to every member of a subset of a committee, a 'Collegium' in Brown's term. The result of relaxing the transitivity axiom is that dictatorial power is spread but does not disappear, and that arbitrariness is increased (Sen 1970, pp. 47–55). In notational form, under quasi-transitivity aIb, and bIc. This can exist along with aPc. When two individuals disagree over a and b, however, the result is aIb. When generalized, this result means that when conflicting interests are encountered in a society or a committee, the *rule of consensus* (Blair and Pollak 1983, p. 79), which describes Sen's formulation of the relaxation of full transitivity, declares collective indifference. This is hardly a useful outcome. Sen himself questions whether the existence of a best alternative (in a and b) in each subset, while itself a sound basis for rational decision-making, (Sen 1970, p. 50) is satisfactory, particularly in view of the fact that as soon as a third option is added, c must be chosen (by the social preference transitivity).

Considering *path-independence*, the assumption here is that the final choice is independent of how the initial decisions were made. This property implies what Sen calls 'property alpha'. It can best be shown by example. If Becker is the best tennis-player in the world, he is also the best tennis-player in Melbourne. Path-independence here requires that his becoming the champion be independent of the ordering of prior matches. Sen complements the 'contraction-consistent' property alpha with an 'expansion-consistent' requirement, beta. It claims that if Becker and Cash tie for Wimbledon then they must also tie for the world championships. While plausible as a constraint on a choice process where candidates are measured in one dimension only, very few social decision processes are this narrow. It is quite feasible that property beta might be violated without thus making the choice process either irrational or unfair. Property alpha is far more convincing. However, it is precisely path-independence, together with alpha, that even in their weakest form lead to dictatorial or oligarchial social preference orderings (see Mueller 1979, pp. 192–193). Finally, the notion of transitivity was inspired by the desire to avoid inconsistency and arbitrariness. But in doing that, Plott (1976) as well as Mueller (1979) argue that Arrow committed the fallacy of composition by illegitimately transferring the properties of an individual to that of a collection of individuals. Both writers draw the conclusion that the notion of a social preference must go. If the transitivity axiom is to stay then, as Mueller argues, it must be shown that the arbitrary outcomes of cyclic preference rankings violate some basic ethical norm. But this is not necessarily the case. Procedures such as flips of a coin or the drawing of straws are often resorted to in small committees to resolve a conflict. The committee may be more interested in a 'fair' decision, rather than a non-arbitrary one. The problem is, however, an obvious one. If the property of fairness were the overriding one then, as Plott (1976, p. 544) points out, we can no longer speak of a social preference ranking. Since the selection process, as was shown above, merely throws up results that are as good as each other, social preference ranking is replaced by what Plott calls a 'social acceptability relation' (1979, p. 544), and Sen a 'Collective Choice Rule' (Sen 1970). In this case, it hardly makes sense to speak of an 'optimum' decision, since the rationality criterion has been abandoned.

Consider next the principle of *universal domain*. This axiom captures in a general sense the liberal idea of freedom of choice and expression. While appealing when considered on its own, conflicts arise quickly, however, since individuals have different, and often opposing, preference rankings. Cyclicity might occur, and if we also consider transitivity as necessary, an impossibility result is unavoidable. To attempt to avoid it two options have been advanced in the literature. The first is to alter the principle so that the types of preference rankings be limited which can be reflected in the social choice. The second is to restrict entry into a society or committee to those who possess the types of preference rankings that make collective choices possible.

It is possible to avoid Arrow's impossibility result on the assumption of a restricted domain if individual preferences satisfy a unimodal pattern called 'single-peakedness' (Black 1971). For example, voters can only vote on the number of text-books; they cannot consider their quality simultaneously. There are some situations that can be decided in this manner by majority rule, but it does not contribute to solving the problems of all those situations that are multi-dimensional and need to be decided in other ways that appear to be prone to Arrow's negative result. Another fundamental problem with single-peakedness is that it already assumes some consensus on the issue to be decided. If voter preferences become more homogeneous the possibility of majority cycles decreases. How one gets such 'similar' preferences in the first place is described in theories of clubs and the voting-with-feet process, which still leaves many questions open. But even if decision-making can be carried out without violating Arrow's axioms that yield a 'best' decision, some decisions cannot be confined to clubs, etc., and the impossibility result looms again. Homogeneity of preferences can only be brought about if individuals adopt or already possess a common set of values. This solution raises more problems still, and since there are few if any issues that are of the requisite unidimensional quality, the option of single-peakedness does little to undermine Arrow's results.

Finally, there is the principle of *independence of irrelevant alternatives*. This has been the most controversial axiom, mainly because Arrow presented it in a confusing way (see Plott 1976). Generally speaking, this principle states that the social choice ranking of any pair of alternatives depends only on individuals' rankings of those alternatives (Blair and Pollak 1983, p. 78). It does not matter whether individuals change their preferences for *infeasible* options, the process outcome will remain invariant. According to Arrow's postulate, every process behaves in this way. It follows that if a process could be found that violates it, i.e. where, upon changes in some infeasible options, the process outcome were also changed, the principle should be given up as invalid. While it appears that one process, the Borda count, indeed seems to violate the axiom, according to some, Plott (1976) advances an interesting argument that even the point system obeys the axiom. To see how this works, consider his example.

Suppose we are choosing from a list of candidates w, x, y, and z who is to receive an offer as full professor. We add four infeasible candidates to our list of four: James Maddison, Thomas Jefferson, J.S. Mill, and Karl Marx. All candidates are ranked according to members' preferences with the lowest ranked receiving the number 1. The feasible candidate with the highest score will receive the offer. But this process can be said to violate the axiom because if you changed your mind about two infeasible candidates and ranked them below the feasible ones, then they would receive a higher score and hence the process outcome would be changed. Also, preferences for the feasible candidates remained the same since only the two infeasible ones were ranked lower. Hence, the argument goes, since this was not supposed to be happening according to the axiom, it must be false. Plott points out that this way of arguing rests on a questionable assumption, i.e. that the reporting of preference ranking was honest (Plott 1976, p. 536). But according to game-theoretic principles about how people behave in such circumstances, it is 'strategic' rather than 'honest' voting that must be assumed. Strategic voting means misrepresenting the actual ranking of the infeasible candidates to gain a more advantageous outcome. What are reported, according to Plott, are *strategic* advantages that have nothing to do with actual preferences for infeasible options. Since those do not change, and are not reported as changed, the process outcome remains invariant. Plott concludes 'that the proposed process could not behave as reported, and that the principle remains intact' (Plott 1976, p. 536). Whether or not individuals always behave in a 'dishonest' manner is not at issue. Rather, the only way in which we could eliminate the principle is if we could be sure that people *never* behave strategically — clearly an impossibility. Does this axiom apply to all processes? The answer given through examination of non-cooperative and cooperative games is that it does. This concludes the overview of attempts of 'Arrow-dodgers' to evade the Impossibility Theorem. The result is, as Plott (1976, p. 553) remarks in his very thorough and careful examination that 'it is reasonable to suppose that the negative results should be taken at face value and that our philosophical positions must be altered accordingly'.

Do the logical inconsistencies Arrow exposed have anything to do with the 'real' world of educational decision-making, or are his considerations merely an exercise in logic?

Arrow's Theorem and Educational Decision-Making

Insofar as school-based decision-making has to be concerned with 'social needs' and 'group wants', it has to be able to determine priorities and the ranking of options by some process. Ranking does not have to be expressed by preferences in the way Arrow does, but can be substituted for the concept 'better'. The technical properties apply equally to the latter concept. It is precisely these properties that pose the problem. In order to demonstrate the educational significance of Arrow's results, let us consider an example of educational decision-making: the selection of a principal. The relevant body for making this decision is the local School Council. For the sake of the present argument, suppose that the Council can potentially avail itself of a number of procedures of selection. In the following, let us consider what the selection process would look like by examining the properties of the most commonly used process, that of majority rule, as well as those of some alternatives. Suppose then that our problem is to choose one candidate from a list of m candidates, where m is greater than 3. Following Mueller's (1979, pp. 59 ff.) description, the procedures are as follows:

(1) *Majority rule*: choose that candidate who is ranked first by more than half the Council members.
(2) *Plurality rule*: choose that candidate who is ranked first by the largest number of members.
(3) *Condorcet criterion*: choose that candidate who defeats all others in pairwise elections using majority rule.
(4) *Borda count*: give each of the m candidates a score of from 1 to m, based on its ranking in a member's preference ordering (as described above).
(5) *Exhaustive voting*: ask each member to indicate the candidate s/he ranks lowest from the list. Remove that candidate who is ranked lowest by most members and repeat the process until a single candidate remains; he or she is the winner.
(6) *Approval voting*: each member casts a vote for all of the candidates on the list of whom s/he approves; the candidate with most votes wins.

The problems with majority rule have already been described. Since a voting cycle results when more than three options have to be considered, this method does not make selection of a winner possible. In practice, however, we do apply this procedure *and* select a candidate, thus doing what the process forbids. The questions that needs to be asked is, what is that choice based on? We return to this problem later. While majority rule is the most egalitarian it is also the most indecisive and the least rational.

Under plurality voting there is always a winner — but not a majority one. For example, in the following set of members' rankings over four candidates (adapted from Mueller 1979, p. 60), x is the winner, but y recommends itself as the 'best' option because it is relatively high on all members' votes:

SCM_1	SCM_2	SCM_3	SCM_4	SCM_5	(SCM = school council member)
x	x	y	z	w	
y	y	z	y	y	
z	z	w	w	z	
w	w	x	x	x	

The problem with plurality voting is that it only takes into account information about the first preferences. The other four procedures do take into account more information about members' preference rankings. In this respect they have an advantage over the first two, but these are purchased at a price too.

According to the Condorcet rule, y, in the above table, is the winner since it defeats z and w in pairwise elections, 4 to 1. It also defeats x, the plurality winner, 3 to 2. Indeed, y is the winner by any of the other three voting procedures. The disadvantage of the Condorcet rule is that it, like majority rule, may not produce a winner. When a cycle over more than 3 candidates occurs, these may form a set of 'top' candidates who can beat all others but who do not allow the selection of a winner from among them. Even where there is a Condorcet winner, a candidate chosen by another procedure may be preferable. Consider the following:

SCM_1	SCM_2	SCM_3	SCM_4	SCM_5
x	x	x	y	y
y	y	y	z	z
z	z	z	x	x

According to this table, x is both a Condorcet and a majority rule winner. However, this situation looks like a 'tyranny of the majority' situation where the first three members can force their selection on the latter two who rank their candidate last. *Y* emerges as a compromise candidate since it is ranked highly on all preference scales and may thus be the 'best' candidate. Using the Borda count for this example makes y the winner *on average*. This introduces a degree of stability or consistency which might be considered desirable on its own account. On the other hand, the Borda procedure, as discussed above, is open to strategic voting which is a potential threat.

Approval voting is an attractive idea when members agree to group candidates into two or three preference sets, as can happen when a large short-list needs to be reduced. Considering the last table, the first three members, although preferring x to y, would accept either x or y's victory over z. The gap between x and y is thus much narrower than that between y and z. These members might then vote for x and y, and y would be the winner. One might argue that the closeness of x to y in the first three members' ranking, plus y's top ranking for SCM_4 and SCM_5, lends intuitive legitimacy to y's being the winner.

The reason for there being agreement between the Borda count, the Condorcet criterion, and other more complete information procedures regarding the winner is that these procedures pick out the candidate who ranks highest *on average*, and thus eliminate 'extremists'. Just how practically important this is can be seen in a study of the 1972 Democratic presidential primaries in the United States, which are run under the plurality rule (Joslyn 1976, as cited in Mueller 1979). Under this rule, the 'extremist' McGovern won overwhelmingly over the middle-of-the-road Muskie: 1.307 to 271 delegates. Joslyn recalculated the final delegate counts under the rules we discussed above, and these show a dramatic increase in Muskie's delegate count in all but the plurality rule procedure. Mueller (1979, p. 65) comments that Muskie arguably *should* have been the Democratic Party's nominee, and that therefore one of the other voting procedures is preferable to the plurality rule. But all this is a matter 'loaded with value judgements' (Mueller 1979, p. 65).

What emerges quite clearly from the above examples is that the technical features of these egalitarian procedures are such that different procedures produce quite different results from the same list of four candidates. Although agreements have also been noted, these came about largely as a result of introducing extraneous considerations into the process such as judging candidates *on average*. On what grounds can these judgements be justified? This raises a number of issues, which can only be listed here. For example, the most urgent one seems to be that, given the advantages and disadvantages of the procedures discussed, how do we select one to begin with? The question is important since results can diverge considerably. Also, what criteria come into play once we have adopted, say, majority rule, when we encounter a cycle? In fact, do we even follow the procedure at all, i.e. are we aware that cycles develop when more than three candidates are on the list? The overriding worry is that since we obviously do make selections of principals in ordinary life, and since according to the technical features of egalitarian procedures, we should not have been able to, we must have made these choices on criteria other than those specified by the rules. This is a question of considerable import which needs to be examined in the future.

References

Arrow K. J. (1951, 1963). *Social Choice and Individual Values.* (New York: Wiley).
Black D. (1971). *The Theory of Committees and Elections.* (Cambridge: Cambridge University Press).
Blair D. H. and Pollak R. A. (1983). Rational collective choice, *Scientific American*, **249**(2), pp. 76–83.
Mueller D. C. (1979). *Public Choice.* (Cambridge, New York: Cambridge University Press).
Plott Ch. R. (1976). Axiomatic social choice theory: an overview and interpretation, *American Journal of Political Science*, **20**(3), pp. 511–596.
Sen A. K. (1970). *Collective Choice and Social Welfare.* (San Francisco, Cambridge, London, Amsterdam: Holden-Day).
Sen A. K. (1982). Social choice theory: a re-examination, in A. Amartya *Choice, Welfare and Measurement.* (Cambridge, Mass: The MIT Press).

13
Explaining and Improving Educational Administration — *Donald J. Willower*

Educational administration has had a long, but chequered relationship with philosophy. In the United States, W. T. Harris, a scholar and practitioner of school administration, founded and for many years edited the *Journal of Speculative Philosophy*. Appointed superintendent of the St Louis schools in 1868 and from 1889 to 1907 US Commissioner of Education, Harris was a prolific writer and an important American authority on Hegel (see Culbertson 1988). Harris is an early example of the connection between philosophy and educational administration. A much more pervasive influence on education and educational administration was John Dewey. Although his work on education was a very small part of his extensive contribution to philosophy, it was thoroughly embedded in his larger view of inquiry and reflective methods which appealed to many educators.

More recently, interest in philosophy was a small but noteworthy feature of the turn toward the social sciences that occurred in educational administration in the mid 1950s. For instance, the important book *Administrative Behavior in Education*, edited by Campbell and Gregg and sponsored by the National Conference (now Council) of Professors of Educational Administration, included a lengthy chapter on values (Graff and Street 1957), and a little later, when the University Council for Educational Administration was up and running in the USA and Canada, it sponsored an array of activities on values and philosophy (see for example Ohm and Monahan 1965). During that time, I presented what I described as a 'general philosophical position concerning educational administration' (Willower 1964, p. 96) and have continued to write on the topic.

However, it was not until the subjectivist and neo- Marxist critiques that marked the social sciences reached educational administration that extensive debates over philosophical issues became part of that field. These debates have been chronicled and extensively analyzed by Evers and Lakomski (1991).

At the outset, I want to make clear that I share with both of those writers and others the aim of advancing in educational administration an alternative philosophic view to positivism, subjectivism, or neo- Marxist critical theory (on [problems of] definition of the latter, see Held 1980; Wallerstein 1986, p. 1302; Evers and Lakomski 1991, p. 139). I share also with Evers and Lakomski the general naturalistic, pragmatist perspective set forth in their landmark book and in some earlier individual papers, despite a few differences in our views, most of which are matters of emphasis. In what follows, some general comments are made on Evers and Lakomski's work, then special attention is given to their consideration of what they call the Theory Movement in educational administration, and finally, some suggestions are made concerning uses of philosophy in educational administration.

Some General Comments

As a philosophical work specifically devoted to educational administration, Colin Evers and Gabriele Lakomski's *Knowing Educational Administration* is perhaps the most comprehensive, technically proficient, and elegantly argued book currently available on that topic. They make abundantly clear the limitations of positivistic thought and of subjectivism and critical theory, both of which attacked positivism and attained new popularity after long periods of intellectual marginality. Rather than dwell on my agreement with Evers and Lakomski's treatment of these views and with much of what they have to say about epistemology and values, I will turn at once to some of the differences in emphasis alluded to before.

Coherence and Empiricism

The first deals with coherentism and empiricism. I am a little less enthusiastic about the former and a little more enthusiastic about the latter than are Evers and Lakomski. Usually coherence means coherence with existing knowledge as a criterion of truth. Evers and Lakomski quite properly emphasize the importance of this kind of fit; however, following Quine, they distinguish between a theory of evidence and theory of truth. For the latter they endorse a kind of scientific realism, in effect a form of correspondence theory, coherentism's traditional enemy. It is their theory of evidence that encompasses what they call coherence criteria. These criteria include testability, simplicity, consistency, comprehensiveness, fecundity, familiarity of principle, and explanatory power (see Evers and Lakomski 1991, pp. 37–44, p. 229). Such criteria are, of course, very familiar in that, for over 50 years, texts in logic and scientific methods have commonly included them (with some variations) in their treatments of adequate theory or sometimes adequate hypotheses. Evers and Lakomski describe these criteria with Churchland as 'superempirical' and omit a criterion usually found in the texts — predictive power — although they aver that a theory of evidence involves empirical adequacy in so far as it can be achieved. Yet, they are chary about empiricism, understandably put off by the many classic examples of naive empiricism, as well as the assumptions underlying Hume's scepticism, and the too confident scientism of logical empiricists and others including Karl Popper.

There is no question that Evers and Lakomski do a thorough job of demolishing foundationalism. One even detects a certain joyful overkill, perhaps made easier by the fact that they have managed to preserve in another category the realism that goes with most foundationalism. My view is that inquiry or science needs no foundation outside of its own self-rectifying processes. If that view involves circularity, which is arguable, then circularity in this instance is a logical bagatelle that pales beside the history of scientific accomplishments. Realism is a reasonable inference from what we know of the world and its priority to humankind, but as with any other inference, however well supported, certainty is not possible.

In any case, science (including social science) is driven by a pervasive practical empiricism that features the interplay of theory and observation. Observation in science is theory laden, but observation that attends to scientific norms can furnish genuine information about the plausibility of the theory with which it is laden. Put differently, science uses explanations to direct observations to determine if the explanations are plausible. Said still another way, observations laden with some theories make more sense and work out better than those laden with other ones. The Deweyan notion that ideas should be judged by their consequences is crucial in science where, of course, those consequences are commonly empirical.

I have followed mainly a Deweyan account of inquiry, while Evers and Lakomski follow mainly a Quinean one. These accounts are not incompatible, but they have different emphases. Dewey is better on science as an active process and its applications to practice and life. Quine is better on the logical posers implicit in scientific methodologies. Accordingly, Evers and Lakomski are superb at the latter as, for instance, in their detailed arguments in refutation of strict and broad forms of foundationalism.

Without detracting in any way from the importance of these careful and compelling arguments for a text on educational administration with an avowed epistemological focus, Evers and Lakomski give limited attention to the concrete, empirical side of knowledge getting in this applied field of inquiry. As a result, a few problems emerge in their treatment of the Theory Movement, which I treat later, and in connection with their chapter on what they call the cultural perspective, examined next.

Field Research and the Cultural Perspective

Evers and Lakomski indicate that the 'cultural perspective' in educational administration is a way of 'conceptualizing organizations such as schools' (Evers and Lakomski 1991, p. 112) that derives its justification from disciplines and traditions that 'consider themselves as alternatives to science . . . such as cultural anthropology, phenomenology, hermeneutics, and interpretive social science' (Evers and Lakomski 1991, p. 112). Lumping all that together paints with too coarse a brush. More than a few cultural anthropologists and sociological field researchers might not welcome the company. Indeed, the implication that field studies should be considered to be phenomenological or hermeneutical applies only to some scholarly endeavours along those lines. My guess is that a minority of those who do field research would accept such labels, although there are some visible writers who would, quite a few of whom are cited by Evers and Lakomski.

Language contributes to the problem. Ethnographers and participant observers commonly think of themselves as engaged in empirical studies because description is a key activity in their work, and one has to define the word empiricism in a very special way to exclude its use in such circumstances. At the same time, a term like phenomenology is often used very loosely by social scientists. For instance, Glaser and Strauss (1967), who produced one of the most influential books on field research, stressed their commitment to the scientific enterprise and argued for the importance of theory derived inductively from data. Their book was about how to develop theory through the systematic and comparative analysis of data. However, because they emphasized the generation of theory from observation rather than from deductive processes, they stated 'our position is not logical, it is phenomenological' (Glaser and Strauss 1967, p. 6), an unhappy and misleading phrase all around in light of the current discussion.

A quite typical statement on field methods is found in Cusick's (1973) oft-cited study of an American high school. After quoting Blanche Geer's well-known comment that a field researcher is at once a reporter, interviewer, and scientist, Cusick characterizes his main methodological work as description and explanation. He chronicles the effort to formulate working hypotheses, which are then sustained or contradicted, discarded or refined, all in the context of his study (Cusick 1973, pp. 229–230). What we have here is a thoughtful social scientist at work, not a philosophical subjectivist dabbling with data.

In any case, my point is again one of emphasis. I wish Evers and Lakomski had more explicitly recognized that a great deal of field research is done under other banners than those they discuss, or is done eclectically without regard to such labels. Their philosophical criticisms are beautifully rendered, but it should have been clearer that they applied only to particular kinds of justifications for field research. The claims made by many such investigators are quite consistent with the kind of science espoused by Evers and Lakomski.

In addition, I find their emphasis on physicalism in connection with cultural studies not particularly useful. It would be nice not to have to infer attitudes, beliefs, and the like from other data, but such inferences are common in science. After all, not everything can be directly observed. Explanation in all of science, especially in the highly complex fields that study individual human beings and their collectivities, is often dependent on just such inferences. That inferences in field research are often quite shaky cannot be denied, nor can the notion that such research is far stronger in generating theory than it is in testing it. One of the genuine delights of doing such research is to craft a plausible explanation, even or perhaps especially when the physical evidence is limited. In making that comment, I hope I can be forgiven for casting myself as a coherentist and Evers and Lakomski as empiricists

The Mirror Of the Social Sciences

A final difference of emphasis concerns Evers and Lakomski's depiction of the changes in thought in educational administration that began in the mid to late 1970s. As we all know, these changes involved the emergence of subjectivism and critical theory as significant strands of scholarship in the field.

My disagreement here is straightforward: Evers and Lakomski do not pay sufficient attention to the parts played by the societal and social science contexts in bringing about change. They rely too heavily on a kind of great lodestar approach that attributes virtually revolutionary powers to a few writers in educational administration as architects of new thinking. While there is little doubt that such writers contributed importantly to the diffusion of certain views in educational administration, they advanced ideas that had become popular in the social sciences using arguments similar to those already made, and with largely comparable reactions and results.

As I have pointed out elsewhere (Willower 1992a), thought and research in the various fields of scholarship are influenced by the social and political forces of their times. The 1960s were marked by a growing distrust of societal institutions, especially on the part of university students and their teachers, with those in the social sciences, and the humanities among the most affected. Government, business, science and an array of social and economic arrangements were targeted by those who were dissatisfied with the status quo. The Vietnam conflict, an ongoing cold war that included the possibility of nuclear disaster, assassinations, Watergate, and struggles for equality by various groups all spurred political activism, especially on the campuses, which in turn fostered the spread of two long-standing, but widely discounted viewpoints: neo-Marxism and subjectivism. The former emphasized the redistribution of power from the dominant classes to the working classes and the powerless, radical economic and social change, and an aggressive political style. The latter, under a variety of labels, rejected conceptions of system and society in favour of the individual and his or her special realities. The notion of the individual against the system appealed to those who were unhappy with current conditions, just as neo-Marxism appealed to those who favoured radical reform and class equity.

Although neo-Marxism and subjectivism had been traditional philosophical enemies, in the 1960s they were allied, at least in their criticisms of societal arrangements and institutions, including science. Both attacked 'positivistic science', which the Marxists saw as serving class interests and the subjectivists saw as detached from the concerns and realities of people.

Without belabouring the point further, the times and social circumstances were made to order for these two views, which attained great popularity and even near dominance in social science fields such as sociology (Smelser 1988). When they finally began to get serious attention in educational administration in the late 1970s and 1980s, they had already passed their peaks and begun to wane in the social sciences (see Alexander 1988). The arguments used in educational administration on behalf of these views and in criticism of them reflected those made in the social sciences, although in educational administration, they had to be tailored to the special conditions and needs of that field. Such applications are an important and often creative undertaking, one that I do not want to derogate or undervalue. At the same time, it is desirable to be clear on contexts and origins.

That propositions and viewpoints can be examined and critically analyzed independently of context and origin is not disputed. That is what Evers and Lakomski have done and done well. My comments on context are of a different order, meant to round out the portrait of intellectual trends in educational administration given by those authors. Their work places subjectivism and critical theory in a philosophical context, but neglects the influence of the times and the consequent impact of those two views in the social sciences, which were in turn mirrored in educational administration.

Theory and Educational Administration

Some of my comments on Evers and Lakomski's treatment of what they and many others call the theory movement are of a similar kind. Here, I have no quarrel with the substantive discussions on particular philosophic issues presented by Evers and Lakomski. However, their history of this period (roughly the mid 1950s to mid 1970s) in educational administration thought and research leaves something to be desired.

They divide the 1950s to 1990s into two segments. One they label the old orthodoxy, associated with some of Herbert Feigl's ideas, logical positivism, closed systems, and Daniel Griffiths' writings on theory. The other, labelled the new orthodoxy, which begins in the 1970s, they describe as oriented to open systems and contingency theory, especially as expressed in the well-known text by Wayne Hoy and Cecil Miskel (1987). Evers and Lakomski's treatment is devoted largely to a finely done analysis of epistemological questions in line with the positions taken in the rest of their book. However, their history of scholarship in educational administration fails to capture important elements of the times. The difficulty is mainly a matter of exaggerating the importance of philosophical considerations in educational administration along with the exclusion of other more salient features. I will very briefly cite some of the major ones (see also Willower 1992b).

I stated at the beginning of this essay that interest in philosophy was a small but noteworthy feature of educational administration's turn toward the social sciences. It was just that, but after one got beyond the notion that writing in educational administration ought to be less hortatory and more scientific, much of the philosophical focus was on values.

The social sciences were the main sources of ideas and the main objects of emulation. Educational administration was ripe for change. The sage advice and practical tips that were a major part of the field's lore seemed increasingly inadequate as recognition grew that the social sciences dealt with the subject matter of school administration. For example, motivation, leadership, behaviour in groups and organizations, polities, and economics were all potentially of great relevance to administration. Diffused and legitimated by professional and professorial associations, the idea that social science knowledge and methods could put educational administration on a firmer intellectual footing was soon widely accepted.

All this led to a great upsurge of research in a variety of areas in educational administration that more or less reflected the different social sciences. Whereas a few writers like Griffiths and Andrew Halpin cited philosophers like Herbert Feigl, the latter's definition of theory, made much of by Evers and Lakomski, was widely ignored or discounted as crafted for physicists. Most researchers in educational administration paid little attention to epistemological issues as formulated in philosophy. They were interested in the norms of science as reflected in their reference group, norms that increasingly mirrored those of the social sciences.

As in most social science, the theories used or developed in educational administration were very seldom presented in hypothetico-deductive form, but there was a great concern that their presentation be clear and consistent, and where possible, build on past work and current knowledge. Methods were rarely experimental, but whether research was done in a survey or field studies mode, it was understood that procedures and results should be specified so that others would know just what was done. Operational definitions, criticized in the abstract but recognized as useful by Evers and Lakomski, were properly valued by practising researchers because such definitions clarified the connections made between concepts and data in particular studies.

A great deal of research was produced in educational administration after the late 1950s, a portion of which was reviewed in Boyan (1988), the most comprehensive single source on scholarship in the field produced to date. A wide array of theoretical frameworks were employed. No single theory was dominant. One of the more popular was a loose version of social systems theory that focussed on norms and values, roles and statuses, informal and formal structures, and other sociological concepts. Theory and research on leadership and organizational climate were also common.

The closed versus open systems dichotomy that Evers and Lakomski and some organizational theorists emphasize was far less a matter of viewpoint than of priority and division of labour. Many of the early studies of educational administration had an internal organizational focus, but these studies usually recognized the impact of external factors, and a substantial number of investigations of school/community relations, school boards, and political forces were completed throughout the period. Put differently, the open–closed system divide was not as sharp as is sometimes depicted. The version of systems theory that uses the analogy of the organism and concepts like homeostasis and equilibrium that Evers and Lakomski discuss in connection with Hoy and Miskel's text was and is very rarely employed in research beyond the occasional use of a concept, for instance, equifinality in its Mertonian version of functional equivalents.

Persons interested in that period of scholarship in educational administration would do well to consult other sources beyond Evers and Lakomski. Their review is about epistemology, which took a back seat to the theorizing and empirical studies that defined those days. The scholarship of that time provided a great many new insights on how schools worked and on their organizational, political, and social contexts. What they write is carefully crafted and is a largely defensible set of arguments which range from the limitations of positivism, naked empiricism, and foundations to those of broad-gauged systems theory. However, I have already pointed out that researchers rarely use the latter. It remains to remark that they did not show substantial interest in the former. Most of them were just not oriented to epistemology.

As Smelser (1988) indicated for sociology, active researchers usually hold to a theoretically eclectic empiricism—even those who may be critical of science. This kind of practical empiricism among those who actually do research has been persistent and is sustained by norms of science that are mainly geared to sceptical openmindedness, impersonal criteria of evaluation, and the communication of results (Zuckerman 1988). That such an approach should be so successful in so many fields is itself an epistemological statement, albeit lacking in philosophical niceties.

While the research got done and increased our knowledge considerably (today's science bashers to the contrary), I believe that greater attention to epistemological questions would have been beneficial. It might have reduced the exaggerated expectations for theory held in some quarters in educational administration and might have made the field a bit less vulnerable than it was to the epistemological attacks launched by subjectivists and neo-Marxian critical theorists. While Evers and Lakomski have quite thoroughly disposed of the arguments underlying those attacks, I am not sure how influential their work will be in the end. Just as many scholars in the field were not oriented to epistemology during the period in question, not many more are today. Others are concerned mainly with practice, eschewing both research and epistemology.

However, those who teach educational administration are well aware that science has been called into question in at least some quarters. In addition, increased interest in field studies has stirred new concerns about the standards that should guide such research and research generally. Epistemological questions will not just go away. What we've had in educational administration is the articulation by a few of a borrowed positivist view, followed by an aggressive subjectivist and critical theory reaction, against a backdrop of epistemological neglect occasioned by preoccupation with other matters and lack of philosophical awareness or interest or both.

All this puts one in mind of the old saw that philosophy will always be needed to counteract bad philosophy. So it is with epistemology, and Evers and Lakomski have done a marvellous job of counteracting, in some detail, many of the flawed views of knowledge getting that have been advanced both past and present. This is a major accomplishment, but much remains to be done, my final topic.

Philosophy and the Invigoration of Educational Administration

I have argued elsewhere (Willower 1992b) that a reinvigoration of the culture of scholarship in educational administration is especially dependent on three things. The first is an adequate conception of inquiry, the second is more explicit attention to values and valuation, and the third is a view of practice that emphasises what the Greeks called praxis, that is, thoughtful practice.

Following Dewey (1922, 1938), I contend that these are interrelated. In Dewey, the process of inquiry can inform science, but also everyday decisions, ethical choices among competing values, and problems of social policy. A distinctive feature of the general philosophical approach taken by Dewey and others using similar frameworks (among whom I would include Evers and Lakomski and myself) is that science and ethics and practice should not be sharply separated and placed in mutually exclusive categories.

Indeed, it is ironic that subjectivists and critical theorists who tout the importance of values but reject science, cut themselves off from the very processes and insights that can inform value choices in concrete situations. As Dewey reminds us, abstraction can be the enemy of morality. Morals have to do with the common problems of living and separating morals from life in favour of an abstract creed creates social problems rather than resolving them (Dewey 1922, pp. 324–330).

Making ethical judgements requires the reflective examination of alternative courses of action and their consequences. An ethical choice is, in effect, one that leads to desirable outcomes in concrete situations. Hence moral choice required that hypotheses be devised concerning probable consequences and outcomes. On this view, the practice of administration is an ongoing exercise in valuation and a profoundly ethical activity.

While good intentions are important in ethical choice, they are not sufficient to ensure desirable outcomes. Administrators who seek to make reasonable value judgements need to be able to make informed estimates of what will happen when particular courses of action are pursued. This in turn requires explanations that are plausible in terms of what is known about individual and collective behaviour in the context of the particular facts of the case. Here, social science concepts and theories can be especially helpful. Put another way, social science can be a substantive aid to the attainment of desirable outcomes.

Ethics and science are, in this view, symbiotic, not antithetical. It is the separation of ethics and the reflective methods of science that allows ritualism and merely verbal commitments to take hold in administration. Hence the importance of being able to explain in order to improve. If administration is essentially about moral choice, and thoughtful moral choice depends on informed explanation and inference, then philosophy and social science both are crucial to thoughtful administrative practice.

In this connection, the critical examination of knowledge- getting and of the strategies and tactics of empirical study and theoretical explanation become important concerns to scholars and practitioners alike. My preferred science of educational administration differs somewhat from Evers and Lakomski's in that I am less sanguine than they about the utility of what appears to be their call for the construction of grand theory. I base my scepticism on the failure of such theory to contribute much so far in the social sciences, while more modest theories have done much better. However, consistent with the view of inquiry to which we jointly subscribe, I intend to keep an open mind on the matter.

Returning to the need for invigorating conceptions of inquiry, ethical choice, and practice in educational administration, it should be clear that such conceptions fit beautifully within the philosophic view suggested here and by Evers and Lakomski. The time for polemics in educational administration is ending and scholars should now devote more energy to a revitalization of scientific work in the field. That such a revitalisation could serve both ethics and practice has, I hope, been made clear.

References

Alexander J. C. (1988). The new theoretical movement, in N.J. Smelser (ed.) *Handbook of Sociology*. (Newbury Park: Sage).
Boyan N. J. (ed.) (1988). *Handbook of Research on Educational Administration*. (New York: Longman).
Culbertson J. A. (1988). A century's quest for a knowledge base, in N.J. Boyan (ed.) *Handbook of Research on Educational Administration*. (New York: Longman).
Cusick P. A. (1973). *Inside High School*. (New York: Holt, Rinehart and Winston).
Dewey J. (1922). *Human Nature and Conduct*. (New York: Henry Holt).
Dewey J. (1938). *Logic: The Theory of Inquiry*. (New York: Henry Holt).
Evers C. W. and Lakomski G. (1991). *Knowing Educational Administration: Contemporary Methodological Controversies in Educational Administration*. (Oxford: Pergamon).
Glaser B. G. and Strauss A. L. (1967). *The Discovery of Grounded Theory: Strategies for Qualitative Research*. (Chicago: Aldine).
Graff O. B. and Street C. M. (1957). Developing a value framework for educational administration, in R F. Campbell and R. T. Gregg (eds.) *Administrative Behavior in Education*. (New York: Harper).
Held D. (1980). *Introduction to Critical Theory: Horkheimer to Habermas*. (Berkeley and Los Angeles: University of California Press).
Hoy W. K. and Miskel C. G. (1987). *Educational Administration*. (New York: Random House, third edition).
Ohm R E. and Monahan W. G. (1965) (eds.). *Educational Administration: Philosophy in Action*. (University of Oklahoma: Norman).
Smelser N.J. (1988). Introduction, in N.J. Smelser (ed.) *Handbook of Sociology*. (Newbury Park: Sage).
Wallerstein I. (1986). Marxisms as utopias: evolving ideologies, *American Journal of Sociology*, 91(6), pp. 1295–1308.
Willower D. J. (1964). The professorship in educational administration: a rationale, in D. J.

Willower and J. A. Culbertson (eds.) *The Professorship in Educational Administration*. (Columbus and University Park: University Council for Educational Administration and The Pennsylvania State University).

Willower D. J. (1992a). Educational administration: Intellectual trends, in M. C. Alkin (ed.) *Encyclopedia of Educational Research*. (New York: Macmillan).

Willower D. J. (1992b). *Educational Administration: Philosophy, Praxis, Professing*. (Madison: National Council of Professors of Educational Administration).

Zuckerman H. (1988). The Sociology of Science, in N. J. Smelser (ed.) *Handbook of Sociology*. (Newbury Park: Sage).

14

The Salvation of Educational Administration: Better Science or Alternatives to Science?—*Peter Gronn and Peter Ribbins*

Evers and Lakomski's *Knowing Educational Administration* (1991) is a tough and uncompromising but important book. It seeks to examine the merits of traditional science and its alternatives as a basis for educational administration as a field of study. As such, a key purpose of this symposium is to enable some of those who have played a major role in the disputations that have taken place upon this issue to respond to the critiques of their views that Evers and Lakomski (1991) make.

Had things turned out differently, Thomas Greenfield would have written this essay but his illness during 1992 made this impossible. Nevertheless, his contribution to the challenge to the traditional science paradigm and, in particular, to its place within the study of educational administration, is widely acknowledged. Griffiths (1988, p. 30), for example, accepts that its 'demise came at the 1974 meeting of the IIP in Bristol'. Indeed, 'the coup de grace was delivered by Greenfield who made an across the board denunciation of every aspect of the theory movement'. Evers and Lakomski (1991, p. 1) also recognize this *implicitly* when they note that 'since the mid 1970s', educational administration, as an area of study, has undergone a fundamental transformation'. And they also acknowledge it *explicitly* when they claim (Evers and Lakomski 1991, p. 75) that since the late 1970s Greenfield has 'broadened and deepened his critique'. 'In an impressive series of papers . . . he has sought to develop a systematic view of social reality as human invention, in opposition to the systems scientific perspective of social reality as a natural system. He has constructed strands of argument on the nature of knowledge, on administrative theory and subjectivity, truth and reality'. Both the magnitude of his effort and elegance of his argument had made Greenfield's work 'the most important theoretical development in recent educational administration'. Given this generous assessment, it is hardly surprising that of the two chapters in their book that focus on the work of a named author, one is devoted to an examination of Greenfield's ideas.

For his part Greenfield acknowledged the seriousness of the critique of his work contained in *Knowing Educational Administration* and the extent of the threat that 'coherentism' represented to the views which he had been advocating over the last 20 years. In a conversation with one of the authors in which they sought to develop a preliminary taxonomy of the kinds of response that his work evoked, Greenfield categorized Evers and Lakomski as 'opponents' and others as 'allies', 'hedgers' and 'enemies'. To set Evers and Lakomski's critique in context it might be worth considering this taxonomy briefly.

In search of Enemies and Hedgers, Allies and Opponents

The years that followed the publication of his paper to the IIP in 1974 were often hard for Greenfield. As he says 'I had not anticipated the bombshell that broke over my head'. The paper was interpreted as 'deliberately challenging, threatening and hostile'. He believed it was being commented upon in unintellectual ways:

> I discovered something about my field: its pettiness, its calcified and limited vision, its conventionality, its hostility to dissenting opinion, its vituperativeness ... I was aware I was being attacked unfairly in an unscholarly manner, that people sought to explain my paper in personalised terms. They began to circulate stories about my administrative competence.

The result was that he 'felt beleaguered and alone' (Greenfield and Ribbins 1993a, pp. 245–247). It was this *ad hominem* style of the attack that led Greenfield to think of those who engaged in it as his enemies.

Reflecting on the hostility that his ideas have sometimes generated in North America, he suggested that 'the observation about the "point of the needle" is explanation of why my critique has not "caught on" in the United States. It is unpopular wherever what [Northrop] Frye describes as "Mercantilist Whiggery" prevails'. A good example is one of his most recent papers, which he referred to as 'phoenix' (Greenfield 1991a), and in which he saw himself attacking 'the ill-effects on education of cultures that accept the excesses of technocratic-pragmatic, systems-empiricist, individualist values as received and unopposed truth'. On the other hand, he concedes that 'phoenix' was 'commissioned by an American group and presented first from an American platform to a small but appreciative audience' (Greenfield and Ribbins 1993a, p. 267). It is worth stressing this last point because ultimately Greenfield reserves his most scathing comments not for his enemies in the United States but for those he thought of as the hedgers in Canada. Hodgkinson, in commenting on the response to his book *Towards a Philosophy of Administration*, has wryly observed that 'as a general rule, a prophet has little honour in his own country. The Canadian perception, as with most things, was vacuous and innocuous' (Ribbins 1993). Greenfield took a similar view. By 'hedgers' he meant those within the community of scholars in educational administration in Canada who, he thought, after 1974, distanced themselves from him in various ways or who reserved their judgement. As he remarked (Greenfield and Ribbins 1993a, p. 248): 'I always felt my Canadian colleagues waited to see how the show would play New York or Chicago before committing themselves'. Eventually they came around, although 'often they would rather put me on some kind of honour roll than ask my views on anything. At one time however some of the colleagues closest to me virtually averted their eyes when they saw the fuss and the furore'. And so 'four years later when IIP '78 was held in Canada, my colleagues ensured there would be no echo of the Bristol error. I was not invited to attend or make a presentation. I watched the IIP caravanserai as it passed briefly through Toronto' (Greenfield and Ribbins 1993a, p. 245).

Given this reception, the support of allies was important to him. Endorsement from outside North America was welcome but was always less significant to him than encouragement from Canada and the United States. Nevertheless, he classified supporters according to their location, their place within 'the great world of theory and accepted thinking about it' (Greenfield and Ribbins 1993a, p. 245) and their contribution to the development of his own thinking. On this basis he was always clear about who the most valued of his allies was. Only Hodgkinson met all three criteria. There is an important study to be written on the ways in which these two Canadian scholars have influenced each other's thinking. Whilst this is not the place to attempt it, something is still worth saying about how Greenfield saw the relationship. Two observations from his conversations with Peter Ribbins illustrate what he felt. In the first he says (Greenfield and Ribbins 1993a, p.247)

> The first person who extended the personal hand of friendship and support to me here was Chris Hodgkinson.... He offered the steel of intellectual argument and the hand of friendship. There were no others like him.

In the second he acknowledged his friend's particular influence (Greenfield and Ribbins 1993a, pp. 264-265):

> what I have taken from Hodgkinson is his argument that a social science of organization can never replace an understanding of administration itself. He deals with the existential reality of the administrative act as virtually no other writer does. He is a fine philosopher, insightful, expressing the power of ancient and modern thought, revealing its essence in pungent human terms. . . . He is relentless that technique can never supplant the wilfulness of human action or release human agents from responsibility for that choice. If there are any ideas from Hodgkinson which have influenced me most, they

are the irreducibility of value choice and the unavoidabilty of human responsibility for that choice.

In the reflection that follows on immediately after this one, Greenfield states the nub of the two Canadians' differences with Evers and Lakomski. 'Free will exists in some measure at least', he wrote, 'and that is where the struggle with Evers and Lakomski begins. They deny mind and free will, reducing everything to matter, arguing as Evers has, that it is easier to physicalise the mental than to mentalise the physical. In opposition to that dehumanising proposition Hodgkinson and I are united' (Greenfield and Ribbins 1993a, p. 265). These are views that we take up below, but we want to stress at this point that if Greenfield felt that some of the ideas advanced by Evers and Lakomski were dehumanizing, he nevertheless cherished them as honourable opponents. As he has written: 'I take great satisfaction from what they have said, not just because it is appreciative... but because they pay attention to the text of what I have written. They pay attention to what I have said, even though their view of it is rather selective. They don't look at all of my writing over the last twenty years and I wish they had looked at some of the other things. But what they look at, they look at squarely and carefully'. He goes on to acknowledge that Evers and Lakomski are critical of those who do not do this 'and that is something which has aggrieved me most over the years. Too often I have been personally attacked rather than attacked on the basis of what I have written'. Accordingly, it was 'satisfying at this juncture to find critics with whom I may disagree, but who will understand what I've said and deal with it' (Greenfield and Ribbins 1993a, p. 263).

Before turning to Evers and Lakomski's critique of his work it might be worth commenting on the view Greenfield expresses above that it is based upon a partial examination of his writings and a selective treatment of his ideas. We doubt if Evers and Lakomski would deny either point. Their discussion of his ideas is based upon eight of Greenfield's papers published between 1975 and 1986. This set deals with many of his most important writings during that period but not all of them. As it happens we know which papers he would have chosen to represent the corpus of his writings between 1974 and 1991 since these are included in *Greenfield on Educational Administration: Towards a Humane Science* (Greenfield and Ribbins 1993b). It would, of course, not be fair to complain that Evers and Lakomski had not included some of his most recent writings since 1986, although future commentators on this work will certainly need to include the two major papers he wrote in 1991 along with the final chapter of his book. Of Greenfield's ten papers (1975, 1979b, c, 1979-80, 1980, 1983, 1986, 1991a, b and c) comprising the Greenfield and Ribbins (1993b) collection, Evers and Lakomski do not discuss two of the papers that Greenfield chose (1979c, 1979-80), but on the other hand they include three papers (1978, 1979a, 1984) that he decided to leave out. Perhaps too much should not be read into this comparison, since these three papers were all seriously considered for inclusion.

What of the claim that Evers and Lakomski's treatment of Greenfield's ideas is selective? Greenfield broadly accepted the account of his critique that Evers and Lakomski (1991, p. 76) set out at the beginning of the chapter they devote to his work, and which we have quoted in full. In the light of this agenda it may be worth citing exactly what Evers and Lakomski promise (1991, p. 76):

> Our discussion of these themes will, of necessity, be selective rather than comprehensive, and is aimed at illustrating the methods of coherentist epistemological critique and defending the possibility of a new, suitably broadened, science of administration. Within this framework our basic position on Greenfield's work can therefore be simply stated. We think that his epistemological objections to science traditionally or positivistically conceived are decisive and cannot be answered from within traditional assumptions.

They continue:

> [On the other hand] we think that Greenfield's objections do not apply to a naturalistic scientific realism justified on holistic or coherentist epistemic criteria.

Whilst we will object to aspects of what Evers and Lakomski have to say on Greenfield's ideas, such as, for example their attempt to lock him into a wretched hermeneutic cycle, or rather to claim that he has locked himself into it, we must acknowledge that the agenda they set out is fair enough. That they have their own agenda, and that this is to press the merits of their coherentist views as against available alternatives, is clearly stated both above and at other points in their book. When reflecting upon what Evers and Lakomski have to say about 'The Greenfield Revolution' it is for the reader to beware.

Reflecting upon 'The Greenfield Revolution'

In their opening statement Evers and Lakomski (1991, p. 1) suggest that 'from following mainly models of theory and research associated with more traditional views of science, the field [of educational administration] has moved to a position of much greater diversity'. This shift was made possible by the development of 'alternative philosophical perspectives on the nature of knowledge which could function as frameworks for rival systematic conceptions of administration' (Evers and Lakomski (1991, p. 1). From such a perspective, it is ideas about the nature of knowledge, or 'the structure of justification, as specified by epistemology', that 'determines much of the overall framework in which theorizing in educational administration can take place' (Evers and Lakomski 1991, p. 3). Evers and Lakomski cite three major developments to illustrate this point. The first and second have already had a considerable impact within the field and the third underpins their own proposed new direction: logical empiricism, the paradigms approach and coherentism.

The discussion of Greenfield's subjectivism to be found in *Knowing Educational Administration* is, in many respects, unsatisfactory. In part this is understandable given that Evers and Lakomski adhere to a theory of the administrative world that is incommensurable with that proposed by Greenfield. As is implicit in the title of their book, as far as Evers and Lakomski are concerned, one can only ever *know* administration. It follows that to ask other kinds of questions about administration or administrators is meaningless. If knowing is all then it is hard to see any point in asking how an administrator feels about the world, or what the right and proper way is for that administrator to act in regard to that world. Nor is there much sense in recounting how actual administrators might construe such issues for themselves.

In particular, we make six main criticisms of Evers and Lakomski's thesis in general and its implications for their discussion of Greenfield's position in particular. The first general point is that in their search for a 'better science', rather than 'traditional science' and its logical empiricist base, they, in effect, leave untouched the core of Greenfield's main position. Indeed, they can be said to jettison it. Unlike Evers and Lakomski, and despite his rather teasing subtitle, 'Towards a Humane Science', Greenfield argues for an alternative to science or, at least, for a definition of science that does not exclude a serious concern for feelings and values. This may, in part, explain why Evers and Lakomski neglect three key aspects of his work. Firstly, the primacy that he accords to values: in particular his claim that all statements about science and values are themselves valuational, that in a sense theories are really moral versions of the world, is nowhere addressed by Evers and Lakomski. Secondly, and arising from this point, whilst they recognize that Greenfield believes organizations are non-natural, socially constructed entities, the substance of his most recently developed view that they entail a moral order is likewise ignored. Thirdly, the idea that such a paramountcy of values for organizations may have implications for the role of the administrator as a kind of entrepreneur of values who wilfully constructs the social world for others, is given no credence at all (Gronn 1994).

There are many ways in which these ideas can be illustrated with reference to educational practice. For example, there are a number of studies that have been undertaken into the English public school and their imperial variants (for example Gathorne-Hardy 1979, Honey 1977) that emphasize its purposes in terms of the notion of character building and the creation of a political and administrative elite through the paraphernalia of the boarding house and fagging and buttressed by conceptions of a headmasterly tradition going back to Arnold of Rugby and Thring of Uppingham. If this represents a particularly striking case in point, it should be stressed that the ideas that underpin the approach can be used to give an account of any school at any time in its history. Thus, for example, Best *et al.*'s (1983) study of continuity and change at Rivendell draws on similar ideas to describe the three regimes of headship that existed at the school between the 1960s and the 1980s. This case study offers an excellent example of the merits of viewing the school as a moral order and the headteacher as an entrepreneur of values.

These are points that Greenfield has developed in many of his papers and most notably in his last, known as 'Science and Service' (1991b), where he concludes that a central issue in management science generally, and in educational administration in particular, is the place that facts and values have in shaping an administrator's action (Greenfield and Ribbins 1993b, p. 221):

> many who struggle in the arenas where theory is still debated can be divided into two camps: those who see the central administrative issue as all fact and those who see it as all value... In a sense both camps offer a pure science, one focused on facts, the other on value.

In the latter camp he sees the critical theorists, the deconstructionists and the post-modernists, and in the former the coherentists. A brief examination of the latter position can be used to show the weakness of attempts to force science to deal only in pure values or pure facts. The problem, Greenfield argues, has been economically presented by Lakomski (1987, p. 71, original emphasis):

> if attending to our values helps us make *better* decisions, then we need specific criteria to help us to decide *between* competing values. In other words, there has to be a way to determine if value X is better than value Y in some specified way. But if values are merely non-cognitive or affective and *subjectivist*, then we cannot determine rationally... which of two conflicting values is better.

It might well be asked: what is Lakomski's answer?

The clue, as Greenfield suggests, lies in the phrase 'if values are *merely* noncognitive [emphasis added]'. The answer, he says (Greenfield and Ribbins 1993b, p. 222), is 'to make values cognitive, to turn them into facts. [Lakomski] notes that even subjectivists will defend their values with arguments . . . defending or acting on personal preference implies that *in practise* not all values are considered equally acceptable or worthy. When people actually defend their preferences, they admit a modicum of rationality and objectivity by admitting that some preferences or values are better than others'. Perhaps. But, as Greenfield asks, what would convince one individual of the validity of another's value position? And why *should* we be rational (Greenfield and Ribbins 1993b, p. 222)?:

> What Lakomski fails to recognize is that an appeal to argument is an appeal to values. Rationality itself is ultimately a value position . . . To be rational

or to decide to look for the 'coherence' of evidence is to make a value choice. Such a choice may be one that many people make or that many scientists make, but consensus about a value does not transmute it from the value realm to the cognitive and rational. If we are to ask what values are better than others, we must look in a domain other than rationality or coherence.

For such reasons, Greenfield chooses to take the middle ground, 'the position that argues the central questions of administration turn on an interweaving of fact and value' (Greenfield and Ribbins 1993b, p. 221).

Our second general point follows from the first. As we see it, one important implication of this refusal or inability to consider Greenfield from within his own terms is that it is just not clear where values and questions of moral choice fit in Evers and Lakomski's coherentism. Their long-term project seems to be to naturalize ethics and to do so in part by attacking such notions as the 'is'/'ought' dichotomy and the naturalistic fallacy. It could well be that these ideas are false. This view has been put forward by Quine and by Alasdair Macintyre in his *After Virtue*? But we do not find the examples Evers and Lakomski give to support their case (for example, either snow is white or we ought not to kill — 1991, p. 168) very compelling. More importantly, we are unclear about the meaning or place that they give to values in their model. Exactly where do right and wrong, justice and fairness, equality and freedom, etc. fit in a pragmatic problem-solving approach geared to finding the 'best' theory? In any case, what does 'best' mean? 'Best' for whom, given that there are usually contending interests involved in the resolution of any value conflict. What might, for example, 'best' mean in the context of disputes that are currently taking place in the United Kingdom with regard to the appropriate levels of pit closures or over the level and structure of pay for teachers during a period of economic recession? And what happens if what is thought to be best is deemed to conflict with other considerations? A contemporary example is the way in which market theories are driving or justifying various macro and micro-level economic reform measures in Western economies. Suppose a coherentist could demonstrate that a particular rational economic doctrine (or 'theory') of, let us say, 'the withering state', is 'best' then what happens if the price to be paid is severe economic dislocation, injustice, and hardship? What does a coherentist position advise an administrator to do?

Our third point refers to what Evers and Lakomski describe as the 'extra-empirical virtues' upon which the coherentist can draw in the resolution of disputes over the determination of what counts as 'best' theory in a specific instance. It is not clear where these virtues come from. Nor is it clear exactly what, in this context, 'coherence' means? Words like 'fecundity', 'simplicity', 'consistency', 'elegance' and 'comprehensiveness' are sprinkled throughout their text, and they seem to be very important to the coherentist position that Evers and Lakomski advance. But how are we to understand them? There are references to the work of Lycan (1988, p. 130) and his five rules for guiding the comparison of two theories that are equal in other respects. What it is for theories to be 'equal' is not discussed further but, following Lycan, they propose that in such cases Theory 1 is to be preferred to Theory 2 if it is simpler, explains more, is more readily testable, leaves fewer messy unanswered questions and squares better with what is already known. Evers and Lakomski (1991, p. 38) accept that 'terms like "simpler" and "explain" await more detailed elaboration' but they claim, nevertheless, that 'at least we have a framework for moving towards critical judgements'. Yet none of the terms that underpin the rules identified above are closely defined in the text. And their meanings are far from self-evident. On the contrary, most of them appear to be 'essentially contested'. Nor are we told if these 'rules' constitute a comprehensive list of coherence criteria and if so, why just these particular rules? And what if in making such comparisons, Theory 1 is better than Theory 2 on some criteria but not on others? What kind of a calculus do we use then? Are some of the rules prior to others and if so how are they to be weighted? Are all of the criteria of equal epistemological status? Again we simply do not know because no answer is given to this issue. These are intriguing, interesting and suggestive features but they have travelled some distance from Greenfield's values.

It could be that there is no way out of the difficulties of applying the coherence criteria discussed above. Where problems exist of choosing between theories on coherence grounds these can be resolved by a process of arbitration against the relevant community of scientists. Greenfield believed that Evers and Lakomski were committed to such a view and he argued that 'to me [this] puts it on a social basis' (Greenfield and Ribbins 1993a, p. 238). If this is the case then to do so might create more difficulties for them than solutions. For example, what is a scientist and who is to be defined as a scientist in this context? In thinking about such questions, an attempt by Carr (1964) to define 'history' and to determine who has the right to be described as a historian might be instructive. In effect, he concluded that history is what historians say it is and historians are those who other historians say are historians. Not only does this entail circular logic but practical difficulties. As Greenfield points out (Greenfield and Ribbins 1993a, pp. 238–239) we need only 'think of the fate of Galileo and other heretics who were right but forced to say their theories were wrong'. And their theories applied to the physical world. If there are difficulties with the application of Evers and Lakomski's ideas in this context then

> their argument becomes shaky indeed when it is applied to the social world. There truth is defined, as Szasz says, not by scientists looking into test tubes and telescopes, but by 'experts' who go not to their laboratories to observe, but to make judgements . . . whether schizophrenia is a disease, a sin or an acknowledgment is a willful and moral choice.

The fourth point concerns the issue of causality. Even if, for the sake of argument, it were to be conceded that Greenfield does not consider the question of the causes of human action, this is not necessarily fatal to his thesis. What causes someone to adhere to and pursue particular values is certainly an important question. This is so because an answer to it might lead us to form judgements about the appropriateness or otherwise of a particular set of social arrangements as, for example, in the case of the socialization for the young. However, it is one thing to point to the significance of causality, another to argue that Greenfield glosses over it, and still another to fail, as Evers and Lakomski surely do, to reveal to their readers what a causal account of an organization might look like. In any case, what does it mean to 'cause' something to occur? Does it mean that if humans are disposed to doing something (that is to say their genetic dispositions propel them to) that they simply learn over time to do so? Or does it mean that they are in some sense innately made to act in spite of whatever they might want to do? How much of what anyone ever does is done because of, as opposed to in spite of, what they themselves choose to do? To what extent is their behavioural repertoire an act of free choice on their part or determined for them?

Fifthly, we want to consider how much credence can be attached to Evers and Lakomski's reduction of explanatory terms to the language of brain states and the central nervous system. Is it just wishful thinking to claim that, because somehow so-called 'folk' explanations elsewhere are crumbling before sophisticated scientific theories, explanations of human behaviour within the field of administration will do likewise? This sort of thing is asserted, in the sense that it simply must follow what is happening in the so-called hard sciences. Beyond this, little evidence and no proof is offered. Instead, the reader is given a promissory note. The assumption seems to be that if the proof does not exist now then it will emerge in due course. Such a claim appears to be 'properly scientific' in the sense that it seems to be potentially refutable. But is it? In this, it is remarkably similar to the claim structural/functionalists make of the notion of 'function'. They hold that the existence and persistence of any part of the system is to be explained in terms of the contribution it makes (its functions) to the needs of the system as a whole. Given this, how can such a theory cope with persistent deviance? As Best (1977, p. 72) points out, 'this highlights a fundamental weakness in the . . . functionalist paradigm'. Indeed,

> there is a circularity here . . . for there appears to be no way we can refute the account it gives of society: to cite instances of persistent and apparently dysfunctional phenomena is merely to invite the answer that, if it persists, it must be functional and the only puzzle is to establish the exact nature of the function.

Best (1977, p. 72) goes on to point out that phenomena like 'ignorance, crime and social inequality have all to be explained as functions' in so far as, and in the final analysis, they all contribute to social cohesion. In both of these cases, coherentism and functionalism, therefore, how long are we going to have to wait for the evidence? As long as it takes, it seems, which might be forever.

However much we might like to believe that the stability of organizations is due to relatively enduring dispositions encoded through learning in each person's central nervous system, can it be proved? Until it can be what we have is merely an interesting hunch like any other. Whilst Evers and Lakomski (1991, p. 94) note 'the lack of any really useful empirical generalizations in social science', they miss their chance to provide their readers with one. Instead their audience has to take them on trust when they say that the causal story 'as it would be currently told, is too complex to be of any general use', and claim that even a simple administrative action like issuing an order presumes 'such enormous networks of causal regularities that we could hardly begin to describe them in engineering terms [!!]' (Evers and Lakomski 1991, p. 94). If this extraordinary sleight of hand is to be believed, it surely invites an amazing leap of faith and, ironically, a very long wait on the part of practising administrators for a pragmatic theoretical standpoint (coherentism), dedicated to the goal of problem-solving, to be able to deliver on its promises to come up with solutions.

Finally, for a book that sets out its purposes so fully and clearly (see p. viii), *Knowing Educational Administration* is, in some respects, curiously incomplete and uneven. Firstly, whilst its authors stress 'the importance which theoretical writers [in educational administration] have attached to epistemology or theory of knowledge' and the significance of these terms to their own thesis, the text does not offer a comprehensive treatment of either concept. This is not, of course, to claim that Evers and Lakomski have nothing useful to say in this area. On the contrary, the opening two chapters offer much that will help the reader to develop a grasp of the epistemological foundations of many of the theories that have dominated thinking in the field over the last fifty years. Furthermore, it might be argued that this is all they have claimed to do in an account that is designed to advance their own coherentist views. As Evers and Lakomski (1991, pp. viii–ix) point out,

> despite our ambitions, this work is not a systematic treatise on the science of educational administration. Neither the critique nor the topics covered offer a comprehensive coverage of the rich and detaned work being done in the field today. Rather, what we have tried to do by constructing a range of quite different theoretical approaches in at least some historical framework of intellectual development, is to indicate the main features of our own perspective on a number of key issues of importance, debate and controversy. The topics we have selected are therefore those most amenable to philosophical discussion.

This may well be so, but the topics selected for inclusion have, in the main, been drawn from the published corpus of work in which Evers and Lakomski have been engaged separately and together over the last decade and more. As such, this book, despite its many and manifest merits, is not the systematic examination of the philosophical roots of educational administration that many of us had hoped for and that these two writers are well equipped to provide.

Postscript

Greenfield, in his final papers and in his extended conversations with one of the authors, acknowledged the importance of *Knowing Educational Administration* Thus he identified (1993b, p. 196) 'a looming shape... in the stimulating work of Evers and Lakomski' and commented (1993a, pp. 268–269) that 'in beginning to write "Science and Service", I knew I had to speak to the challenge from Evers and Lakomski'. He concluded:

> as I see it Evers and Lakomski's great contribution is to remind us of the world of fact, of the error of valuing too much. My contribution and Hodgkinson's is perhaps to remind us of the mystery beyond fact, of the error of valuing too little.

Had he seen *Knowing Educational Administration* earlier then he might, as he said (1993b, p. 196), 'have spoken to its challenges', 'admiring its depths of scholarship, and warning of and attempting to refute the false byways opened by their theory, which notably says much of epistemology and little of administration. That task of rejoinder must await more thought than I can give it now'. Greenfield's untimely death means that this was one task he was unable to complete. We hope that what we have said above might give some indication of what his rejoinder might have looked like had he lived to write it, and also emphasized just how much he valued dialogue with two people he regarded as cherished opponents.

References

Best R (1977). Sketch for a sociology of art, *British Journal of Aesthetics*, 17(1), pp. 68–81.
Best R., Ribbins P. and Jarvis C. with Oddy D. (1983). *Education and Care*. (Oxford: Blackwell).
Carr E. H. (1964). *What is History*? (Harmondsworth: Penguin).
Evers C. W. and Lakomski G. (1991). *Knowing Educational Administration: Contemporary Methodological Controversies in Educational Administration Research* (Oxford: Pergamon Press).
Gathorne-Hardy J. (1979). *The Public School Phenomenon*. (Harmondsworth: Penguin).
Greenfield T. B. (1975). Theory about organization: a new perspective and its implications for schools, in Greenfield T. B. and Ribbins P. (1993b) (eds.) *Greenfield on Educational Administration: Towards a Humane Science*. (London: Routledge).
Greenfield T. B. (1978). Reflections on organization theory and the truth of irreconcilable realities, *Educational Administration Quarterly*, 14(2), pp. 1–23.
Greenfield T. B. (1979a). Ideas versus data: How can the data speak for themselves? in G. Immegart and W. Boyd (eds.) *Problem-Finding in Educational Administration*. (Lexington: Lexington Books).
Greenfield T. B. (1979b). Organisation theory as ideology, in Greenfield T. B. and Ribbins P. (1993b) (eds.) *Greenfield on Educational Administration: Towards a Humane Science*. (London: Routledge).
Greenfield T. B. (1979c). Organisationen als Rede, Zufall, Handlung und Erfahrung in, *Die Psychologie des 20. Jahrhunderts: Lewin und die Folgen*, Band VIII, Annelise Heigl- Evers and Ulrich Streeck (eds.). (Zürich: Kindler Verlag). (Title in English: Organizations as Talk, Chance, Action and Experience), in Greenfield T. B. and Ribbins P. (1993b) (eds.) *Greenfield on Educational Administration: Towards a Humane Science*. (London: Routledge).
Greenfield T. B. (1979–80). Research in educational administration in the United States and

Canada, in Greenfield T. B. and Ribbins P. (1993b) (eds.) *Greenfield on Educational Administration: Towards a Humane Science*. (London: Routledge).

Greenfield T. B. (1980). The man who came back through the door in the wall: discovering truth, discovering self, discovering organizations, in Greenfield T. B. and Ribbins P. (1993b) (eds.) *Greenfield on Educational Administration: Towards a Humane Science*. (London: Routledge).

Greenfield T. B. (1983). Against group mind: an anarchistic theory of organization, in Greenfield T. B. and Ribbins P. (1993b) (eds.) *Greenfield on Educational Administration: Towards a Humane Science*. (London: Routledge).

Greenfield T. B. (1984). Leaders and schools: willfulness and nonnatural order in organizations, in T. J. Sergiovanni and J. E. Corbally (eds.) *Leadership and Organizational Culture: New Perspectives on Administrative Theory and Practice*. (Chicago: University of Illinois Press).

Greenfield T. B. (1986). The decline and fall of science in educational administration, in Greenfield T. B. and Ribbins P. (1993b) (eds.) *Greenfield on Educational Administration: Towards a Humane Science*. (London: Routledge).

Greenfield T. B. (199la). Reforming and re-valuing educational administration: Whence and when cometh the phoenix?, in Greenfield T. B. and Ribbins P. (1993b) (eds.) *Greenfield on Educational Administration: Towards a Humane Science*. (London: Routledge).

Greenfield T. B. (1991b). Science and service: the making of the profession of educational administration, in Greenfield T. B. and Ribbins P. (1993b) (eds.) *Greenfield on Educational Administration: Towards a Humane Science*. (London: Routledge).

Greenfield T. B. (199lc). On Hodgkinson's moral art, in Greenfield T. B. and Ribbins P. (1993b) (eds.) *Greenfield on Educational Administration: Towards a Humane Science*. (London: Routledge).

Greenfield T. B. and Ribbins P. (1993a). Educational administration as a humane science: conversations between Thomas Greenfield and Peter Ribbins, in Greenfield T. B. and Ribbins P. (1993b) (eds.) *Greenfield on Educational Administration: Towards a Humane Science*. (London: Routledge).

Greenfield T. B. and Ribbins P. (1993b) (eds.) *Greenfield on Educational Administration: Towards a Humane Science*. (London: Routledge).

Griffiths D. (1988). Administrative theory, in N. J. Boyan (ed.) *Handbook of Research in Educational Administration*. (New York: Longman).

Gronn P. C. (1994). Subjectivity and the creation of organizations: Australian educational administration and the writings of T. B. Greenfield, in C. W. Evers and J. Chapman (eds.) *Educational Administration: An Australian Perspective*. (Sydney: Allen and Unwin).

Honey J. R. de S. (1977). *Tom Brown's Universe*. (London: Millington Books).

Lakomski G. (1987). Values and decision-making in educational administration, *Educational Administration Quarterly*, **23**(3), pp. 70–82.

Lycan W. (1988). *Judgement and Justification*. (Cambridge: Cambridge University Press).

Ribbins P. (1993). Conversations with a condottiere of administrative value: some reflections on the life and work of Christopher Hodgkinson, *Journal of Educational Administration and Foundations*, **8**(2), pp. 29–45.

15

On Knowing: Cultural and Critical Approaches to Educational Administration—
Richard Bates

Evers and Lakomski in *Knowing Educational Administration* (1991) have presented us with several philosophical analyses of various approaches to educational administration. Such analyses are to be welcomed as a contribution to the development of the field inasmuch as they both clarify and provide evaluations of particular positions. Such analyses should provide useful ground for the further development of more adequate accounts of administrative theory and action.

Such analyses do, however, need to be grounded in accurate representations of the arguments they purport to critique. In this respect Evers and Lakomski's presentation of at least some of the arguments they critique is rather less straightforward than it appears. For instance in reference to some of my early work (Bates 1980, 1983) they assert that I claim that '... a science of administration is *essentially* manipulative and concerned with social control' (Evers and Lakomski 1991, p. 12, emphasis added). In fact I said something quite different:

> ... as *currently conceived* by professional and professor alike, educational administration is a technology of control Moreover, it is a technology of administrative control that systematically ignores both educational issues and those social and cultural issues that lie at the heart of people's commitment to, or alienation from, educational institutions (Bates 1983, p. 46).

Nothing here or in the associated argument asserts that a science of administration is *essentially* or necessarily a manipulative form of social control. My point is a simple and different one, one concerned with the social construction and purposes of the particular conception of educational administration that dominated the field during the 1960s and 1970s – a conception that separated educational from administrative concerns and privileged the latter. Indeed, the point I was making was not concerned with the *epistemological* foundations of science or administration *per se*. It was rather, an observation based upon the *historical* association of three social movements that were partly responsible for this state of affairs:

> This separation of administrative from educational concerns was shown to have its roots in three social movements: the municipal reform movement, occupational professionalism, and the cult of efficiency (Bates 1983, p. 47).

My conclusion was certainly that

> The result has been the establishment in education of hierarchical structures of authority and control that both mirror and reproduce the systematic inequalities of the wider society (Bates 1983, p. 47).

My point, however, was not that this result was a necessary and inevitable concomitant of 'a science of administration', but that it was an historically contingent product of particular social forces and interests at a particular time. Educational administration *could* possibly have developed differently, *could* possibly have represented different interests and *could* possibly have been built on an alternative conception of science to that to which it had in fact appealed.

It is true that I made passing reference to inadequacies in the model of science to which traditional theorists in educational administration constantly appealed but not on the basis of a 'robust and systematic indictment of poor philosophy' that Evers and Lakomski claim (Evers and Lakomski 1991, p. 13). My grounds for complaint were quite different and rested on observations drawn briefly from the sociology and anthropology of science — particularly those traditions concerned with processes of the production of science and scientific knowledge. A further difficulty is that Evers and Lakomski represent the argument of articles they quote as being based upon a particular reading of Habermas:

> In developing this claim (that a science of administration is essentially manipulative and concerned with social control) Bates draws upon a reading of the early work of Habermas for an understanding of science: particularly the epistemological theses of Knowledge and Human Interests and the 'General Perspective' lecture published as the appendix to the English translation (Habermas 1972, pp. 301–387) (Evers and Lakomski 1991, p. 12).

In fact my 1980 article makes no reference to Habermas (I don't believe I had even *read Knowledge and Human Interests* at that date) though Evers and Lakomski provide an extensive quotation that suggests that not only had I read the volume in question but that my whole argument was based upon it!

Similarly, my 1983 paper makes a single brief reference to Habermas in the context of introducing Watkins (1983) discussion of the distorted structures of communication that arise from the inappropriate exercise of power in hierarchical organizations. But Evers and Lakomski get the reference wrong: it was in fact to Habermas's 1975 publication of *Legitimation Crisis* and not the 1972 *Knowledge and Human Interests* as they suggest.

Again, they represent my paper as being fundamentally concerned with an exposition of Habermas's thesis as the basis for a *philosophical* attack on the traditions of educational administration. In fact, both papers are conceived within the tradition of the New Sociology of Education. Concerns with philosophical issues arise en passant and are neither grounded in nor articulated in the terms suggested by Evers and Lakomski.

However, even had I couched my argument in terms of the Habermasian thesis I could not possibly have agreed with their absurd conclusion, which supposedly finishes off critical theory as a serious contribution to the debate. However, since my work is taken as the starting point for their supposed exegesis I feel bound to address it. Evers and Lakomski claim that

> There are a number of things that are puzzling about (Habermas's) account of science and administration, especially in view of the fact that Bates thinks traditional empiricist accounts of the practice and conduct of science are mistaken. For if the traditional view of science is wrong, and we know that it is thanks to the work of Quine, Kuhn, Feyerabend, Hesse and others, then the story Habermas tells of empirical science being constituted by technical interests of control and manipulation is also wrong (Evers and Lakomski 1991, p. 13).

This is, of course, not necessarily so. For such a claim to be sustained it would need to be shown that the grounds on which the traditional (essentially positivist) view of science was successfully challenged by Quine *et al.* were precisely those on which Habermas had based his claims. In fact the grounds on which the challenge was mounted by Quine *et al.* were essentially *epistemological* (grounds which are further explicated by Evers and Lakomski in their assessment of the Theory Movement in educational administration and with which Habermas largely agrees). But the grounds on which Habermas makes his claim concerning the role of science in modern societies were *sociological*. That is, Habermas claims that

> From the very beginning the pattern of human sociocultural development has been determined by a growing power of technical control over the external conditions of existence . . . We know how to bring the relevant conditions of life under control, that is, we know how to adapt the environment to our needs culturally rather than adapting ourselves to external nature (Habermas 1989, p.260).

This knowledge has resulted from a modern conception of science and an alteration in its purpose:

> In this connection modern science assumes a singular function. In distinction from the philosophical sciences of the older sort, the empirical sciences have developed since Galileo's time within a methodological frame of reference that reflects the transcendental viewpoint of possible technical control. Hence the modern sciences produce knowledge which through its *form* (and not through the subjective intentions of scientists) is technically exploitable knowledge, although the possible applications generally are realised afterwards (Habermas 1989, p. 249) .

Or an even more pointedly social claim:

> ... the cognitive function of modern sciences must be understood in connection with the system of social labor: they extend and rationalise our power of technical control over objective or (what amounts to the same thing) objectified processes of nature and society (Habermas 1989, p.36).

On the basis of a confusion of their epistemological concerns with the social concerns expressed by Habermas, Evers and Lakomski suggest that Habermas would therefore claim that

> If traditional views of science were true then technical control and manipulation would occur. Therefore if everyone (professor and practitioner alike) acted as though they were true then technical control and manipulation would occur (Evers and Lakomski 1991, p.14).

Habermas, of course makes no such claim. This is a mischievous invention of Evers and Lakomski, as is their attribution of an extreme subjectivist view to Habermas:

> The missing premise in this (Habermas's) argument is a subjectivist claim that the effect of my having a particular theory of the world somehow makes the world that way, or brings it into line with my theory (Evers and Lakomski 1991, p. 14).

But Habermas explicitly rejects a purely subjectivist theory of meaning. His theory of communicative action

> ... relies on a cooperative process of interpretation in which participants relate simultaneously to something in the objective, the social and the subjective worlds even when they thematically stress only one of the three components in their utterances. (Habermas 1987, p.120).

To dismiss critical theory as subjectivist clearly involves a careless reading of Habermas. To suggest, as Evers and Lakomski do, that critical theorists would agree that water flows uphill if that was the 'dominant orthodoxy' is simply mischievous.

I shall return later to Evers and Lakomski's failure to treat the social claims of critical theory seriously and with due care. Suffice it to note here that there is a major problem inherent in their insistence that social arguments be treated solely as epistemological arguments. Their exclusion of the social is a fundamental flaw in their approach to *Knowing Educational Administration*. This flaw is particularly evident in their treatment of *The Cultural Perspective* (Chapter 6) and *Administration for Emancipation* (Chapter 7), to which I now turn.

The Cultural Perspective

Evers and Lakomski choose two major representatives of the cultural perspective: Geertz from the field of anthropology and Sergiovanni as the champion of such a perspective in educational administration.

Choosing Geertz as the sole representative of the complex and contested terrain of anthropological debate over the concept of culture runs the danger of presenting a very partial view of the field. For their purposes, however, Evers and Lakomski have chosen well for Geertz does indeed argue that cultures are only to be understood in their own terms — that they are essentially incommensurable as systems of thought. Moreover, in his 1983 address to the American Anthropological Association Geertz mounted a strong attack on the notion of a universal social science and against the universalising of common sense, arguing that social practice and common sense itself were only amenable to study in their particularity (Geertz 1984). Indeed, as Austin Broos argues,

> Geertz suggests we should commit ourselves to an 'ethnography of thought' recognising that different cultures like different academic disciplines constitute different 'forms of life' (after Wittgenstein) or different ways of being in the world (after Heidegger) (Austin-Broos 1987, p.155).

But Geertz's point of view is highly controversial among anthropologists, notwithstanding the fame of his analysis of the Balinese Cockfight. The relativism of his position is by no means a universally agreed position among anthropologists. Indeed, the foundations of his relativist 'ethnography of thought' are strongly challenged by others in his field. In commenting on his most famous ethnographies of Modjokuto and Balinese societies Austin-Broos comments that

> There are real differences between (these) societies and it appears that in some the aesthetic component is more pronounced, or more pronounced in some areas of life. Yet Geertz has not stopped here. He has taken up this discontinuity between cultures and made it a uniform and total breach

between each culture and any others we currently identify. For all the boldness of this move, it has been carried forward on the basis only of some very vague ruminations on common sense, rationality, and the nature of semantics and cognitive symbols that do not withstand rigorous investigation (Austin-Broos 1987, p.157).

Choosing Geertz as the champion of the contemporary debate over culture as a concept is therefore a very partial representation of a debate in which many players have opposed and denied the relativism of his position. 'Turtles all the way down' is not a proper characterization of current anthropological thought in the area of culture. Such a characterization reduces a complex and sophisticated argument to the level of parody — which, of course, is easier to dismiss, as Evers and Lakomski do.

Part of this parody relies again upon the exclusion of qualifying arguments. In their discussion of the cultural perspective the cultural is equated with individual subjectivity. That is, the hermeneutic attempt to understand as well as simply describe human interaction is reduced to a discussion of subjectivity:

> This, means, then, that the validity of understanding an act or event is a matter of subjective assessment of an individual, or a group, given prevailing attitudes in that culture or time (Evers and Lakomski 1991, p. 127).

Or, again:

> What makes us 'human' in the cultural perspective, is that we possess 'inner' phenomena such as motives, purposes, and intentions (Evers and Lakomski 1991, p. 128).

What such an interpretation of cultural anthropology suggests is that culture is reducible to internal psychological states that originate somewhere inside the individual in much the same way as hunger and thirst. This is a complete denial of the vigorous assertion by anthropologists such as Bourdieu that culture is a result not of our appetites or desires, but of our socialness (as indeed may hunger and thirst be social in their origins — though that is another argument). Indeed:

> Bourdieu is concerned with our socialness, but with a socialness conferred by the complex of dispositions which we as social strategists, employ in our everyday life... For Bourdieu, to concentrate on individuals and actual populations is to fail to transcend the ideology of individualism which dominates our society; it is to fetishise our personalities at the expense of our sociological knowledge, to succumb to subjectivism (Miller and Branson 1987, p. 215).

Evers and Lakomski fail to take this socially oriented interpretation of culture into account. Preferring instead a version that is indeed prone to accusations of relativism, subjectivism and reductionism. Once again, their exclusion of the social has led them to trivialize an important argument, dismissing it on grounds that do not engage some of its most important characteristics which do indeed address their complaints (see for instance Bates 1987).

Their treatment of Sergiovanni is hardly more serious. While they, quite accurately, represent Sergiovanni as emphasizing that

> ... the cultural perspective in educational administration is concerned more with understanding than explaining, with sense-making rather than description, and with 'community' and shared meanings and values (Evers and Lakomski 1991, p. 125).

They also go on to eliminate the social from Sergiovanni's argument, suggesting that what he is doing is a form of reductionism that posits inner states that are unobservable and therefore cannot contribute to our understanding of the processes of understanding.

Their response to this problem however is a further reduction: one that reduces psychological phenomena to physical ones.

> In our view, interpretation requires us to draw on all the theory that is relevant to human action. Since this manifestly includes powerful scientific physicalist theories of human behaviour, a coherentist view of evidence for interpretation, for assigning motives, beliefs, desires, and intentions, will place a premium on reducing the mental to the physical, at least at the level of substance (Evers and Lakomski 1991, p.133).

Or, even more explicitly,

> ... while we do not deny the existence of inner phenomena as causing human behaviour, it is not intentions which do the causal work but fine grained, neuro-physiological mechanisms. Intentions, as referred to in ordinary talk, are more fruitfully conceived of as 'promissory notes' for these underlying mechanisms which, unlike the former, are, and will be, identifiable by our best science (Evers and Lakomski 1991, p.134).

Once again, there is a total denial of the social: a reduction of the social to the psychological, the psychological to the biological, and the biological to the physical. Yet we are given no reasons why such an extraordinary assertion would be accepted. We are given no examples or illustrations of the power or utility of such an argument. We are simply told that 'a better ontological candidate for "inner" states would be the fine-grained physical properties of human subjects' (Evers and Lakomski 1991, p. 132). Turtles all the way down indeed.

Administration for Emancipation

Evers and Lakomski's discussion of neo-Marxist and Critical theory in relation to educational administration is curious, to say the least. The claim of their book is that it examines post-positivist theories of educational administration in the light of their philosophical claims. In their first chapter they acknowledge the critique of traditional educational administration theory made by critical theorists as a major post-positivist challenge. They recognize that 'critical theory approaches to administration are complex and multi-faceted, covering ethical, political, social, linguistic, and personal dimensions' and promise a 'systematic treatment' of these issues in Chapter 7. What is delivered in Chapter 7 is, however, an analysis that eliminates any consideration of the ethical, political, and social; which makes only brief reference to Foster's development of Critical theory in educational administration and which completely ignores any textual analysis of my work, which they claim in Chapter 1 to be central to the critical theory tradition in educational administration.

Instead, we are presented firstly with an analysis of neo-Marxist organizational theory which has in fact not been fundamental to the emergence of critical theories of educational administration and an analysis of Habermas's early work, which is also wrongly claimed as the basis of my own early work (which in fact grew out of the New Sociology of Education) as well as that of Giroux (whose work in fact appealed to earlier critical theorists of the Frankfurt school and who is quite critical of Habermas (see Giroux 1992) and Apple, who has, in fact, paid little attention to Habermas's work and who is mainly interested in curriculum).

Rather than deliver its earlier promise, Chapter 7 avoids the quite extensive work in the Marxist/Critical theory tradition in educational administration (see for instance Bates 1985). Despite this, it also claims to have dispensed with such work on educational administration on the basis of an analysis of Habermas's work of the earlier period. This, however, is not dealt with in its political, social and historical aspects, but only in its epistemological aspects — ones that it is acknowledged that Habermas has turned aside from.

Yet it is these aspects that provide the foundation of a critical theory of educational administration. This theory has derived almost wholly from Habermas's concern with the political economy of late capitalist societies — especially those issues articulated in *Legitimation Crisis* (Habermas 1975) and more recently reformulated in *The Theory of Communicative Action, Vol. 2* (1987).

These concern the encroachment of administrative rationality via systems theory and the mechanisms of power and money on the 'lifeworld' within which communicative rationality arises and through which society is symbolically reproduced. The encroachment of the 'system' on the 'lifeworld' produces a series of crises, which can only be resolved through communicative rationality, which is itself only possible within the context of the lifeworld. The 'communicative rationality' of Habermas's later work is a substantial revision of his earlier work to which Evers and Lakomski refer. It is certainly not the case that Habermas's notion of communicative rationality is simply a 'consensus theory of truth (which) rules out the possibility of statements being true by virtue of empirical reality' as Evers and Lakomski suggest (1991, p. 161). Indeed, Habermas is at pains to emphasize that communicative rationality addresses objective, subjective and inter-subjective domains of reference simultaneously. Empirical referents are therefore at least implicit in all instances of communicative rationality.

In choosing 'to examine the original version of Habermas's critical theory, rather than the reading given to it by writers in educational administration or organizational analysis' (Evers and Lakomski 1991, p. 163), Evers and Lakomski therefore avoid discussing the social and political roots of critical educational administration theorists' interest in Habermas, and Habermas's reformulation of his earlier interest in the philosophy of consciousness which he recognizes as imposing significant limitations on his original formulation.

Evers and Lakomski can therefore claim to have dealt fairly neither with the work of critical theorists of educational administration nor with Habermas's reformulation of critical theory.

Conclusion

It has not been the purpose of this chapter to articulate fully either a cultural perspective or a critical theory of educational administration. The beginnings of each of these projects I have suggested elsewhere (Bates 1984, 1985, 1987, 1988). Nor has it been to deny the worth of philosophical analyses of theories of educational administration. Indeed, I welcome the attention of philosophers in the field. But the minimal conditions they should observe are care in the treatment of the arguments they examine, a proper reticence in representing social arguments as directed towards solving philosophical problems, and avoidance of taking single instances as representative of the field under examination.

References

Austin-Broos D. J. (1987). *Creating Culture.* (Sydney: Allen and Unwin).
Bates R. J. (1980). Educational administration, the sociology of science and the management of knowledge, *Educational Administration Quarterly*, **16**(2), pp. 1–20.
Bates R. J. (1983). *Educational Administration and the Management of Knowledge.* (Geelong: Deakin University Press).
Bates R. J. (1984). Education, community and the crisis of the state, *Discourse*, **4**(2), pp. 59–81.
Bates R. J. (1985). A marxist thaw of educational administration? Paper presented at the Annual Conference of the American Educational Research Association, Chicago, USA.
Bates R. J. (1987). Corporate culture, schooling and educational administration, *Educational Administration Quarterly*, **23**(4), pp. 79–115.

Bates R. J. (1988). Is there a new paradigm in educational administration? Paper presented at the Annual Conference of the American Educational Research Association, New Orleans, USA.

Evers C. W. and Lakomski G. (1991). *Knowing Educational Administration: Contemporary Methodological Controversies in Educational Administration Research.* (Oxford: Pergamon Press).

Geertz C. (1984). Anti-antirelativism, *American Anthropologist*, **86**(2), pp. 263–278.

Giroux H. (1992). *Border Crossings.* (New York: Routledge).

Habermas J. (1972). *Knowledge and Human Interests.* (London: Heinemann, translated by J.J. Shapiro).

Habermas J. (1975). *Legitimation Crisis.* (Boston: Beacon Press).

Habermas J. (1987). *The Theory of Communicative Action. Vol. 2: The Critique of Functionalist Reason.* (Cambridge: Polity Press).

Habermas J. (1989). Technology and science as 'ideology', in S. Seidman (ed.) *Jürgen Habermas on Society and Politics: A Reader.* (Boston: Beacon Press).

Miller D. and Branson J. (1987). Pierre Bourdieu: culture and praxis, in D. Miller and J. Branson (eds.) *Creating Culture.* (Sydney: Allen and Unwin).

Watkins P. E. (1983). Scientific management and critical theory in educational administration, in P. E. Watkins (ed.) *Educational Administration and the Management of Knowledge.* (Geelong: Deakin University Press).

16

The Epistemological Axiology of Evers and Lakomski: Some Un-Quineian Quibblings—
Christopher Hodgkinson

The contribution of Evers and Lakomski to an emergent discipline of philosophy of administration is without question; they have endowed this putative subset of educational administration with a basic text (Evers and Lakomski 1991). The quality of their scholarship is rigorous and the sincerity of their intent beyond question. Nevertheless, their text is set down into the swirling contentious maelstrom of academic quibbling and the authors are, in Nietzsche's phrase, 'human, all too human' (Nietzsche 1989). I am sure they will allow me this reservation, especially as they have so kindly devoted a chapter to 'Hodgkinson on Humanism'. Given the constraints of time and space, I shall confine my comments not to this excellent chapter, with which I can hardly find fault, but instead to quibblings of the academic variety about Chapter 8 'Ethical Theory and Educational Administration' and Chapter 9 'Policy Analysis: Values and Complexity.' The former is, I think, more directly attributable to Colin Evers, and where authorship is concerned will hereinafter be referred to as EL (for Evers-Lakomski), while conversely the latter will be referred to as LE (for Lakomski-Evers).

Between them, in these two chapters, EL and LE lead us a merry dance. 'About it and about' as Omar Khayyam (Fitzgerald 1955) might say: EL treading a dainty minuet about the value–fact distinction and LE leading us in a stately pavane through that entangled web of conceptual confusion known as policy analysis. It is all educative, informing, entertaining, but, yes, there are difficulties. It is these difficulties that I shall expatiate upon, but first, I think, one must consider some more general difficulties with EL and LE. The largest is that they are more interested in epistemology than axiology, in truth rather than goodness, and their orientation is perhaps inversely related to the distribution of interest in practice. Most practitioners are concerned with what is good and bad or right and wrong; what is true or false seems to them self-evident. And this leads directly to the distinction between value and fact; a distinction that EL/LE wish to eliminate, or at least fudge over. More about this below. EL and LE do emerge finally with an axiology but it is a weak and watery Deweyism set of Type II values that might help the educational administrator through the day but would hardly inspire him. The 'touchstone' of consensus, especially scientific consensus, is after all a sort of lowest common denominator. One is unlikely to die, or go to the stake, for it. And yet administrators may have to make those 'job-on-the-line' decisions from time to time. When they do, I submit, they need more than epistemological axiology to guide them (Hodgkinson 1982, 1983, 1991). But let us scrutinize the arguments.

EL

The central problem in ethics for EL is straightforward. They simply wish to overturn the conventional wisdom. The latter, from antiquity through to Wittgenstein maintains a distinction between values and facts; between, in other words, science and ethics, and asserts that one cannot *logically* get from the one to the other. Or, more precisely, '... that evidence for one cannot function as evidence for the other' (Evers and Lakomski 1991, p. 166). This is how EL launch their case. They are going to show by their holistic epistemology that science and ethics can be unified, that the claims of the one (science) can serve as 'evidence' for the other (ethics). A bold, and some would say overweening, ambition, but a gallant one — and, after all, the evolutionary function of all orthodoxies would seem to be their capacity to incite the intellectual effort that will ensure their ultimate overturn and replacement by something fitter and better.

Let us, however, note at the outset that this relationship between science and ethics is, even in the EL canon, non-reciprocal and intransitive. Science is to serve as 'evidence' for ethics but not vice versa. One cannot claim, as some North American educational administrators might like to, that blacks are the academic equals of Asiatics, and then go seeking scientific evidence to substantiate this 'ethical' claim. Science might prove it wrong. And that very possibility precludes any scientific engagement on *political* grounds. Worse, even if science were to prove it right, the distinction between value and fact would remain immaculate, unsullied; for the initial ethical claim was, upon analysis, not truly that blacks *were* equal (this would have been a scientific claim) but that blacks *ought to be* equal (the unspoken ethical claim). Certainly ethics will limit scientific activity by setting the bounds of investigation as it did in Galileo's time and as it does today through the sanctions of political correctness, but it cannot function as *evidence* for science, only, at best, as motive. On the other hand no one can dispute that science and empiricism can provide 'evidence' for ethics. That it is proven that the earth moves (*eppur si muove*) occasions eventually, 360 years later in this instance, an ethical apology from the Pope. In this sense science sets the bounds for ethics. Moreover, anything that *social* science can tell us about *what* our values are and *how* we acquire, maintain, and change them is grist, that is, *evidence* for the ethical mill. But if this is all that EL maintain, then it is merely conventional wisdom. The basic claims of orthodoxy would remain and these are what EL labour to overturn.

The Is–Ought Dichotomy Still Stands

EL rightly take as their starting point Hume's assertion that the quality of *ought* is entirely different from the quality of *is*. Note the modifier *entirely*. What is being remarked here is that the value experience is utterly, that is, logically, psychologically, and ontologically different from that of objective empirical reality, fact, or logical truth. To say, 'That is a stone before me', or 'that $1 + 1 = 2$', is really *quite* different from saying, that ought to be a stone, or one plus one ought to equal two. How to challenge this?

The EL challenge consists in three moves. The first is to shift the argument to one of formal logic (ignoring the ontological and empirical bases of Humean analysis). This leads to an invocation of the 'vocabulary criterion' (Evers and Lakomski 1991, p. 168), which invalidates logical conclusions if they contain 'non-logical' words (such as 'ought') if such words are not contained in the premises of the argument. So far so good for Hume. But then comes the second move: to refute the vocabulary criterion by citing the work of A. N. Prior (1960) and an arbitrary rule of squiggle logic (If p then either p or q) that permits such patent (but logical) absurdities as: Snow is white. Therefore: Either snow is white or we ought not to kill. And presumably: Pigs have wings and either pigs have wings or logicians ought to keep well away from administration.

EL go on to cite even more extravagant examples of logic chopping *pace* Peano, Black and Kurtzman (Evers and Lakomski 1991, p. 169), some of which might persuade a professional logician and all of which would surely mystify the administrator behind the desk. They do concede that 'it is possible to argue that these deductions are trivial' (Evers and Lakomski 1991, p. 169), and in an intermediate sequitur allow that 'non-trivial deductions of "ought" from "is" would be invalid by definition' before proceeding to their third move, which is contained in the sentences

> We do not propose go press the point further. It is sufficient for our purposes *merely to note* that there is an *awkward complexity* involved in specifying a version of the 'is/ought' dichotomy *once it is admitted* that one can validly derive statements containing 'oughts' from statements containing only 'is's' (Evers and Lakomski 1991, p. 169, my emphasis).

Indeed, this conclusion is a masterpiece of caution. And a composition of error. Wisely EL propose not to press the point. But they do. They do. It is *not* sufficient to terminate the attack this way. 'Awkward complexities' cannot be brushed off. 'Version' needs clarification. And 'once it is admitted' does not constitute a proof. Or a refutation of Hume. The dichotomy stands.

A psychoanalytic test for discriminating between a psychotic and a neurotic is said to be that while the former might *know* that 2 + 2 = 5, the latter knows that 2 + 2 = 4 but *just can't stand it*. But perhaps one might excuse a little neuroticism on EL's part considering that their axiological edifice depends so much on obviating the distinction between value and fact.

The Naturalistic Fallacy Still Stands

The fact–value distinction is inseparable from the naturalistic fallacy. They stand to each other as obverse and reverse, both sides proclaiming that one cannot get an *ought* from an *is*. There is a further subtlety to Moore's (1903) great contribution to philosophy and that is the notion of good being a 'primitive' or atomic term. It represents, to each subject using the term, a bedrock of desire that cannot be further analyzed. Good (to the subject) is just that. It is not complex like pleasure or happiness, nor is it natural since its origin is phenomenological, nor is it definable. It is a simple non-natural property; a Type III preference in terms of the value paradigm.

EL's treatment of the naturalistic fallacy is nice until it ascends from description to prescription. There then follows a confusing exposition based upon the authority of Quine (Evers and Lakomski 1991, p. 171), in which we are sidetracked into a debate about the justification of the analytic–synthetic distinction, apparently denied by Quine, and illustrated by the rather tortuous and tenuously related assertion that the term 'bachelor' is not amenable to analytic (i.e. tautological) definition as an unmarried adult male but must instead be referred to empirical or 'touchstone' evidence in ordinary language (i.e. synthetic definition). Frankly, I do not understand how this refutes the naturalistic fallacy and perhaps the error is mine, the subtleties being beyond my grasp. But from an axiological standpoint there would appear to be a confusion between values (Type III) and ethics (Types I or II) in the EL line of argument. They assert, *pace* Quine, that Moore does no more than ask

> whether a particular theory of good is really justified. It does not show that the mere asking of this question demonstrates that such a theory can never be justified. In fact, not only is the matter of justification methodologically open, but in our view a particular naturalistic theory of good is more *coherent* than its major rivals, and hence to be *preferred*. Such a theory *could be construed* as offering a reduction of good to some natural quality (Evers and Lakomski 1991, p. 172, my emphasis).

But there are *two* things here: good and *right*. The good is the axiological preference, the justification of it the ethical claim. That Guinness is good for you does not entail that one ought to drink Guinness. To assert this is surely to commit the naturalistic fallacy as indeed EL themselves unconsciously do when they assert that to be coherent is to be preferred. Coherence just happens to be their preference. *De gustibus non est disputandum*.

To their credit EL modestly note that they have not examined other versions of the naturalistic fallacy and its consequence, the autonomy of ethics. Instead they settle for

> expressing the *belief* that the whole strategy of defending partitions in knowledge is extremely difficult given *our* holistic theory of knowledge (Evers and Lakomski 1991, p. 172, my emphasis).

Indeed it is, but the difficulty is theirs not ours. Meanwhile, ethics remains autonomous for the infidels of orthodoxy. And incoherent for reductionists like EL. I would merely add that the distinction claimed by the human species to render itself ontologically other than the rest of the animal kingdom is that, in the language of myth, it has eaten the fruit of the tree of knowledge of good and evil. Mankind *has* moral capacity. It can do ethics. Cats and dogs can't. Automata can't. Administrators can. And *should*. A most peculiar 'non-natural' capacity when one reflects upon it.

The Value Paradigm Still Stands

The disaffection of EL with what they call my moral theory seems to centre about the possibility of such leaders as Adolf Hitler being exponents of Type I values, the highest in the value hierarchy. Let me take the opportunity of declaring that my analysis of value is not intended to prescribe moral theory but rather to *describe* its infrastructure. It is axiological rather than ethical. And it cannot be discounted by the demonizing of Hitler and the Nazi ideology (Evers and Lakomski 1991, p. 176). 'But moral consciousness and its informing folk theory is not so fragile as to be distorted by alternative moral pronouncements', say EL (Evers and Lakomski 1991, p. 176). Would they consider, however, that if the Nazis had won the war the 'informing folk theory' would then have been a *Volkstheorie* that could have led to our all favouring brown shirts. Touchstone here depended upon who won the war and, as German philosophers have often observed, Macht macht Recht. Might makes right and 'Power is the first term in the administrative lexicon'. For seventy years and three generations, Marxist-Leninist informing folk theory gave Soviet citizens a very coherent touchstone Type IIb frame of value-reference, often inspired by authentic Type I aspirations to create the New Soviet Man and a new world moral order. They also gave us the Gulag.

Given the historical record one can sympathize with EL's discomfiture at a paradigm that *coheres* with that record. It is not in accord with their preferences. They would like the world (and its administration) to be nicer. What remedy then do they offer? Why, a faith (Type II not Type I, I fancy) in that good old stand-by, science. Against the 'banality of evil' (Arendt 1963) they raise the standard of science. '... there are gains to be had in being able to speak scientific truthfulness' (Evers and Lakomski 1991, p. 177). One's moral posture is to derive from one's scientific grounding or touchstone.

Ignore that EL are here embarking on the naturalistic fallacy. Consider only that science itself is not necessarily coherent (see the next section) and that *social* science is always constrained by mores and ethos. Philippe Rushton at the University of Western Ontario could not investigate racial differences unhindered, nor could Jencks at Harvard nor Eysenck at London; their science was not politically correct. *Per contra*, the floodgates today opened wide to an outpouring of pseudo-science on politically correct women's studies or women's issues. Science alas, is fickle, handmaiden to the consensus (IIb) of the day. And, even more alas for EL, where science is 'hard' and 'pure' it has historically funded only the value theory of logical positivism. In terms of the paradigm the best it can do for us is enhance our sophistication at the IIa level of the paradigm. To the extent that it causes wonder in our souls at the architecture of the universe, it leads us beyond science: to the leap of faith and, possibly, to Type I value commitments.

What Doesn't Stand

EL most clearly reveal their empiricist–rationalist biases in the following passage:

> ... the only evidence we have for our moral beliefs is evidence which is *on a par* with the evidence we have for our *so-called empirical* beliefs. [Ethical theories] are justified according to the virtues of system that accrue to the global theories of which they are part ... Because humans are complex physical systems living in a physical world, we assume the most parsimonious accounts of our moral beliefs will be *naturalistic* accounts, since these will *cohere* most *economically* with our best *physical* science. Attempts within biology to use evolutionary theory to explain our current folk moral dispositions are a case in point (Evers and Lakomski 1991, p. 185, my emphasis).

This of course is coherent with the EL epistemology. It is pro-science — hard science, that is — but it is wrong-headed, and therefore not *good* science. Too much is left out in the EL–Quineian worldview. Essentially the axiological path traced by EL is one from Skinnerian behaviourism, which reduces values to affect and ethics to programming, through to Wilson's bioethics, which reduces everything to evolutionary process and ultimately to Professor Evers's own favourite reductionism: brain chemistry and neuronal electro-mechanics. But this trajectory rapidly becomes unscientific: it omits too much and it assumes too much: for example, that the mind–brain problem in philosophy has been solved; that the value–fact distinction is destroyed, leaving us only with facts (But what are facts?); that quantum mechanics is reconciled with general relativity; that there are no Chomskyian depth structures in language; that we can explain what is meant by a belief; that all metaphysics will be reduced ultimately to physics (i.e. coherent naturalistic accounts) (the hoary ploy that 'science' will give us pie in the sky by and by); that science itself coheres (or that we should take out a promissory note that it will some day cohere); that the Turing-machine problem has been solved; that non-equilibrium thermodynamics can be reconciled with Newtonian physics; that the anthropic principle coheres with scientific epistemology; or that Jungian synchronicity does; or Sheldrake's morphogenetic thesis; or the Gaia hypothesis; or that, in short, Truth, Beauty, and Goodness are not teleological ideals but mere epiphenomena of sub-molecular activity.

Within the welter of 'many divergent perspectives' that EL do admit to, how then is one to achieve a parsimonious base for axiology? EL are modest about this '... considerable further work seems to be required to argue for a preferred normative ethic for educational administration' (Evers and Lakomski 1991, p. 185). Well, yes, that would seem about right.

What also would seem right is the EL dismissal of Rawls and Harsanyi (Evers and Lakomski 1991, pp. 180–183). Yet these versions of liberal–rational utilitarianism are not so far removed from the rehabilitated Deweyian pragmatism that EL and Willower (*qv.*) seem to advocate. The problem with this from my standpoint is that the faith in democratic problem-solving (and its implicit codicils of welfare-utopianism and social engineering) ignores the problem of evil — banal or otherwise. It exalts reason over the passions that really govern men. It leads ultimately and proximately to hyper-relativism and neo-nihilism. Its end state is worst than its first. It ought not to be recommended as touchstone to the educational administrator without being hedged about with more cautions than EL provide. Nor is it quite right to suggest, as EL do in their summary, that the value–fact distinction has been disposed of and that the solution to ethics lies, though distantly, within rational and epistemological sight. Such solution, or even resolution, is not upon the horizon. Their reach extends not a little beyond their grasp in this chapter, which remains, nonetheless, a most impressive account of profound and intractable problems. Their work may not flood the darkness of administration with light but it may serve to show that there is still darkness where before was thought to be illumination. And it does thread pathways through the black.

LE

From the admittedly abstruse plane of ethical analysis surveyed with undeniable skill by EL we move now to the more concrete realm of practice as treated by LE. Here the authors weave a stately pavane in their survey of the complexities associated with policy formation; here we touch upon not so much practice as *praxis*, that is to say, the conjunction of value systems with administrative planning and managerial implementation. In short, with philosophy-in-action.

An immediate merit of this chapter is that it deals succinctly and concisely with a vast and contentious literature. In this alone it would reward the student or practitioner who wishes for whatever reason to join in the pavane. The dance is patterned, however, and the philosophical line of the previous chapter, the EL–LE 'party line', is enunciated clearly at the outset as presenting an argument for a 'fallibilist, coherentist view of values within a much broadened scientific view of policy and decision' (Evers and Lakomski 1991, p. 193). One refrains from italicizing only with difficulty, but the LE axiology is now apparent. Values can be grounded (sort of, by touchstone) in an empiricist epistemology and scientific worldview which, presumably, is evolving towards the best of all possible worlds. It has to be the best because that's what evolution does, *n'est-ce pas*? Should we go wrong somewhere, science will show it up and set us aright on our way towards the Omega point.

LE structure their discussion about two questions: How to integrate values into policy analysis? and How to improve our knowledge of complex human systems? These are good questions, the first clearly axiological, the second directly epistemological and indirectly axiological. I shall return to them at the end of this chapter but we should already note that in seeking to answer such questions we need to visit not just philosophy but also psychology, sociology, social psychology, and political science. Not to mention art, literature, and history. LE are correct to include the word Complexity in their chapter title. There is nothing simple about policy-making unless it be to refute the notion that rational-synoptic strategies are viable — in educational administration or any other kind of administration. This is a dead horse, which, however, LE are still inclined to flog. Their energies could be spared.

LE's specific example of dead horse is Lasswell. He was, of course, doomed from the outset because his definition of value (not entirely inconsistent with my own) was 'a category of preferred events'. The problem here is that 'preference' goes unanalyzed at the psychological level; it is assumed merely that policy makers can observe and aggregate in some quantitatively respectable manner the component preferences of a complex policy context. Yet not even this can be done *pace* Arrow (1963). We are also up against the fact/value dichotomy which LE like no more than do EL. Their motivation in dismantling the rational-synoptic approach must at least in part be to indicate their epistemological axiology, viz. 'maintaining a sharp distinction between fact and value does our capacity to learn *new* values from experience a disservice' (Evers and Lakomski 1991, p. 197) (my italics). LE would prefer that the distinction were not sharp. They would rather have it blurred, and they look expectantly to the future for our *new* values since experience (i.e. the growth of scientific knowledge) teaches us that tomorrow is always better than yesterday. As if evolution endows humanity with an arrow of time so that each generation can gasp in wonderment at the folk-theoretic follies of their predecessors. This, after all, is the democratic tradition in which policy makers follow policy makers across the stage, entering to the applause of the demos and eventually exiting to cat calls, whistles, and abuse.

Of course values and facts never *appear* separately any more than do substance and quality, or time and space. The world of the administrator is a world of mixed values and facts, inseparably intertwined, but this is not to say that the distinction *ought* to be blurred and *ought not* to be sharp. There is an inescapably *sharp* distinction here, a *break*, which no epistemology has yet managed to bridge. Despite EL's and LE's best efforts, values and facts continue to form not a continuum but a dichotomy.

If then rational-synoptic large-scale planning will not do, what are we left with? Well, strictly speaking, only incrementalism or muddling through, about which LE have little to say, or mixed scanning about which they have nothing to say. Instead they concentrate upon policy analysis as interpretive social science or counsel. Wildavsky and Jennings are the favoured exponents, especially the latter to whom several pages (Evers and Lakomski 1991, pp. 199–203) are devoted. Jennings is indeed worth lingering over because he lays claim to a certain 'post-positivistic objectivity' about values. This objectivity permits the analyst to determine the preferred policy option

> in the light of the more general vision of the good of the community as a whole as well as the more discrete interests of the policy makers themselves (Evers and Lakomski 1991, p. 202) .

Post-positivist Platonism this might be called since it seems to elevate the analyst at once to the rank of Guardian in the new Republic. And how is the objectivity warranted? By categories

> drawn from a common, intersubjectively meaningful set of cultural norms, traditional values, and serviceable common-sense understandings of what human beings need and how they react to various circumstances (Evers and Lakomski 1991, p. 203) .

In terms of the value paradigms these are Type IIb values (while Wildavsky's legal adversarial emphasis is Type IIa). At best they are 'folk-theoretic' or touchstone, at worst mere opinion polling. Presumably, however, the Jennings approach finds favour with LE because it is 'fallibilistic'. But the fallibility here is what I have labelled elsewhere as the homogenetic fallacy, the notion that values are not hierarchically discriminable amongst themselves, as in the value paradigm, but are logically homogenetic . . . a value is a value is a value. This fallacy is committed not just by Wildavsky, Jennings, and the other authorities cited by LE, but also by LE themselves. The point that is continually overlooked is that *any* value can be held at *any* level, from the preferential to the transcendental. Thus, for example, the value of education itself (that education is desirable) can be a mere preference (III); a result of peer group pressure and conformity (tradition or folk-theoretic — IIb); a conclusion of cost–benefit analysis (IIa); or a searing passion commanding lifelong sacrifice and commitment (I). The policy maker needs to know which is which, and LE do not make these distinctions.

In fact they tend to compound this error. Against Weber and Habermas they seek to blur the distinction between *Naturwissenschaft* and *Geisteswissenschaft*; between, that is, the different kinds of science and again, between value and fact, because this distinction 'relies on an unnecessarily narrow and now rejected instrumentalist conception of natural science', (Evers and Lakomski 1991, p. 203). Rejected by whom? one is tempted to ask. LE wish to redefine or undefine the concepts of value, fact, and science. And to do it in such a way that the colours will all run into each other. When one does this with paint, one gets grey or black. A mess. When one does it with concepts important to our understanding of the world, its perception, and its administration, one gets less light and more darkness. Confusion. To call this semantic holism will not relieve them of this problem. For example, one *can* say that values are themselves facts, but this cannot be allowed to blur the distinction, because values are a different *order* of facts. Distinct logically, ontologically, and epistemologically.

Furthermore, if the conception of science is to be broadened so as to include the social sciences, the arts, the humanities, all knowledge and all experience, then indeed it really would subserve the policy-making endeavour. It really would fund administration as LE would like, but it would not be science, it would be philosophy. LE's Humpty Dumpty approach to semantics doesn't help us here.

What, in the end, does the LE (and EL) axiology look like and what kind of map can it offer the policy maker and the practitioner–navigator? Tentatively and conditionally the touchstone values are 'tolerance, freedom of speech and thought, openness of debate and discussion, and a promoting of learning and scholarship' (Evers and Lakomski 1991, p. 206). These, it seems, are their preferences (Type III) and one would assume that they are held, given LE's philosophical credentials, at least at the IIa level. But preferences they are, and one man's touchstone is another man's brimstone. LE's liberal values were once espoused by Islam, to the everlasting benefit of science, when they were denied by Christianity. They are denied today by the grievance groups who embrace the ideology of political correctness. Feminism, Marxism, Fascism, radicalism of all sorts adhere to different touchstones. LE express their own preferences and argue for them with what they consider to be the true criterion of policy values: epistemic dialectic. But this ignores the range and gamut of Type I and Type III possibilities and is in error, at least *axiologically*, in assuming that sweet reason alone can solve value problems. That extra-rational factors are integral to the resolution of value conflict does not discount reason, it merely reiterates that the policy analyst is obliged by his calling to reach beyond the limits of what Simon (1976) calls 'bounded rationality. The implications of this for administrative philosophy are profound but cannot be fully entered into here.

Per contra, it seems that LE's solution to the intractable problems of value underlying all policy making is to accept the 'sheer usefulness' and 'empirical adequacy' of folk theory (i.e. Type IIb values) despite the fact that 'there are anomalies, of course' (Evers and Lakomski 1991, p. 209). In effect this would be to accept the black box of bounded rationality and decline to look beyond it.

This is not without merit. It would be convenient for administrators and policy makers. It would be the line of least resistance. It would be pragmatic. It could be justified, or at least defended, on incrementalist grounds. It would represent a 'rational' epistemological axiology. But it would fall short.

It would fall short and fail because it discounts, neglects, elides, overlooks man's passions. It takes into account neither the depths nor the heights of these passions, neither evil nor transcendental glory. Aristotle also fell short when he declared man to be the rational animal. Nietzsche was closer to our modern condition when he declared man to be the beast with red cheeks, the animal that blushes. Perhaps it is just more modest to say that man has the capacity to reason and exhibits through his history and culture a commitment to the notion of rationality as a concept of the desirable. A current manifestation of this value is the very idea of 'policy science'; the faith, falsely founded or not, that by the exercise of intellectual energy the 'best' or the 'right' or the 'good' can be discriminated from less desirable options, contingencies, and alternatives. But this remains a faith, that is to say, it is grounded in belief, in the hypothetical, in the irrational, and it is to LE's credit that at the end of the pavane we are returned to a condition of ignorance — but an ignorance that may be enlightened and sophisticated as a result of scholarly disquisition. The questions asked at the beginning of the chapter remain, and they assume sharper focus in the light of LE's attempts at their answering. Dancing is, after all, not only aesthetically pleasing but also good exercise.

Conclusion

The work of Evers and Lakomski will surely stand as a major contribution to the literature of administrative philosophy. That that literature is itself so limited only enhances the value of a serious scrutiny of the epistemological arguments underlying, as their subtitle has it, 'contemporary methodological controversies in educational administration research'. Such an examination, from a *philosophical* standpoint, was indeed necessary and, so far as epistemology goes, it will serve to sophisticate the emergent discipline. But Evers and Lakomski are perhaps over-ambitious when they attempt to exceed their warrant and propound a new systematic doctrine of 'coherentism'. Such coherentism is, if not incoherent, then perhaps inchoate, as evidenced in their concluding remark in this issue that

> coherentism in general is a shift towards holism, with an attendant blurring of distinctions between observation and theory, fact and value, brute and non-brute data, foundational and derived knowledge.

My fear is that the lust for 'holism', a lust with which I sympathize, will depreciate rather than advance administrative philosophy. Between cognition and affection there is a gulf, as there is also between cognition and volition. These gaps cannot be bridged by blurring or by fuzzy logic. Light itself is discontinuous. The universe is not necessarily coherent, while the art of administration is itself largely the manipulation of incoherence, contingency, and flux.

In my view, and I submit it only as a respectful footnote to the critical examination of but two chapters, the philosophy of administration calls for a lesser emphasis on epistemology and a greater emphasis on axiology. Ultimately the administrator wishes to know about what is right and what is good; about the nature of evil and irrationality; about consciousness and belief and will; about the strange *Verwaltungsmensch*; and about the way in which things go wrong. On these things our disciplinary texts are wondrously silent (Hoy and Miskel do not even have values in their index) and one wonders, along with Saul Bellow, that, 'with everyone sold on the good how does all the evil get done?'

One endorses then the invitation put out by Evers and Lakomski for interdisciplinary participation in large-scale theory constructions and one invites them in turn to engage their formidable talents in the exploration of the passions and the will — in short, in the study of *homo administrativus*. And one commends them to an axiological epistemology.

References

Arendt H. (1963). *Eichmann in Jerusalem: The Banality of Evil.* (London: Faber and Faber).
Arrow K J. (1963). *Social Choice and Individual Values.* (New York: Wiley, second edition).
Evers C. W. and Lakomski G. (1991). *Knowing Educational Administration: Contemporary Methodological Controversies in Educational Administration Research* (Oxford: Pergamon Press).
Fitzgerald E. (1955). Myself when young did eagerly frequent/Doctor and Saint, and heard great Argument/About it and about: but evermore/Came out by the same Door as in I went. (London: Ward, Lock, translation of *The Rubaiyat of Omar Khayyam*).
Hodgkinson C. (1982). *Towards a Philosophy of Administration.* (Oxford: Blackwell).
Hodgkinson C. (1983). *The Philosophy of Leadership.* (Oxford: Blackwell).
Hodgkinson C. (1991). *Educational Leadership: The Moral Art.* (Albany: State University of New York Press).
Moore G. E. (1903). *Principia Ethica.* (Cambridge: Cambridge University Press).
Nietzsche, F. (1989). Menschliches, Allzumenschliches, in G. T. Martin. *Nietzsche to Wittgenstein.* (New York: Lang).
Prior A. N. (1960). The autonomy of ethics, *Australasian Journal of Philosophy*, **38**(3), pp. 199–206.
Simon H. A. (1976). *Administrative Behavior.* (New York: Free Press).

17
Response to Commentaries

In this chapter we take the opportunity to respond to the commentaries offered in the preceding four chapters. In doing so, we would like to express our gratitude for the care and thought that has gone into each commentary. In such a short space we can take up only a very few of the many useful points that have been made.

Matters of emphasis aside, Willower is largely in agreement with our analysis of traditional science of educational administration (the Theory Movement), its problems and its prospects. Happily, even differences in emphasis appear to be benign. For example, Willower wishes we had been more discriminating in aggregating views into the 'cultural perspective'. We are glad to oblige, for it is foundational patterns of justification in cultural studies that are the main target of criticism. Jim Walker's *Louts and Legends* (1988) is a good example of a non-foundational, naturalistic, ethnographic study of youth cultures, which we have warmly endorsed (Evers 1989).

Willower is concerned that our focus on the philosophical context of intellectual trends in educational administration occurs to the neglect of social circumstances. For us this emphasis is partly a matter of filling the gaps that exist. However we agree that changes in the social realm can often force revisions on the most basic, nay philosophical, assumptions of received wisdom, especially if social change is theorized from the perspective of alternative philosophical assumptions. Indeed we would want to blur the distinction between philosophy and social theory since the interpretation of change always invokes some philosophical perspective. This suggests that those who discount the importance of philosophy are discounting the influence of their own philosophical assumptions.

Gronn and Ribbins find our discussion of Greenfield's subjectivism unsatisfactory, though this is understandable, they claim, in view of the incommensurability of our view with Greenfield's. The curious reader will find our attack on the notion of incommensurability in our Chapter 10, 'Against Paradigms'. Getting down to detail, they have six main criticisms.

The first concerns our neglect of the primacy Greenfield accords to values and our ignorance of his recently developed views on organisational moral orders.

Here we plead guilty, our only excuse being that Greenfield's key publications on these topics appeared *after* our book was published. Greenfield's preliminary treatment of values is, of course, contained in his 1986 'Decline and Fall' paper, but on most matters of importance he cites the work of Hodgkinson. We therefore refer interested readers to our Chapter on Hodgkinson, and also chapter 8 on 'Ethical Theory and Educational Administration'.

Their second criticism is really a series of questions about the structure of our ethical theory and how we might respond to particular issues, for example pit closures, teachers' pay, and micro-economic reform. But what is the point of the examples? Most ethical theories worth their salt *can* offer a moral perspective on these issues (and lots of others), and ours is certainly no exception. But we think the task is difficult because we deny there is any separate realm of moral reality that special persons of great insight can appeal to. In denying the separation of fact and value we claim moral judgement cannot escape addressing the complex empirical reality of issues, the social, historical, economic, and cultural contexts. Worse still, we think all moral judgements are fallible. We therefore urge the maintenance of a social context for improving knowledge, including moral knowledge. But this social context itself embodies substantive moral values that can be used to adjudicate issues — values such as tolerance, freedom, respect for persons, equality, and justice. Our point is that if Gronn and Ribbins wish to argue for alternative values, they will find themselves tacitly relying on the sort of moral framework we think is required to sustain the social conditions of reasoned argument.

In writing a book whose central methodological thesis was that theories of knowledge shape the structure and content of administrative theories — hence the term 'Knowing' in the title — our biggest fear was of overindulging our taste for epistemology. We are therefore relieved to discover that Gronn and Ribbins, in their third and sixth points, feel we have understated these matters. They want more on coherence, knowledge, simplicity, comprehensiveness, and the like. Happily more is provided, but of a non-technical nature and embedded in the various applications of our epistemological method given throughout the book. Readers who wish to pursue the technicalities will find Chapter 2 'Knowledge and Justification' a useful guide to the relevant philosophical literature.

Criticism four amounts to the demand for a theory of causality, and when one is suggested for mental events in terms of brain states, criticism five demands it be proved. But does the onus of proof shift when, for example, a child triumphantly demands to know how the toys could possibly have found their way into a Christmas stocking if there were no Santa Claus? We think Gronn and Ribbins need to do a bit more than just ask questions. They need to supply better answers. And if they wish to champion a species of methodological dualism then they might like to start by answering how the so-called mental phenomena of believing, desiring, and knowing relate to associated physical phenomena like talking, walking, and doing.

Our debate with Hodgkinson is an enduring one (Evers 1985, 1993, Evers and

Lakomski 1991, Hodgkinson 1986, 1993), so we shall take up only a few issues here from his very detailed critique, beginning with his claim that we 'wish to overturn the conventional wisdom' of a distinction between facts and values. Well, we do. And we would start by denying that it is a conventional wisdom, at least where it matters, namely in philosophy. Following the Harvard logician Hilary Putnam (1981, p. 127), we would call it a 'cultural institution', a term Putnam prefers because he thinks belief in such a distinction is both mistaken and that it gets transmitted uncritically 'regardless of what philosophers may say about it'. So when it comes to a choice of pleasing the logician or mystifying the administrator behind the desk, on the technicalities of deriving value from fact, we aim to please the logician.

We also dispute the claim that ethics cannot function as evidence for science. The reason is that we have a much broader, post-positivist, view of evidence, including what counts as a plausible basis for rationally persuading people to revise their beliefs. We think that what counts as rational is influenced by larger world views, including a view of what promotes the conditions for human flourishing, a familiar ethical concern (Evers and Lakomski 1991, pp. 182–183).

On the question of values in policy-making, we do indeed acknowledge the sheer usefulness of any empirical adequacy enjoyed by folk theory, or commonsense. However, we do not decline to look beyond the black box, as Hodgkinson suggests. Rather, our proposal is to look *inside* the black box to where the human brain is, to model the processes of learning, cognition, and decision-making according to our best accounts of how neural networks operate. Against our urging the collapse of fact and value, it is ironic for Hodgkinson to suggest that we account for the rational by overlooking passions — because from a physicalist perspective, the passions are as much a part of the total cognitive process as the rational. Indeed, we see a naturalistic, coherentist approach as offering the best prospect for illuminating the cognitive dynamics of will, choice, and decision.

Lastly, we turn to Bates, who begins by interpreting us as essentialists and then attacks us thus construed. Lest readers be misled, in acknowledging the influence of Quine on our work, we are acknowledging the influence of perhaps the most anti-essentialist living philosopher. For 'essentially' in our book read 'in the main' rather than 'necessarily' as Bates does.

He then moves on to query our scholarship, surprisingly arguing that he was less well read, when he wrote his 1980a and 1983 pieces, than we supposed him to be. Fortunately, Bates is being excessively modest in describing his own reading habits. His 1980a paper cites Bates (1980b) which does refer to Habermas, notably *Theory and Practice* (1974). This work is particularly helpful for readers ignorant of the epistemological theses of Habermas's *Knowledge and Human Interests* (1972), since the translator thoughtfully provides a long critical introduction to the main themes of both books. Similarly, Bates (1983) cites Bates (1982) [later published as Bates (1984)] which not only cites *Knowledge and Human Interests*, but also commends it (Bates 1984, p.270).

In his final criticism, Bates chides us for focussing on Habermas's epistemology rather than his political, economic, and social critiques of the legitimation crises of late capitalism, claiming these have been much more influential in educational administration. Waiving our point about the close connection between philosophy and social theory, we can agree with this claim. Indeed, one of the strengths of Critical Theory has been its willingness to embrace substantive empirical theories about the structure and dynamics of societies. The bad news — political, economic, and social — for Bates in particular and Critical Theorists in general, therefore, is not the demise of late capitalism at the end of the 1980s, but the collapse of its principal rivals across Eastern Europe and the old Soviet Union. We see no evidence in his critique of our book that he has come to terms with the consequences of capitalism's new international legitimacy.

References

Bates R.J. (1980a). Educational administration, the sociology of science, and the management of knowledge, *Educational Administration Quarterly*, 16(2), pp. 1–20.

Bates R.J. (1980b). New developments in the new sociology of education, *British Journal of Sociology of Education*, 1(1), pp. 67–79.

Bates R.J. (1982). Towards a critical practice of educational administration, Paper presented at the Annual Conference of the American Educational Research Association, New York.

Bates R.J. (1983). *Educational Administration and the Management of Knowledge*. (Geelong: Deakin University Press).

Bates R.J. (1984). Towards a critical practice of educational administration, in T.J. Sergiovanni and J.E. Corbally (eds.) *Leadership and Organizational Culture*. (Urbana, Chicago: University of Illinois Press).

Evers C.W. (1985). Hodgkinson on ethics and the philosophy of administration, *Educational Administration Quarterly*, 21(4), pp. 27–50.

Evers C.W. (1989). Theory competition and intercultural articulation: methodological reflections on *Louts and Legends*, *Educational Philosophy and Theory*, 21(1), pp. 78–82.

Evers C.W. (1993). Hodgkinson on moral leadership, *Educational Management and Administration*, 21(4), pp. 259–262.

Evers C.W. and Lakomski G. (1991). *Knowing Educational Administration: Contemporary Methodological Controversies in Educational Administration Research*. (Oxford: Pergamon Press).

Greenfield T.B. (1986). The decline and fall of science in educational administration, *Interchange*, 17(2), pp. 57–80.

Habermas J. (1972). *Knowledge and Human Interests*. (London: Heinemann).

Habermas J. (1974). *Theory and Practice*. (London: Heinemann).

Hodgkinson C. (1986). Beyond pragmatism and positivism, *Educational Administration Quarterly*, 22(2), pp. 5–21.

Hodgkinson C. (1993). Foreword, in T.B. Greenfield and P. Ribbins (eds.) *Greenfield on Educational Administration*. (London: Routledge).

Putnam H. (1981). *Reason, Truth and History*. (Cambridge: Cambridge University Press).

Walker J.C. (1988). *Louts and Legends*. (Sydney: Allen and Unwin).

18

Three Dogmas of Materialist Pragmatism: A Critique of a Recent Attempt to Provide a Science of Educational Administration—
Trevor H. Maddock

The predominant contemporary conception of an administrator is of someone who gives directions. Administration, on this view, is to do with control, whether it is of people, of events, or of objects, even when this control is conceived of as rational control. With few exceptions, it is this view of administration that predominates in contemporary educational thought, and the current concern with educational administration as a science is hardly meaningful without it. It is the objectivity of science that is important if rational control is to be established, but this objectivity presents a problem, casting doubt on the notion that educational administration can be adequately conceived of as a disinterested, objective and rule-governed activity.

Current attempts to present the general activities of administrators of education as scientific have been conditioned historically by the apparent failure of traditional empiricist philosophy to provide justification for particular actions and, so, for the scientific management of administrative direction. Because administration is a kind of practice or action, a theory of administrative direction must allow administrators to determine the facts of the matter while remaining objective, but the traditional empiricist view of science has difficulty in providing such a conception. For the advocates of scientific administration, the problem is to find a way of justifying their view in light of recent criticisms of traditional empiricism.

This article deals with *materialist pragmatism*, an influential recent attempt to resolve this problem. Numerous publications by its advocates, Evers, Lakomski, and Walker, have appeared in the educational administration literature for well over a decade (for example Evers 1979, 1987a, b, Lakomski 1985, 1987, Walker 1987, 1988, Evers and Lakomski 1991). For example, Willower has described works by Lakomski and by Evers as 'gems'. In a recent book review, he wrote that two papers by Evers and Lakomski 'alone make this booklet a worthwhile acquisition' and he suggested that these writings are 'a welcome addition to the effort to furnish a non-positivist, non-subjectivist, and non-Marxist alternative in educational administration'. There is further support from within the literature aimed less at an academic readership. Recently the *Practising Administrator* published a series of articles based on materialist pragmatist claims and ideas, and their practical application. The first article in the series is Duignan and Macpherson (1990).

Since its inception, materialist pragmatism has been involved in controversy, a characteristic that is diagnostic of the contemporary development of the philosophy of educational administration generally, but which also relates to the specific *modus operandi* of its practitioners. The technique of *Knowing Educational Administration*, by Evers and Lakomski (1991), is a paradigmatic example of this. In this book the technique employed is briefly to set up a theoretical framework that provides the basis for the evaluation of contending theories of educational administration. These critiques are also intended to fill out the sketch of the theoretical framework. The basic argument against many of these contending theories is that, in *materialist pragmatist terms*, they fail to reach an adequate standard of coherence. (Some notable critiques of materialist pragmatism are Hodgkinson 1985, 1986 and Eisner 1992).

The materialist pragmatists have attempted a synthesis of a number of philosophical perspectives to provide a different account of educational administration as a science. They draw on Quine's holistic epistemology and on Lakatos's development of Popper's falsificationism. The result is taken to be a philosophical system that has its own epistemology, metaphysics, and ethics, and it provides the basis of critique of many of the standard dichotomies — e.g. between analytic and the synthetic, and between facts and values — which have plagued Humean and Kantian philosophy, and which are shared by logical positivists, like Simon (1965), and subjectivists, like Greenfield (1978).

The notion of a coherentist criterion of theory evaluation is taken seriously in this article in order to criticise materialist pragmatism on its own terms. (This kind of approach is discussed in detail in Adorno 1973.) The question considered is whether there are adequate reasons for supposing that materialist pragmatism is superior to alternative philosophical accounts of educational administration according to coherentist standards. It will be argued that this claim to superiority is not convincing, that materialist pragmatism, like all the other philosophical theories, is both enlightening in certain respects and problematic in others.

The kind of pragmatism being considered here is associated with the contemporary *neo-pragmatist* orientation in philosophy, perhaps the dominant academic approach over the last decades of the twentieth century in the USA. As with neo-pragmatism generally, materialist pragmatism involves a rejection of what are called 'foundationalist justifications' for a justification in terms of coherence and empirical and/or practical success. The analysis presented here suggests that Quine's holist epistemology does not meet the materialist pragmatists' need for a method of theory evaluation, and that remarks by Lakatos emphasize the problems of such comparisons. The concept of coherence is intended to overcome the difficulties associated with evaluating theories solely in terms of their practical success, but it is questionable whether it does this. In fact, the concept of coherence, which is the linchpin of materialist pragmatism, is vague and tends to gloss over considerable philosophical problems. The programme to naturalize ethics is also fraught with difficulties. The critiques of Moore's naturalistic fallacy and Hume's dichotomy of facts and values are unconvincing. While the idea that philosophical foundationalism should be avoided is a good one (see Adorno 1977) an examination of the critical literature on Quine suggests that the problems of foundations have not been successfully bypassed, for traces of both empiricist and transcendental foundations are still in evidence. Despite the neo-pragmatists rejection of any operative distinction between empirical and philosophical theories, it is doubtful that the latter theories can be straightforwardly assessed in terms of their empirical success or adequacy. In this sense, materialist pragmatism does not move far from the position of logical positivism that initially motivated the Theory Movement. These problems suggest that, in terms of the coherentist criterion, materialist pragmatism has its difficulties and, to this extent, it is in keeping with the other attempts to provide a philosophy of educational administration. All such accounts have contained their insights *and* their problems. It is doubtful that any conclusive selection between them can be made on the basis of their internal coherence, and even if this could be done there is little substantial evidence to show that materialist pragmatism is superior to its competitors in this respect.

Quine's Version of Neo-Pragmatism

The term 'neo-pragmatism' is used to refer to a range of epistemological perspectives sharing some common characteristics. Generally, these approaches offer some kind of coherence alternative to foundationalism, they are naturalist with regard to the relation between philosophy and the natural sciences, and they are weakly realist, rather than instrumentalist or idealist with regard to the nature of knowledge. Walker suggests that basic to neopragmatism is the idea of 'criteria for solving problems arising in the pursuit of our goals and satisfaction of our needs' (Walker 1987, p. 4).

The term 'foundationalism' refers to the belief that humans possess a privileged basis for cognition, that, for example, direct sensory experience may be taken as the most reliable cognitive act, or that transcendentally determined preconditions provide a reliable foundation for cognition and action. These examples, empiricism and transcendental idealism, are the most common forms of foundationalism, and have been the subject of some criticism. Margolis (1986, p. 166) writes of foundationalist theories that, in some way, they all 'hold to a fixed, certain, or strongly privileged cognitive access to the real world, on which all other cognitively relevant claims depend, however weakly'. Materialist pragmatism is said to provide a coherence alternative to foundationalism, an epistemological approach that denies that there are any privileged cognitive insights. Instead of drawing on metaphysical speculation, criteria such as simplicity, comprehensiveness, elegance and explanatory power are taken to coalesce as coherence, which combines with practical success to provide evaluative criteria for assessing cognitive insights.

Within the neo-pragmatist tradition, the concept of coherence is frequently associated with the notion of *naturalism*, the idea that, as Roth (1978, p. 350) states, 'theories in natural science provide the regulative principles for determining all that can be known. Natural science is not one species of knowledge; rather it is all there is to epistemology'. Naturalism goes hand-in-hand with positivism, restricting all questioning to a line of development that conforms to the methodological criteria of scientific inquiry. It is perhaps because, in the neopragmatist tradition, the term 'positivism' is usually restricted to the logical positivism of the Vienna Circle that neo-pragmatism is not more generally described as a kind of positivism.

In this tradition, the term 'pragmatism' refers to non-epistemic strategies in realist epistemology, this latter being the view that humans have some knowledge of the actual world that is structured independently of human inquiry, or at least that there are some facts about this world of which humans have some knowledge. The idea is that scientific knowledge is a true description of at least some aspects of an actual, material world. Traditionally, pragmatism is instrumentalist and anti-realist. Coherence and practical warrant seem, on the face of it, to have more to do with operational control than with establishing the truth about independent reality.

Neo-pragmatists legitimate their commitment to realism indirectly, in terms of the survival and viability of species, or in terms of the actual practices of societies, which are taken as signifying the possession of suitable cognizing powers. The success of science in terms of the survival and viability of the species and as an actual practice of humans is taken as sufficient reason for believing that any scientific assertion about independent reality is true. In essence, non-epistemic criteria are taken to resolve epistemic problems. Epistemic legitimations of realism are generally associated with some form of foundationalism, and the move to non-epistemic accounts is motivated by the need to avoid such foundations by turning to coherence and success in practice.

Quine's philosophy is the version of neo-pragmatism favoured by the materialist pragmatists. Quine came to prominence in post-war Anglo-American philosophy largely through a celebrated critique of the influential doctrine of his teacher Carnap, who was perhaps the major representative of the logical positivist movement. Carnap's concern is with the logical reconstruction of scientific languages. His doctrine, which is critical of speculation on metaphysics and ethics, and based on empiricist foundations, advocates a *verificationist theory of meaning, physicalism*, and the methodological unity of science.

The major contrast between Carnap's doctrine and Quine's lies in the latter's preference for a non-foundationalist empiricism, for an empiricism without certainty. In his critique of Carnap, Quine's philosophical objective is to find a basis for favouring or for confirming the relative superiority of physicalism or extensionalism as first-order empirical disciplines in relation to contending theories. *Extensionalism* and *physicalism* involve the view that the description of human behaviour can be achieved in terms restricted to those necessary to account for purely physical phenomena and immediate experience, with this latter ultimately expressible in purely physicalist terms. Quine sees this narrowness of expression as the source of the unity of science, which he takes as the only source of objective knowledge.

Carnap is noted for arguing that, while the extensionalist reduction of non-observation statements to complexes of observation statements cannot be achieved, it could be determined just which observation statements were specified by non-observation statements. Although there could be no reduction of non-observation statements to observation statements, concepts of material objects could nevertheless be reduced to concepts concerning sensory states. For Carnap, this weakening of extensionalist requirements meant that statements may be determinately false, when a consequence is observed to be false, but they cannot be determinately true.

Quine does not agree that the consequences of non-observation statements can be specified, so, he argues, it cannot be determined whether such statements are true or false. Different non-observation statements can have the same consequences and, what is more, statements do not stand alone but are a part of complexes of statements. When a consequence is observed to be false, any part of the complex may be altered and there will always be alternative means of doing this. Quine's response to Carnap in many ways matches Lakatos's response to Popper, which is discussed in the next section.

Quine subscribes to a particular version of the verificationist theory of meaning, where meaning is associated with complexes of statements rather than single statements. Because different complexes of statements may have the same empirical consequences, like individual statements, the empirical significance of complexes of statements cannot be determined. As Quine (1961, pp. 41, 42) argues, 'our statements about the external world face the tribunal of sense experience not individually but only as a corporate body. The unit of empirical significance is the whole of science'. Quine's point is that science *as a unified whole* must be accepted on the basis of its practical consequences, or it must be rejected. No specific claim or bunch of scientific claims is determinately true or false. The choice is to accept the lot or to accept nothing

In Quine's critique of Carnap a major point of focus is the *analytic–synthetic distinction*, a cornerstone of post-Kantian philosophy. Distinguishing between logical truths and definitions, Quine focusses attention on this second class of analytic statements and he argues that the notion of analytic truth requires other notions, such as synonymy, which presuppose analyticity, and thus the account of analyticity is circular. He suggests that the point of calling a statement 'analytic' is to justify why it cannot be revised in the light of experience, and he charges that no statements are immune to revision, not even logical truths. Quine's holism is incompatible with anything other than a pragmatically justified distinction between the analytic and synthetic. All statements are revisable and, in the light of consequent experience, any statement may be revised. As long as science outstrips its competitors in its successes, this instrumental attitude to language is justified.

Quine's holism has been the object of criticism, some of which is directly relevant to the analysis of materialist pragmatism. In particular, Margolis has argued that Quine's holism is incoherent or irrelevant to Quine's own objectives. He argues that, generally, pragmatism lacks the resources to evaluate the claims made on behalf of specific scientific theories, and, in particular, that perspectives like behaviourism and physicalism, to which both Quine and the materialist pragmatists subscribe, cannot be justified. This represents a serious problem. If, say, for the theory of behaviourism, there can always be a different theory with the same empirical consequences, why subscribe to behaviourism? As Margolis (1986, pp. 151–152) notes,

> Quine clearly commits himself to a very strong form of extensionalism. . . But there is no way to make sense of such commitments, except in terms of some thesis about the legitimation of distributed claims. Quine, therefore, is obliged to reconsider the viability of some operable distinction between analytic and synthetic criteria.

Given that it is the whole of science that faces the tribunal of experience and given that no distinction can be forged between analytic and synthetic claims, Quine cannot explain how to decide rationally between alternative sets of distributed claims within first-order science. It is for this reason that Margolis feels that Quine's holism may be either ultimately incoherent or irrelevant to his own realist objectives. Quine's rejection of intentionality and his commitment to extensionalism and physicalism are dogmas and there is no compelling reason to favour them rather than other alternative perspectives.

The problem of explaining which theories should be favoured is not just seen as applying to Quine's theory but is taken to be a general one. The move to pragmatism arguably precludes any privilege for extensionalism and physicalism, and it is a move against the unity-of-science programme pursued by the logical positivists. The pragmatist approach to epistemology is based on the notion that the biological viability of the human species is a consequence of the cognitive intercessions of science. However, non-epistemic legitimations provide no differential grounds for appraising particular parts of science. It simply cannot be justifiably said that this or that theory is preferable or superior in terms of the viability of the species. In this sense, pragmatism fails to signify an unearned cognitive privilege. It needs concrete epistemically focussed guidelines for any evaluation of the realist status of science. Holistic and anti-foundationalist pragmatist treatments of knowledge and meaning abandon the quest for commensuration, and so are relativist and can provide no justification for any particular theory. The peculiarities of materialist pragmatism as a philosophical approach result largely from an attempt to resolve this problem.

The Significance of Falsificationism

If the material pragmatist characterisation of educational administration is to be of value, it must provide a means of selecting between alternative solutions to problems and, in this way, guide practice. Quine's version of neopragmatism requires modification if it is to play a role, and, in particular, the thesis of the indeterminacy of theories must be discarded. Evers, Lakomski and Walker seek an objectifiable scale of achievement based on the twin criteria of coherence and comparative success of application, and the development of Popper's sophisticated methodological falsificationism by Lakatos is seen to provide the key to this problem. Two elements of Lakatosian theory are taken up by the materialist pragmatists: his account of *series of theories* (*research programmes*) as holistic structures, and his account of *touchstone theory*, which is seen as providing a way of adjudicating between rival theories.

Lakatosian theory is the result of an attempt to improve on Popper's falsificationism and so save it from the criticisms of Kuhn, Feyerabend, and others, Quine among them (for details of this debate, see Lakatos and Musgrave 1970). Stated briefly, falsificationism is based on the logical principle of *modus tollens*; if p then q, not q therefore, not p, where p is a set of non-observation claims and q is an observation statement. Carnap had a similar idea in arguing that theoretical claims could be falsified but not verified. Lakatos noted that Popper has a more sophisticated methodological understanding of this principle than many critics recognize, and he aimed to develop this understanding still further. Popper augmented the falsification principle with an account of *verisimilitude*, this latter understood in terms of an increasing imbalance of corroborations over falsifications, what amounts to a *progressive problem-shift*. To these principles Lakatos added the constraint that they only come into play in the advent of theory-competition. When this occurs, certain non-observation claims may be protected, by fiat, as a *hard core*. Only certain components of the theoretical armature can be altered during the competition between theories. This competition occurs on the basis of some touchstone, a joint acknowledgment of problems and, perhaps, some commonality over ways of conceiving of these problems.

Both Lakatosian and Quinean holism are associated with attempts to adjudicate over rival theories but, in this respect, Lakatos appears to provide the more advantageous alternative, for his account has a concept of the structure of theories, and a concept of their relationship to testing and empirical matters, and this gives content to the notion of the adjudication of competing theories. However, whether or not Lakatosian theory is adequate for the task required by materialist pragmatism is a moot point. For one thing, Lakatos has argued that 'anomalies, inconsistencies, and ad hoc stratagems can be consistent with progress, and that research programmes should be allowed 'to outgrow infantile diseases'. If this is the case then, despite any touchstone, criteria for adjudication cannot be simply and unproblematically applied. As Lakatos (1978, p. 149) remarks,

> it is difficult to decide, especially if one does not demand progress at each single step, when a research program has degenerated hopelessly; or when one of two rival programs has achieved a decisive advantage over the other.

In fact, when all is said and done, it is only with the hindsight of divorce from the issues and their social milieu that judgements of comparative worth can be made, and even granting Lakatos this concession is generous because it attributes a standard of historical objectivity that is probably unattainable. For Lakatos, there is no instant, let alone mechanical, means of adjudication between rival theories. The materialist pragmatists make a stronger claim than Lakatos, maintaining that on-the-spot comparative assessment is possible.

The Concept of Coherence

For the materialist pragmatists, theory choice does not hinge solely on the question of practical success but also turns on what is claimed to be the *extra-empirical* virtue of *coherence*. As Evers and Lakomski remark, 'theory choice needs to be guided by a consideration of the extra-empirical virtues possessed by theories. These virtues of system include simplicity, consistency, coherence, comprehensiveness, conservativeness, and fecundity' (Evers and Lakomski 1991, p. 4). Walker and Evers had identified simplicity, comprehensiveness, elegance and explanatory power as properties of coherence in an earlier work (Walker and Evers 1984), but more recently these extra-empirical virtues are referred to collectively as *considerations of coherence*, and they underlie what is called a *coherentist account of justification* (Evers and Lakomski 1991, p. 33, Walker 1991, Evers 1991). The most coherent theory, on this view, is 'the one that enjoys more than any other the extra-empirical virtues of system' (Evers and Lakomski 1991, p. 8). These extra-empirical virtues include consistency, comprehensiveness, a near absence of counter-examples, relatively few anomalies that cannot be deductively or systematically explained, simplicity and elegance.

The coherentists argue that their approach to theory evaluation is an alternative to selection processes relying on epistemological foundations. A coherentist evaluation is based on a complex association of success in practice and internal coherence, of empirical and extra-empirical virtues. However, there are a number of problems with this concept of coherence, as well as reason for thinking that coherence, as specified by the materialist pragmatists, should not be valued so highly. The criteria of coherence, the extra-empirical virtues, are generally vague and it is not clear that they need always do the job claimed of them. Some of the virtues seem not to be epistemological, while others appear to be at least partly empirical virtues. There is also no indication of how the extra empirical virtues are to be aggregated, or of what criteria can be brought into play to evaluate different aggregations comparatively. The point that one may be wise only well after the event when comparing theories seems also to apply to the determination of coherence. Finally, it needs to be explained why the specified conditions of coherence are so important, why a theory of educational administration which met these conditions must be preferable to an alternative with less internal coherence.

Simplicity and elegance are aesthetic conditions, and it not clear why they are relevant in epistemology. (Davson-Galle 1990a, b makes this point in two somewhat different contexts but it is relevant to the claims of materialist pragmatism.) Little is said about the concept of elegance in the materialist pragmatist literature but simplicity receives some attention. Simplicity, a hallmark of modernism in art, is said to unite with comprehensiveness 'in the sense of using the least amount of explanatory apparatus to account for the largest range of phenomena'(Evers and Lakomski 1991, p. 9) but why is this condition so desirable? Why must a simple explanation be better than a complex one? If the world is complex, simplicity would be something imposed upon it that misrepresented the nature of things. Parsimony is perhaps an economic virtue, but that does not make it an epistemological one.

The concept of comprehensiveness is also unclear. Statements that are true come what may are maximally comprehensive, because they explain everything, but they are generally considered to be of dubious empirical import, and it also needs to be shown why it is *always* preferable that a theory deal with a wide range of situations rather than a narrow one. It is arguably the case that administration is at least equally well served by narrow-range theories. Adorno (1977, pp. 6-8) has argued that the idea that there can be one unified, over-arching theory that explains everything is a pipe-dream, for the world is fundamentally contradictory, driven by opposing political forces and ideas, economic needs, religious views, neuroses, petty jealousies and differing normative convictions (poststructuralists and post-modernists have also made this general point; for example Lyotard 1984, Baudrillard 1975, Deleuze 1988). This volatile mixture does not produce a coherent, rational world but a situation in which even rationally determined ideas come to clash with other rationally determined ideas, and the outcome need not be rational in the sense that the ends reflect some reasoned coherence. Given this situation, an attitude of maintaining a distance from grand theory is desirable and narrow-range theories appear more palatable than comprehensive ones (Adorno 1977, pp. 18-26). In such situations, consistency becomes a regional condition belonging to narrow-range theories, and cannot be expected either of the social totality or of the grand unified theory which depicts it.

Quine (1974, p. 130) has suggested that conservativeness and simplicity are *the* guiding considerations when modifying a theory to accommodate empirical findings. The notion that there is such freedom to modify theories is questioned in a later section. However, the idea of balancing simplicity with 'a less drastic departure from the old theory' is also in need of justification. Evers and Lakomski (1991, p. 38) also feel that it is preferable to adopt a theory that 'squares better with what you already have reason to believe', a similar point to that raised by Quine, but in neither case is any argument advanced for why this should be so. It is conceivable, perhaps in a situation of recalcitrant difficulties, that a radical departure from previous beliefs and understandings could be in order. Bachelard (1968, 1984) has argued that such epistemological radicalism is fundamental to scientific change and advancement. Conservativeness may be a prejudice that hampers scientific advancement. In any case, the possibility that a radical response may sometimes be the most appropriate cannot and should not be ruled out before the fact.

Even if it was the case that simplicity, comprehensiveness, conservatism and fecundity are desirable attributes of the most accepted scientific theories, it begs the question to argue their superiority in all theory choice. The comparative superiority of science cannot be assumed but must be established. The neo-pragmatists' conception of the success of science has been increasingly brought into question as this century has developed, so that it is difficult to say at present whether or not science is like a pathological growth that blossoms immediately between bringing about the destruction of its host and itself. Musil (1953, pp. 360–361) has remarked of the scientific virtues that, 'if one investigates what qualities it is that lead to discoveries...one sees nothing but the old hunter's, soldier's and merchant's vices, simply transposed into intellectual terms and re-interpreted as virtues'. What is a virtue in terms of science need not be a virtue *per se*. In the broader scheme of things, scientific virtues may well be vices.

The notion that all of these coherentist virtues are extra-empirical should also be questioned. Comprehensiveness entails being 'able to explain more phenomena rather than less, and with fewer anomalies, counter-examples, and falsifying instances rather than more' (Evers and Lakomski 1991, p. 9). This is, on the face of it, an empirical virtue. It is to do with the successful application of the theory, with its operational success. Comprehensiveness makes sense as a virtue in terms of increasing operational control, which is its standard scientific meaning, but whether operational control has particular epistemological significance is another question.

Further, even if the extra-empirical virtues ware unproblematic, a fundamental problem of weighing the different virtues of theories, and of comparing the aggregated virtues of competing theories, remains to be addressed. It is precisely this latter problem that led Lakatos to argue that such evaluations could only be made well after the event. In the case of contemporary competitors, what appears as a virtue to one may seem a vice to another. No neutral procedure has been proposed to resolve this problem to date. Even within a scientific framework, what counts as a virtue may depend on the camp to which one belongs.

The intuition that criteria, like fruitfulness, consistency, explanatory power, etc., are important properties of theoretical accounts is a sound one. The problem in the materialist pragmatist account lies not in the simple valuing of such criteria but in their over-valuing. There are no abstract formulae for determining coherence. While coherence, conceived of as an aggregate of virtues, is an aspect of theorizing that is important, and even aesthetic constraints have a role to play, it is one that must always be assessed in terms of the specific concrete circumstances. Coherence is one virtue among many, it need not always be so virtuous, and it needs to be weighed up against others when dealing with specific instances, and this holds in the administration of education and elsewhere. It is an error to see coherence as an abstract virtue divorced from concrete circumstances, according to which all theories may be judged, and when this abstraction occurs the concept itself loses coherence.

Coherence, which includes empirical comprehensiveness, makes sense when applied to first-order empirical theories, that is, scientific theories, where the question of increasing the scope of operational control is important. However, it is questionable that philosophy should meet essentially the same operational constraints — while it is characteristic of neo-pragmatism to deny that there is an operative distinction between first- and second-order theories, essentially, between science and philosophy, it remains to be demonstrated that first-order evaluative criteria are appropriately applied to philosophical theories, whether it is possible to say that one theory is more quasi-empirically virtuous than another. If, as Lakatos has argued, this is difficult to do in the case of first-order theories then the problems are compounded when comparing second-order theories. The criteria of coherence are not only vague and of differing and changing value, it is questionable that the criteria can be readily applied to philosophy. Why should philosophy be constrained to meet the requirements of system? The answer to this question is necessarily a concrete one.

Quine and Wittgenstein on Language

Subjectivist accounts of educational administration, such as those of Greenfield and Hodgkinson, resemble the logical positivist view in that they too acknowledge a categorical distinction between facts and values. In opposition to such views, the materialist pragmatists' project is to weaken this distinction, to suggest that facts and values can be components of a single system capable of producing factual and normative judgements. In particular, the critique of Hodgkinson's work has provided the basis for this project. Hodgkinson's subjectivist and elitist conception of administration, resting on the radical separation of facts and values, leads ultimately to a view of administration in which acts of will play the most significant part (Evers and Lakomski 1991, pp. 109–110). The problem for Evers (1985, p. 71) in Hodgkinson's approach lies in the latter's acceptance of Moore's *naturalistic fallacy* and Hume's *is/ought distinction*. Moore's defence of the naturalist fallacy — the confusing of non-natural things, like good, with natural things, like pleasure — depends on the analytic/synthetic distinction, which Quine has shown to be suspect. Hume's defence of the fact/value distinction rests on the contention that practical conclusions cannot be based on solely factual premises, and Evers attempts to cast doubt on the status of this argument. If these critiques were decisive, a qualitatively different conception of administration as a science would become a real possibility, while accounts which separated the factual and the normative would be unnecessary. Neither Quine's nor Evers' critiques are convincing, however. In this section, Quine's and Wittgenstein's conceptions of analyticity and necessity are considered, a comparison that puts the Quinean view into critical relief, while, in the next section, the arguments by Evers against Hume's distinction are subjected to scrutiny.

The intention here is not to propose Wittgenstein's philosophy as an alternative theory of educational administration, but simply to focus on a well-known response to Quine's thesis on analyticity and his general conception of language in order to show that the matter is not as clear-cut as it appears. Dilman (1984, p. 81), following Wittgenstein, argues that Quine's criticism of the definition of analyticity for being circular is too formal. As he notes,

> From the impossibility of such an explanation ... it does not follow that the notions in question are unclear, nor that there is nothing that philosophy can do to throw light on them in so far as they give rise to philosophical difficulties.

Grice and Strawson (1956) argue that it does not follow from the failure to explain analyticity in the formal way that analyticity cannot be explained at all. It is true that there is no understanding of analyticity independent of the notion of necessary truth that can clarify the matter, but this kind of argument misses the point that such accounts perform an elucidatory function.

For Quine, necessity cannot be found in the structures of understanding, and it cannot be found in language, so, there cannot be any necessity. As Dilman (1984, p. 105) remarks, 'Quine, having recognized that propositions can only be necessary within a system, concludes that since the system itself cannot be necessary, the necessity of any proposition within it cannot be real necessity'. On the other hand, for Wittgenstein, it does not follow from the fact that no law or rule is inexorable in itself that it cannot be inexorable for us. Dilman (1984, p. 101) paraphrases Wittgenstein as claiming,

> that the possibility of any concept-formation demands that we should take such an attitude towards certain things, and to certain moves or transitions within the language-game, that we should regard certain propositions in which these moves are generalised as unassailable.

Humans do not hold things as unassailable simply because it gives them the most convenient concept-formation. In this respect, Quine sees language users as individuals that have the single purpose of understanding nature in the same way as physical scientists. On the other hand, Wittgenstein sees humans as many-sided language users who develop in a two-way interaction with the language they employ. As Dilman (1984, p. 102) remarks of language users, 'By and large their contribution is not an exercise of their rationality — as it is for Quine'.

While both philosophers regard language as an instrument, Quine thinks of persons as wielding this instrument to satisfy their pre-existing purposes and he has no sense of the dependence of people's purposes on the language they speak. Consequently, he gives no consideration to language having an independent life into which people are born, grow, develop their ability to reason, and find themselves. For Wittgenstein, language is not simply a human instrument for understanding nature, and human aims are largely those of the language they speak. Humans serve their spoken language as much as the language serves those who speak it.

Wittgenstein (1969, proposition 44) remarks that 'it belongs to the logic of our scientific investigations that certain things are in deed not doubted', and he rejects the idea that propositions, such as postulates and assumptions, can be readily manipulated for the purposes of theory-construction. Dilman agrees with Grice and Strawson that truth values cannot be readjusted, as Quine proposes, without altering the meaning and identity of statements, without bringing about a change in concept-formation. Although the distinction between the analytic and synthetic is not as Kant or the logical positivists envisaged it, there is nevertheless a difference between giving up propositions believed to be true and giving up necessary truths. For Wittgenstein, propositions regarded as necessary truths are different instruments of language to falsifiable propositions. While he agrees with Quine that logic is not immutable, this does not mean that the distinction between necessary and contingent truth must be rejected. Humans cannot do what they like in the language game, for there would be no language without its rules, and it is these rules that safeguard the integrity of the analytic. In playing the language game, humans operate on the basis of established analyticity.

From a Wittgensteinian perspective, language games can only be understood in relation to forms of life, and it is in terms of this last concept that questions of competition and evaluation become meaningful. Wittgenstein denies that forms of life can be evaluated, while Quine argues that forms of life can be evaluated in terms of their comparative success. Quine's argument is not compelling, however. If the present system was pathological to other forms of life, it would tend to dominate, to become the only form of life. Weber made this point (see for example Gerth and Mills 1958). Whether domination makes a particular form of life the most fitting or the best is nevertheless questionable. Just because cancerous growths come to dominate over other tissue formations, they are not considered the most fitting. Quine's argument for science begs the question in a crucial way, and fails to distinguish between pathological and genuine flourishing.

Whereas, from Quine's perspective, science is conceived of as the most successful means of achieving indeterminate ends, Wittgenstein situates the perspective within a network of gestures, expressions, and activities, and it is in terms of this network that science is evaluated as the most successful. Wittgenstein's approach remains detached from history in a crucial way, but it has the comparative merit of allowing both science and administration to be conceived in relation to the form of life within which they are meaningful. However, it is necessary that forms of life are subjected to scrutiny, and, to do this, what is required is an investigation of the concrete instances of administrative failure — an activity that puts forms of life into critical relief. Quine's proposal for evaluating forms of life is inadequate because it fails to consider the historical circumstances in which science has developed, but it is important that criticism is carried out.

Hume's Problem

The second component of the materialist pragmatists' argument for ethical naturalism is based on a critique of Hume's problem. Hume denied the possibility of determining normative pronouncements on the basis of factual considerations alone. Evers (1985, p. 71) aims to show that such a view can be sustained. He notes that propositional calculus accepts as valid:

(1) snow is white, therefore snow is white or killing is wrong; and
(2) vivisection causes gratuitous suffering to animals, so, if any act that causes gratuitous suffering ought not be done then vivisection ought not be done.

Both arguments are examples of valid reasoning where the premise is a factual statement while, in the case of (1), the conclusion is the disjunction of a factual and a normative statement — which may, for purposes of exposition, be called a 'mixed' statement — and, in the case of (2), the conclusion is a normative statement. On this basis, normative claims and mixed claims can be validly derived from factual claims.

This argument is questionable, however. The conclusion validly derived in (1) provides neither directions nor justifications for acting, and it provides no warrant for accepting the proposition that killing is wrong, while it is this proposition that is needed to derive the normative or action-guiding assertion that 'killing ought to be opposed'. The following argument is needed to justify opposition to killing:

(3) snow is white or killing is wrong, so killing is wrong, and any act that is wrong ought to be opposed, so killing ought to be opposed.

For a disjunction to be true only one of the disjuncts need also be true, so 'killing is wrong' does not follow from the disjunction 'snow is white or killing is wrong'.

In the case of argument (2) validity is again established but the conclusion is a conditional with no warrant for unconditional acceptance, and yet it is precisely unconditional acceptance that is required to justify action. In argument (2) if both the antecedent and the consequent of the argument's conclusion were false then this conditional would be true and the argument valid. However, in this example, the falsity of the claim that 'vivisection ought not be done', which is the consequent of the conclusion, would make it difficult to act on the basis of the validity of the argument. If we followed the direction of the argument on the basis of its validity, we would find ourselves always opposing vivisection despite the fact that we believed the claim 'vivisection ought not be done' to be false. More than validity is required if arguments like (2) and (3) are to provide acceptable normative guides; as well, the antecedent of the conclusion must also be true. Like (1), argument (2) fails to provide a guide for action simply on the basis of its validity. If arguments providing action-guiding assertions that have been validly derived only from factual propositions can be given, Evers has not demonstrated this.

There is at least one case where normative conclusions can be validly derived from factual claims and this is the case of contradictions. The following argument is valid:

(4) snow is white and snow is not white, so killing ought to be opposed.

Argument (4) is valid because the premise is a contradiction and anything follows from a contradiction. There are only two alternatives for dealing with this result. On the one hand, it could be argued that there is a difference between logical derivation and justification. On the other hand, to argue that contradictions can function in the justification of actions is to risk admitting arguments like the following as acceptable justifications for action:

(5) If God is supremely good and all powerful there is no evil in the world. God is supremely good and all powerful but there is evil in the world. Therefore, the Church should be suppressed (taken from Newton-Smith 1985, p. 65).

Certainly, if logical derivation was all there was to justification then intuitively undesirable or even abhorrent administrative pronouncements, like (5), would be just as justified as intuitively desirable pronouncements. This suggests that practical justification cannot be reduced to logical derivation. Hume's claim can be represented as two distinct arguments:

(1) an evaluation cannot be *solely* derived from a set of facts.

(2) no evaluation can be *wholly justified* in terms of a set of facts.

Examples to the logical derivation of normative claims from factual claims have been provided, but Evers has not addressed the justification argument.

There is also a distinction between *practical derivation and logical derivation*, which Evers fails to recognize. When, for example, Popper (1965, p. 64) writes that 'it is impossible to derive a sentence stating a norm or a decision or, say, a proposal for a policy from a sentence stating a fact', he is not denying that a policy claim cannot be logically derived, for, as we have seen, it can be simply derived from a contradiction. What Popper is suggesting is that policy claims cannot be practically derived. The policies that guide public administration are reasoned policies in the sense that these claims do depend on other claims, but the derivations are more than formal and turn on a particular content. The derivation of policy is not simply logical derivation. Even if it cannot be made to conform completely with systems of formal deduction, the practical derivation of administrative claims intuitively corresponds to informal ideas of deductive validity, which involve the assessment of norms, facts and perhaps other kinds of claim. Both Hume and Popper are concerned with practical derivation and not just logical derivation.

Just as it is necessary, when thinking about the practical problems of educational administration, to distinguish between logical derivation and justification, and between logical and practical justification, it is also important to consider *ethical justification*, for the question of whether some administrative action is morally correct is more fundamental than the question of whether it is practically feasible. In fact, the practical and the ethical make a dialectical pair. What *should* be done is always mediated in terms of what *can* be done, while practical possibility is always restricted by what is acceptable. The idea of professional ethics, which is of particular significance in the theory of educational administration, may be understood dialectically as being motivated by the practical need for *durable agency* (on this concept, see Scheffler 1985) — a mode of behaving over time that is identifiable to the client group, the community, and the need to pursue what is right in education, even when this leads to antagonisms. Administrative decisions require not merely logical but also practical and ethical justification.

The Question of Foundations

Margolis (1986) not only argues that to the extent that pragmatist approaches lack epistemic foundations they cannot achieve their realist objectives, he even questions whether such approaches really are without foundations. As well as Margolis, Cornman (1977), Dancy (1985), and Boorse (1975) have raised objections to the claim that Quine's theory is non-foundationalist. The latter three feel they detect an asymmetry between Quine's concept of an observation sentence and his concept of a non-observational sentence that leads to empiricist foundationalism. Walker and Evers note that Quine's tempering of his views on theory dependence and semantic holism have led critics, such as Cornman *et al.*, to construe Quine's epistemology as foundationalist but, they argue, Roth (1978) is able to show that this is not the case (Walker and Evers 1984, p. 27).

Margolis makes a different point: he proposes that, in seeking necessary preconditions for scientific practice, Quine resorts to transcendental analysis and thus to an idealist version of foundationalism. To date, the materialist pragmatists have not responded to the accusation that Quine is committed to transcendental foundationalism.

Cornman (1977) argues that Quine can be a foundationalist even if he proposes that no particular set of statements is immune to refutation and so none is irrevocably at the foundation of empirical science. This, he argues, is because the 'rejection of the given is not a rejection of foundationalism'. Even if they are not given, 'Quine seems to hold that observation statements are individually verifiable and are the tests of explanatory systems' and he takes the claim that '[t]he observation sentence, situated at the sensory periphery of the body scientific, is the minimal verifiable aggregate; it has an empirical content all its own and wears it in its sleeve', from Quine's (1969) book as evidence of an empiricist foundationalism. Quine's is simply a peculiar kind of empiricism that seems '... to hold a foundational theory of justification that utilises explanatory coherence but avoids "the given" '.

Dancy's criticism is rather similar to Cornman's. Dancy claims that Quine's theory contains an asymmetry between observational and nonobservational which is characteristic of the foundationalist. From the same pages of *Ontological Relativity* that Cornman draws his quotation, Dancy (1985, p. 99) shows that, for Quine, the notion of an observation sentence is:

> fundamental in two connections... Its relation... to our knowledge of what is true is very much the traditional one: observation sentences are the repository of evidence for scientific hypotheses. Its relation to meaning is fundamental too, since observation sentences are the ones we are in a position to learn first... they afford the only entry into language.

Observation sentences do not report private events, such as the occurrence of a sensation but, rather, the occurrence of certain sensory stimuli that can occur in more than one person. Examples of single-word observation sentences are 'Red', and 'Rabbit'. Because, for Quine, observation sentences afford the only entry to a language, there is a semantic asymmetry in his theory, and because these sentences can be individually verified and their acceptance be justified one by one, and because they constitute the evidence on which the nonobservational must rest, there is also an epistemological asymmetry. The contrast between holism at the non-observational level and a putative atomism at the level of observation establishes, for Dancy, that Quine is a foundationalist.

Roth, however, disputes these judgements by Cornman and Dancy. While he recognizes Quine's distinction between observation and non-observation sentences, Roth claims that, for Quine, observation sentences do not have the philosophical significance attributed to them by Cornman and Dancy. To demonstrate this, Roth invokes Quine's distinction between *occasion sentences* and *standing sentences*. Occasion sentences are those sentences the verification of which involves some empirical factor when they are uttered. Observation sentences are a subset of occasion sentences which have stimulus meaning, that is, as Roth (1978, p. 359) describes, sentences having a 'non-verbal correlate that makes the sentence learnable as an individual sentence'. As well as occasion sentences, Quine identifies as standing sentences those utterances the verification of which need not require any empirical factor on each utterance. Roth argues that both occasion and standing sentences require information from within the theory for their verification. Observation sentences simply have their verification linked to what Roth calls some 'passing perceptual scene'. So, Quine's recourse to observation sentences need not have the foundational consequences asserted by Cornman and Dancy.

Clearly, Roth is right that the concept of observation statements does less work in Quine's theory than it does in Carnap's. However, whether or not this concept does sufficiently little to avoid the accusation of foundationalism is arguable. Certainly, Quine and Ullian (1970) claim that theories are arrived at speculatively from the data of observation, and they assert that observation statements are nearly infallible. Even if these statements require theoretical information for their verification, there seems at least a prima facie case for arguing that Quine is a foundationalist, even if his foundations are less secure than those of Carnap. Walker and Evers (1984, p. 29) argue that naive falsificationism is a form of foundationalism 'because ultimately, what counts as a knowledge claim will always be settled by some observation', while Quine and Ullian (1970), like the naive falsificationists, acknowledge that observations will always demand some response in any related theory.

Even if Roth's rejoinder to Cornman and Dancy saved Quine's theory from their accusation of foundationalism, this theory is not necessarily cleared of all such accusations, for Margolis makes a different point, which is not affected by Roth's rejoinder. Based on his argument that pragmatism is inadequate for the task of theory appraisal, Margolis suggests that either Quine is a crypto-foundationalist or he is unable to defend his own extensionalist and physicalist programmes, and he thinks that there is a sense in which Quine is a foundationalist, albeit of a different kind. He paraphrases Quine as arguing that foundational or transcendental questions are internal to every practice of science and are reflexively projected by attending to what are taken to be cognitive achievements and querying what could serve as preconditions for the practice at stake. This, argues Margolis, is a foundational question that does not arise in a foundationalist way. Questions of preconditions for the possibility of human knowledge, no matter how weak these preconditions may be, raise transcendental and foundational issues. To provide an answer to these questions is to provide foundations, and this is precisely what Quine attempts to do (Margolis 1986, p. 166).

There is less certainty attaching to direct sensory experience for Quine and for the materialist pragmatists than for the logical positivists. However, their discussions have seldom lost that a-historical quality, characteristic of empiricism, a quality that attaches to the naive reception of immediate experience. Empiricism, including that of Quine and the materialist pragmatists, abstracts immediate experience from actual, concrete circumstances and, in this sense, remains an abstract theory that fails to address circumstances as the sum total of their determinants. It fails to avoid the interpretivists' criticism of scientific approaches to educational administration that they fail to take into account the richness and complexity of actual circumstances (see Greenfield 1978, 1979–80, 1986).

In keeping with logical positivism, Quine and the materialist pragmatists employ Kant's transcendental method of establishing preconditions, an approach that, once again, leads to a focus on an abstract object, a scale of evaluation that is then imposed on actual events. However, interpretivists have argued, the application of abstract criteria of assessment is not an appropriate way of administering education, which requires a concrete focus on the actual determinants of particular circumstances, which abstract evaluation is not able to provide. Adequate administration demands that all determinants of a problem be taken into account, while the abstract character of empiricism and transcendentalism, by its very nature, results in the tendency systematically to reduce determinants to a minimum according to operational criteria. The neo-pragmatist rejection of foundations does not lead to a concrete focus but to an abstract approach with different foundations, practical success and accommodation within a rational system.

Materialist Pragmatism and Positivism

Willower's (1986) claim that materialist pragmatism provides a non-positivist view of administration must be addressed. Quine's theory is certainly a movement away from the original expression of logical positivism, as indeed is the more falsificationist view of the later Carnap, and materialist pragmatism represents a similar move. It is more correct to see what is known as 'logical positivism' as a stage in the development of contemporary empiricist and analytic philosophy rather than as a fixed doctrine. Like Quine's philosophy, materialist pragmatism involves a modification of the basic ideas of logical positivism, such as a weakening of the verifiability criterion, while certain traditional Humean and Kantian conceptions are also rejected. Nevertheless, materialist pragmatism remains in keeping with the philosophy of the logical positivists, in a fundamental sense, in that it shares the prejudice that scientific knowledge is all there is to knowing. Positivism does not establish the superiority of science but assumes it.

In the case of materialist pragmatism, the principles by which evaluative criteria are sought are essentially scientific in character. They involve what Margolis calls 'first-order investigations', or, at least, the distinction between orders is not recognized. This is the meaning of the naturalisation of epistemology. This investigation is said to have discovered that success and coherence are the required evaluative criteria, and that, on comparison with alternatives, this newly conceived scientific philosophy turns out to be superior. The doubtfulness of this claim has already been discussed, but the procedure of assuming the superiority of science in order to establish it must also be questioned, for it is a prima facie case of begging the question. Positivism in general begs the question in this way. It is not surprising that, from the perspective of science, the scientific management of affairs appears superior to other management practices, but, from a perspective that is independent of science, it is not clear that scientific theories are the most coherent or the best in practice.

In the case of administration, Greenfield (1978) has argued against positivism, but he is not the only critic to have rejected this view. Scheffler (1985) has argued that, while there are important aspects of administration that are appropriately conceived in scientific terms, because they are technical, administrative activity cannot be confined to dealing with such problems. Additionally, administration demands an awareness that problems do not arise from the segregated domains of special inquiries but from the fullness of daily existence, and so scientific approaches are not sufficient to orient the response of administrators. In such contexts, these people must ask critical questions of specialists and make decisions that are not necessarily technical. Administration requires a self-conscious, critical attitude. Scheffler identifies four components of such an attitude: awareness of individual agency, of cultural context, of habit, and of contemporary knowledge. Social situations represent many potentials, and administration is to do with choosing which to attempt to realize, and which to minimize, to counter, or to ignore. As well, for Scheffler, administrators make decisions of *principle*, that is, decisions creating and reflecting a *normative style*, providing an identity for both administrators and the community by promoting a sense of durable agency. Actions need to be assessed by administrators not just in terms of successes brought but also in terms of precedents wrought. In a contradictory world, durable agency takes on added significance.

The Argument against Materialist Pragmatism

Materialist pragmatism presents itself as nothing less than a complete philosophy, containing a specific epistemology and related ethics, a theory of human nature, and general theories of education and administration, all of which are integrated within a comprehensive web of belief. It is diagnostic of this philosophy that no firm distinction is made between first- and second-order theories, although it is impossible to eliminate the distinction completely. Thus the epistemology is said to be in keeping with the best empirical theories, while the best empirical theory, behaviourism, is identified according to epistemological criteria (Evers and Lakomski 1991, p. 8). Ethics and psychology are interrelated in a similar way, and, in general, there is a tendency to blur distinctions between the different philosophical concerns.

This chapter has attempted to show that the epistemological theory lacks in coherence in the sense identified by the materialist pragmatists. Margolis has argued that an operative distinction between first- and second-order theories is necessary if the claims regarding the superiority of physicalism and behaviourism are to be established (Margolis 1986, pp. 259–60). However, materialist pragmatism contains no such distinction, but instead maintains that second-order, philosophical theories can be evaluated in the same way as first-order, scientific theories. Lakatos has suggested that such evaluations are problematic even in the case of first-order theories. Such difficulties are exacerbated in the case of second-order theories.

The superiority of behaviourism as a first-order theory is assumed rather than established, and, as Chomsky (1959) has shown, it is extremely doubtful that such an assumption is warranted. In fact, Chomsky (1975) has noted that by the late 1960s Quine had acknowledged that behaviourist theory is incapable of providing an adequate account of language, and he proposed that innate structures were needed to explain language learning. As Chomsky notes, Quine's continued adherence to behaviourism is difficult to reconcile with such admissions, although Evers and Lakomski (1991, p. 183) see no difficulties with this reconciliation. Chomsky describes behaviourism as just another name for weak verificationism. Certainly, if Evers and Lakomski are correct that epistemology should 'embody our most powerful and sophisticated theories of knowledge acquisition,' (Evers and Lakomski 1991, p. 8) then materialist pragmatist epistemology is suspect.

Even if it were possible to choose between competing theories on empirical or instrumental, and coherentist grounds, a most doubtful proposition, it would still remain to be shown how normative pronouncements could be determined and justified in terms of a chosen theory. The rather fragmentary and unconvincing critiques of Moore and Hume contribute no support to this contention, nor does their argument that behaviourist metaphysics and pragmatist epistemology must underscore an adequate ethical theory (Evers and Lakomski 1991, p. 183). Even if it could be determined how people think and learn, which the materialist pragmatists have yet to do, Quine's biologistic account of human preferences is unconvincing and provides the most flimsy basis for a normative theory (Quine 1978). The main evidence that Evers and Lakomski (1991, p. 189) can marshal in favour of Quine's theory is its comparative systematicity and parsimony — criteria that, as has been argued, amount to no more than dogmas of their own choosing. The problems of their undifferentiated holism carry over into their account of morals and effectively restrict adequate consideration. The substantive theories of society required by theorists such as Rawls, which Evers and Lakomski (1991, p. 189) find problematic, are exactly what is needed for an adequate theory of ethics. In fact, the absence of any theory of society is perhaps the major lacuna in materialist pragmatist theory. More than anything else, it is this gap that renders materialist pragmatism inadequate as a theory of educational administration.

While the intuition that all philosophical foundations are suspect is a sound one, it is not clear that the materialist pragmatists have done any more than lost sight of the foundation upon which their philosophy is built. The flawed character of all philosophical systematics should not be allowed to obscure the fact that it is not possible to proceed without some kind of framework. It must be frankly acknowledged that humans seek the truth with faulty implements, for these tools are all that is available. The materialist pragmatists would provide a greater service to educational administration if, instead of seeking certainty or superiority where none exists, they acknowledged the flawed character of all conceptual thought, for it is in this recognition that the core of genuine opposition to foundationalism lies. Rather than in the vain search for the superior system, it is through the relentless criticism of all conceptualization that philosophy provides its service to educational administration.

References

Adorno T.W. (1973). *Negative Dialectics*. (London: Routledge and Kegan Paul).
Adorno T.W. (1977). The actuality of philosophy, *Telos*, **31**, pp. 120–133.
Bachelard G. (1968).*The Philosophy of No: A Philosophy of the New Scientific Mind*. (New York: Orion).
Bachelard G. (1984).*The New Scientific Spirit*. (Boston, Mass.: Beacon).
Baudrillard J. (1975).*The Mirror of Production*. (St Louis: Telos).
Boorse C. (1975). The origins of the indeterminacy thesis, *Journal of Philosophy*, **72**, pp. 369–387.
Chomsky N. (1959). Review of B.F. Skinner, Verbal Behaviour, *Language*, **35**, pp. 26–58.
Chomsky N. (1975). *Reflections of Language*. (New York: Pantheon).
Cornman J.W. (1977). Foundational versus non-foundational theories of empirical justification, *American Philosophical Quarterly*, **14**, pp. 287–297.
Dancy J. (1985). *Introduction to Contemporary Epistemology*. (Oxford: Blackwell).
Davson-Galle P. (1990a). Scientific progress: truth, knowledge and coherence, *Metascience*, **8**, pp. 2–5.
Davson-Galle P. (1990b). Some clarifications and cautions essential for good philosophy of science teaching, *Educational Philosophy and Theory*, **2**, pp. 25–28.
Deleuze G. (1988). *Bergsonism*. (New York: Zone).
Dilman l. (1984). *Quine on Ontology, Necessity and Experience*. (London: Macmillan).
Duignan P.A. and Macpherson R. J. S. (1990). Educative leadership and how to get it right, *The Practising Administrator*, **12**, pp. 28–30.
Eisner E.W. (1992). A reply to Gabriele Lakomski, *Curriculum Inquiry*, **22**(2), pp. 205–209.
Evers C.W. (1985). Hodgkinson on ethics and the philosophy of administration, in F. Rizvi (ed.) *Working Papers in Ethics and Educational Administration*. (Geelong: Deakin University Press).
Evers C. W. (1987a). Editorial comment, *Educational Philosophy and Theory*, **19**, pp. 1–11.
Evers C. W. (1987b). Naturalism and Philosophy of Education, *Educational Philosophy and Theory*, **19**, pp. 11–21.
Evers C. W. (1979). Analytic philosophy of education: from a logical point of view, *Educational Philosophy and Theory*, **12**, pp. 1–15.
Evers C.W. (1991). Towards a coherentist theory of validity, *International Journal for Educational Research*, **15**, pp. 521–535.
Evers C. W. and Lakomski G. (1991). *Knowing Educational Administration: Contemporary Methodological Controversies in Educational Administration Research*. (Oxford: Pergamon).
Gerth H.H. and Mills C.W. (1958). *From Max Weber: Essays in Sociology*. (Oxford: Oxford University Press).
Greenfield T.B. (1978). Reflections on organisation theory and the truths of irreconcilable realities, *Educational Administration Quarterly*, **14**, pp. 1–23.

Greenfield T.B. (1979-80). Research in educational administration in the United States and Canada: an overview and critique, *Educational Administration*, **8**, pp. 207-245.
Greenfield T.B. (1986). The decline and fall of science in educational administration, *Interchange*, **17**, pp. 57-80.
Grice H.P. and Strawson P.F. (1956). In defence of dogma, *Philosophical Review*, **65**, pp. 141-156.
Hodgkinson C. (1985). The value bases of administrative action, *Journal of Educational Administration and Foundations*, **3**(1), pp. 20-30.
Hodgkinson C. (1986). Beyond pragmatism and positivism, *Educational Administration Quarterly*, **22**, pp. 5-21.
Lakatos I. (1978). *The Methodology of Scientific Research Programs.* (Cambridge: Cambridge University Press).
Lakatos I. and Musgrave A. (1970) (eds.). *Criticism and the Growth of Knowledge.* (Cambridge: Cambridge University Press).
Lakomski G. (1985). Theory, value and relevance in educational administration, in F. Rizvi (ed.) *Working Papers in Ethics and Educational Administration.* (Geelong: Deakin University Press).
Lakomski G. (1987). Critical theory and educational administration, *Journal of Educational Administration*, **25**, pp. 85-100.
Lyotard J.-F. (1984). *The Postmodern Condition: A Report on Knowledge.* (Manchester: Manchester University Press).
Margolis J. (1986). *Pragmatism without Foundations: Reconciling Realism and Relativism.* (Oxford: Blackwell).
Musil R. (1953). *The Man without Qualities*, Vol. 1. (London: Secker and Warburg).
Newton-Smith W.H. (1985). *Logic: An Introductory Course.* London: Routledge and Kegan Paul).
Popper K. (1965). *Conjectures and Refutations: The Growth of Scientific Knowledge.* (New York: Basic Books).
Quine W.V. (1961). *From a Logical Point of View.* (Cambridge, Mass.: Harvard University Press).
Quine W.V. (1969). *Ontological Relativity and Other Essays.* (New York: Columbia University Press).
Quine W.V. (1974). *The Roots of Reference.* (La Salle: Open Court).
Quine W.V. (1978). On the nature of moral values, in A.I. Goldman and J. Kim (eds.) *Values and Morals.* (Dordrecht: Reidel).
Quine W.V. and Ullian J.S. (1970). *The Web of Belief.* (New York: Random House).
Roth P.A. (1978). Paradox and indeterminacy, *Journal of Philosophy*, **75**(7), pp. 347-367.
Scheffler l. (1985). *Of Human Potential: An Essay in the Philosophy of Education.* (London: Routledge and Kegan Paul).
Simon H. A. (1965). *Administrative Behavior.* (New York: Free Press).
Walker J. C. (1987). Democracy and pragmatism in curriculum development, *Educational Philosophy and Theory*, **19**, pp. 1-10.
Walker J. C. (1988). *Louts and Legends.* (Sydney: Allen and Unwin).
Walker J. C. (1991). Coherence and reduction: implications for educational inquiry, *International Journal for Educational Research*, **15**, pp. 505-520.
Walker J.C. and Evers C.W. (1984). Towards a materialist pragmatist philosophy of education, *Educational Research and Perspectives*, **11**, pp. 23-33.
Willower D. J. (1986). Review of Rizvi F. (ed.) Working papers in ethics and educational administration, *Educational Administration Quarterly*, **22**, pp. 133-143.
Wittgenstein L. (1969). *Philosophical Investigations.* (Oxford: Basil Blackwell).

19
Three Dogmas: A Rejoinder

In educational administration, much of the debate over administrative theory since the early 1950s has been shaped by philosophical considerations, particularly views on the nature of knowledge and scientific theory. For example, traditional behavioural science of educational administration owes much to logical empiricism; and varieties of subjectivism, critical theory, and cultural theory, have drawn on standard anti-empiricist philosophical arguments.

In his article 'Three Dogmas of Materialist Pragmatism: A Critique of a Recent Attempt to Provide A Science of Educational Administration' (Chapter 18), Trevor Maddock does not disagree with us regarding the problems of scientific administration and the need to justify such a view in light of the by now well documented problems of logical empiricism and positivism as the foundation for a science of administration. Our recent work in educational administration, especially our book *Knowing Educational Administration* (Evers and Lakomski 1991), has sought not only to show what the nature of these problems is, but also to provide the beginnings of a more coherent post-positivist account of a science of administration. We have located the above debates within a current philosophical context which, in Trevor Maddock's words, is 'the contemporary *neo-pragmatist* orientation in philosophy, perhaps the dominant academic approach over the last decades of the twentieth century, in the USA'. As is clear from our work, we obviously concur with and endorse his judgement. Maddock further describes our approach as 'an influential recent attempt to resolve this problem [the problem of justifying scientific administration]', points to our numerous publications (including the earlier work of J. C. Walker), which have appeared for well over a decade in the educational administration literature, and also notes that our approach has been recommended by 'well-known figures in the discipline', as well as being supported by more practice-oriented publications such as the journal *The Practising Administrator*. In addition, Maddock rightly notes that our approach is also controversial. He wishes to criticize our 'materialist pragmatism' on its own terms by taking seriously 'the notion of a coherentist criterion of theory evaluation'. Given such a weighty and considered introduction that describes the impact and appraisal of our work in the educational administration literature to date, both internationally and nationally, and noting Maddock's acceptance of our coherentist principles as his methodological guide, his conclusions, to say the least, are surprising.

Despite our willingness to employ possibly the most powerful, widely accepted, contemporary English language philosophy available for developing a new perspective on educational administration, Maddock finds almost all of our philosophical machinery to be 'unconvincing', 'without justification', 'not compelling', 'inadequate', and 'with no warrant'. A partial explanation for his conclusions is that, regrettably, Maddock has made the task of critique especially hard by failing to engage directly the most systematic recent presentation of the arguments we advance for our position (Evers and Lakomski 1991). Instead, he chooses to give close attention to Margolis's criticisms of Quine, Dilman's and Grice and Strawson's criticisms of Quine, Dancy's and Cornman's criticisms of Quine, and Chomsky's criticisms of Quine. Maddock's justification for dwelling on all this criticism of Quine is because he thinks we accept many of Quine's most distinctive doctrines. And so we do, though not without reasons, some modifications here and there (see Evers and Lakomski 1991, ch. 9), and some accommodation for new developments (P. M. Churchland 1989, P. S. Churchland 1986). The modifications and accommodations sometimes mean that a criticism of Quine fails to carry over as a criticism of us. Most of the time, however, Maddock's main concern is that he just feels 'unconvinced' by what we say. Since he agrees that the literature we have produced 'is large', and since the thriving Quine industry (Davidson and Hintikka 1968, Hahn and Schilpp 1986, Barrett and Gibson 1990) in philosophy contains discussions and replies to all the main points he raises, we are naturally cautious about seeing this reply as a matter solely concerned with convincing Maddock. A further reason for our caution is that we do not accept Maddock's strategy of framing issues in a way that systematically shifts the onus of proof in advance of any actual argument or evidence that he might provide for his own position. In what follows we shall give a brief account of our research programme in educational administration, and how it relates to our philosophy. After that we shall indicate where we stand on a number of Maddock's more basic criticisms.

Philosophy of Educational Administration

Our research programme is easily summarized. We are attempting to develop a new science of administration, one able to incorporate values and human subjectivity. To do this we are working with a new view of knowledge and its justification. We call our theory of knowledge, or epistemology, 'naturalistic coherentism', although in some earlier writings we used what turned out to be the somewhat misleading expression 'materialist pragmatism'. Although developed primarily in the philosophy of education literature (for more recent overviews see Evers 1987, 1993), the theory has been applied and substantially extended in a number of areas of educational studies, for example sociology of education

(Walker 1988), educational research methodology (Lakomski 1991), and, of course, educational administration.

One major motivation for pursuing our view is the same as that advanced by critics of traditional science of administration. Logical empiricism's theory of knowledge is too narrow to be usefully applied to any systematic account of educational administration. While the statistical apparatus of quantitative research methodologies will always be able to extract patterns from survey and other data if the patterns are there to be found, the complexity of social phenomena rather sharply limits the scope for making sound inferences from these data. Generalizations are hard to come by. Moreover, much knowledge of organizations is acquired informally, not by processes associated with the hypothetico-deductive testing of hypotheses, but by processes more akin to socialization and enculturation. Traditional behavioural science, with its emphasis on testability, operational definitions, and the exclusion of values, turned out to be more an attack on knowledge than a general way of advancing it. (Not even physics can meet the demands of logical empiricism.) As a methodology applied to social phenomena, it actually *loses* information, discounting much of what is known to be important about organizational life.

From the rise of the Theory Movement in the early 1950s, behavioural science dominated the scene in educational administration both methodologically and substantively (through the incorporation of systems theory into administrative theory building) for the next twenty years. The first systematic challenge, mounted by Thomas Greenfield in his 1974 International Intervisitation Address (Greenfield and Ribbins 1993, ch. 1), took advantage of a number of the most important philosophical criticisms of logical empiricism. The story is complex, but essentially critics noted that the relationship between empirical evidence and theory is much more problematical than had been believed (Evers and Lakomski 1991, ch. 1, 2 and 4). For example, there is no such thing as experience free of interpretation. All that we see is filtered by our theory-driven expectations. The language we use to describe phenomena is a theoretical language, reflecting the role terminology plays in articulating a conceptual scheme. This problem causes difficulty when it comes to giving operational definitions, for a theoretical term being defined cannot be specified in some purely observational way; that is, all the purported observation terms have a theoretical loading as much in need of operational definition as the original term. And so a vicious regress sets in. A problem also arises over the status of data for correcting theory, since the two are now intertwined; it becomes plausible to propose revisions to data in the light of better theory. Additionally, since data are always finite, any number of theories will fit the same data set. But confirmation now becomes troublesome, since all these theories seem to be equally confirmed.

Given the problematic relationship between theory and empirical evidence, Thomas Kuhn's contribution (Kuhn 1970) is of fundamental importance. For Kuhn suggested that the nature of this relationship is given primarily by the theory side; in particular, by large-scale theories, or paradigms. Thus a paradigm provides its own determination of how empirical evidence may function for or against that paradigm. Greenfield (Greenfield and Ribbins 1993, ch. 3) used this idea to great effect in his attack on traditional science and in developing his subjectivist alternative. Where theories are construed in terms of the beliefs people hold, then what a person sees, feels and experiences is very much a matter of their beliefs, their inner subjective assessments and evaluations of the world. In fact, on this view, the worlds people inhabit are largely determined by inner subjective factors, not by some objective appeal to the empirical evidence.

Since the early 1970s the paradigms perspective has taken root in educational research methodology where it now dominates. Research methodologies, which are organized ways of determining the nature of evidence and its inferential relations to theory, are on a more equal footing for fear of mistakenly imposing the canons of one paradigm onto the subject matter of another (for an overview see Walker and Evers 1994). Further developments in educational administration beyond both traditional science and Greenfield's subjectivism have benefited from paradigms epistemology and the assumption that there is more to knowledge than scientific knowledge. Critical theorists have been able both to attack science and to defend new models of research, such as action research; developments in cultural theory and qualitative methodologies have surged; and a rather more radical version of subjectivism in educational administration is now emerging in the form of postmodernism (for example Chapter 21; Maxcy 1994). Values perspectives are also getting a wider hearing (Hodgkinson 1991).

The upshot is that educational administration now has a much richer and more varied fund of knowledge than at any previous time. It is informed by traditional quantitative methodologies and a range of qualitative methodologies, all cohering to some extent with substantive theories of administrative phenomena. Perhaps it is ironic, then, that one of our concerns with more recent work in educational administration is that these newer approaches still result in knowledge being lost to the field. For in discounting, or excluding, science, these approaches detach themselves from what is the most prolific source of knowledge available today.

Our diagnosis of the central problem is that science is still being equated with one or other version of empiricism, usually referred to generically as positivism. Many of the arguments against science have really been arguments against positivism, which is merely a theory about science, and one we think is wrong. However, our proposal is to embed science within a postpositivist view of knowledge and its justification, notably a naturalistic coherentist view. We favour coherentism because we draw a different lesson from worries over the relationship between theory and *empirical* evidence. These worries imply an attack on objectivity only if it is assumed that all the evidence there is for a theory is *empirical* evidence. Since empirical evidence is never sufficient for determining whether a theory should be accepted or abandoned, it is mistakenly assumed that theory choice is a subjective matter, depending on preference, will, or lifestyle. The conclusion we draw from criticisms of positivism, however, is that there is more to evidence than empirical evidence. Because some theories plainly help us get around in the world much more successfully than other theories, and because these successful theories may not be distinguishable from the others on grounds of empirical adequacy, we conclude that additional, or *superempirical*, criteria of theory choice are doing valuable epistemological work. These additional criteria include consistency, comprehensiveness, simplicity, and explanatory unity, and, when taken together with empirical adequacy, are known as *coherence* criteria. We think that these criteria are, in fact, used by all the so-called different paradigms when it comes to justifying knowledge claims, so we do not accept a sharp division among research methodologies (this argument is elaborated in Walker and Evers 1994). Indeed, we argue that the case for a paradigms epistemology itself makes use of these criteria in attempting to be persuasive (Evers and Lakomski 1991, ch. 10).

Our argument for naturalism in epistemology follows from the fact that theories of knowledge make all sorts of assumptions about the cognitive powers of humans: how we learn and process information. In the history of epistemology, these assumptions have often been the result of armchair theorizing, where philosophers sit and reflect on their own cognitive processes. The findings, ranging from the extreme rationalist claim that all real knowledge is innate (as Plato argued) to the extreme empiricist claim that nothing exists but sensory experience, all look a little threadbare nowadays when they are contrasted with the findings available from empirical psychology and the neurosciences. This is why we follow Quine in freely using our best natural science to help determine how knowledge is acquired. For us, philosophy (or epistemology) does not come before scientific knowledge; rather, it is continuous with science.

One advantage of this kind of approach is that it counsels against less reliable knowledge in epistemology being used to exclude more reliable bodies of substantive knowledge; for example, the use of sense data empiricism to exclude relativity theory. We protest the parallel practice in educational administration of using modest epistemologies to attack the possibility of administrative science. We are open, instead, to the possibility of using the neurosciences as a tool for explaining human cognition, and human subjectivity in general. Attempting to explain administrative decision-making as a species of pattern processing is a case in point (Chapter 8). And so, too, is the use of information theory to illuminate accounts of the nature of knowledge and theory (Chapter 10; Lakomski and Evers 1994).

Responding to Maddock: Some Preliminary Issues

Since Maddock takes exception to many of the distinctive philosophical features that drive our view of educational administration, we begin by clearing up some smaller points that may mislead the unwary reader. Maddock thinks that 'naturalism goes hand-in-hand with positivism...' and identifies one respect in which this might be so. However, this is a confusing use of terminology since our naturalism, like Quine's and Dewey's, is systematically opposed to all major varieties of positivism. An advantage in being more careful over philosophical usage is that one can avoid being mistaken in thinking that arguments against positivism function equally well as arguments against naturalism.

Maddock says that our commitment to realism 'is legitimated in terms of the survival and viability of species'. This is false. Our argument is the familiar one that supposing a real world exists is the best (most coherent, non-random) account of why a system of beliefs would yield accurate predictions over the long run (BonJour 1985, pp. 169-179).

Maddock claims that we subscribe to behaviourism. Again, this is simply false. In our published work on these matters (for example, 'Educating the Brain', here reprinted as Chapter 9), we have championed neural network models of cognition and cognitive explanation, not behaviourist accounts. A more recent example of our use of neural network models to account for administrative phenomena can be found in 'Administrative Decision-Making as Pattern Processing' (here reprinted as Chapter 8). We suspect that many of the 'peculiarities' Maddock purports to find in our position arise not so much from misreading us as, perhaps, not reading us at all. Certainly in his comments on behaviourism, his primary source is Quine rather than our work. So why do we rationally choose neuroscience over behaviourism? Well, given alternative sets of distributions of sensory experience, that is, alternatives that are of equal empirical adequacy, we choose the most coherent set, which in our view (and Margolis's) is the one containing the neurosciences. However, contrary to Maddock's assertion, we do not believe that 'on-the-spot comparative assessment is possible'. (Does anyone?) Instead we say: 'Coherence justification, because of its global character, is just a more intricate and difficult business than foundational justification' (Evers and Lakomski 1991, p. 9).

Not only does Maddock freely conflate our views with Quine's, he also expects us to have published replies to Quine's critics. Thus he writes: 'To date, the materialist pragmatists [that's us] have not responded to the accusation that Quine is committed to transcendental foundationalism.' To the best of our knowledge Quine has not responded either, although this may reflect a provisional ranking of accusations worth responding to. Or it may reflect our own practice of responding to arguments ahead of accusations.

Maddock on Coherence

We are gratified to see that, despite his many concerns over the detail of our coherentist criteria of theory choice, for Maddock 'the intuition that criteria, like fruitfulness, consistency, explanatory power, etc., are important properties of theoretical accounts is a sound one'. So where do we go wrong? Evidently in 'over-valuing' these criteria. Maddock cautions that 'there are no abstract formulae for determining coherence'. We agree, and argue that applying coherence considerations 'is always a matter of detailed critique of particular issues' (Evers and Lakomski 1991, p. 9). However, he gives two examples where application of coherence criteria of theory choice may result in error, or the wrong choice being made.

The first concerns describing complexity in the world — a point that would be of special interest to those concerned with social theories. Thus, according to Maddock, 'If the world is complex, simplicity would be something imposed upon it that misrepresented the nature of things.' This sounds plausible, but would Maddock also say that if the world is contradictory, incoherent, and confusing then *our theories of it* should be contradictory, incoherent, and confusing? Or even, if the world is ninety percent green then our theories of it should be ninety percent green? Maddock's mistake seems to lie in making the assumption that all properties which apply to the world should also apply to good theories of the world. Pending an adequate defence of this assumption, Adorno's observations notwithstanding, we shall go on thinking it is sound epistemic practice to work towards developing simple, clear, consistent, and coherent theories of phenomena that appear complex, confusing, inconsistent, and incoherent (for further debate on this point see Chapters 21, 22, and 23).

Maddock's second example is not so troublesome for educational administrators, unless they be philosophically inclined. He thinks that while coherence criteria may apply to 'first-order theories', presumably theories about the world such as those developed by scientists, educators, and administrators, they do not apply to 'second-order theories', presumably those developed by philosophers that deal with first-order theories. Since our work in educational administration has mainly been about first-order, administrative, theories the amount of 'over-valuing' we are guilty of is relatively modest. We are even less troubled because, as he rightly notes, we think that neither he nor anyone else has succeeded in drawing a principled distinction between first and second order theories. However, if he wishes to exempt philosophy from the demands of our epistemology, he is of course welcome to develop an inconsistent, incoherent, fragmented, and narrow philosophical position. Our only query concerns what epistemic principles he might use in persuading anyone of its virtues.

Maddock on Values

One of the assertions we defend is that ethical claims are justified in the same way as any other knowledge claim: by considerations of coherence. We also deny that there is any sharp distinction to be drawn between fact and value (for a detailed discussion of our view on ethics see Evers and Lakomski 1991, ch. 8). We defend the latter position by taking a stand on two issues. First, as noted by Maddock, we challenge Hume's point that one cannot derive an 'ought' from an 'is'. Maddock seems to accept our argument that Hume is mistaken about logical derivation — after all, the deductions we give are all familiar from the literature, and are all valid. However, he protests that 'Evers has not addressed the justification argument'. But this protest is an odd one since coherentist justification does not have to proceed by the deduction of conclusions from premises. An ethical claim is justified if its inclusion into the system makes for greater systemic coherence.

Maddock's alternative model of ethical justification appeals to 'practical derivation' not 'logical derivation', a distinction 'that Evers fails to recognize'. But Evers fails to recognize the distinction because there is nothing substantive to recognize. Not even Maddock is able to spell out the nature of practical derivation: 'Even if it cannot be made to conform completely with systems of formal deduction, the practical derivation of administrative claims intuitively corresponds to informal ideas of deductive validity.' Earlier standards of clarity, so austere when it came to the terminology of coherentism, are remarkably relaxed in dealing with the intuitions and informal ideas required here.

The second issue on which we take a stand is the naturalistic fallacy. We think it is possible to define ethical terms naturalistically, as some natural quality like human happiness, or the growth of knowledge, or human flourishing. Moreover, it is not clear how our position can be ruled out unless it can be shown that ethical properties are essentially non-natural; that they logically *could not be otherwise*. Maddock almost recognizes the challenge but then slips back to placing the onus of proof on us — as though ethical properties were obviously non-natural unless proved otherwise. Our view of ethics (and politics) is in the Deweyan/liberal tradition. We argue that knowledge has touchstone advantages over ignorance, but that all knowledge is fallible. Maintaining conditions for the growth of knowledge would therefore cohere well, as a heuristic, with a range of other substantive theoretical claims comprising a global theory. But the contingent material conditions for knowledge growth require some ethical infrastructure to prescribe the relevant distribution of resources and the social relations of inquiry. Our arguments in defence of the values of tolerance, freedom, and respect for the welfare of others owe much to Popper's defence of the Open Society against Plato and Marx (Evers and Lakomski 1991, pp. 207–209).

In administrative theory this ethical perspective coheres best with the organizational learning tradition of administration, a view of leadership as educative, that is, as conducive to promoting individual and organizational learning, and a democratic, participatory vision of society (see for example Chapter 5). Despite our writings on these themes, Maddock complains that 'the absence of any theory of society is perhaps the major lacuna in materialist pragmatist theory'. A less patchy reading of our work might have led Maddock to reflect on the matter less polemically.

Conclusion

For Maddock 'it is through the relentless criticism of all conceptualizations that philosophy provides its service to educational administration'. It is true that some philosophers see themselves as the 'thought police' of administrative studies. However, this is a shallow view of philosophy, reflecting belief in a mistaken distinction between conceptual matters and empirical matters. Philosophy, at its best, is informed by the theories it engages, its critical function as much in the service of theory development as theory criticism. These tasks are best carried out by paying closer attention to systematicity and coherence than has been the case in Maddock's discussion of our work. As central and vital ingredients of disciplined debate and argument in educational administration theorizing, as in theorizing anywhere, they have much to recommend themselves in the pursuit of better theory.

References

Barrett R.B. and Gibson R.F. (1990) (eds.). *Perspectives on Quine*. (Cambridge. Mass: Blackwell).
BonJour L. (1985). *The Structure of Empirical Knowledge*. (Cambridge, Mass: Harvard University Press).
Churchland P.M. (1989). *A Neurocomputational Perspective*. (Cambridge, Mass: MIT Press).
Churchland P.S. (1986). *Neurophilosophy*. (Cambridge, Mass: MIT Press).
Davidson D. and Hintikka J. (1968) (eds.). *Words and Objections*. (Boston: Reidel).
Evers C.W. (1987). Naturalism and philosophy of education, *Educational Philosophy and Theory*, **19**(2), pp. 11–21;
Evers C.W. (1993). Analytic and post-analytic philosophy of education, *Discourse*, **13**(2), pp. 35–45.
Evers C.W. and Lakomski G. (1991). *Knowing Educational Administration: Contemporary Methodological Controversies in Educational Administration Research*. (Oxford: Pergamon Press).
Greenfield T.B. and Ribbins P. (1993). *Greenfield on Educational Administration* (London: Routledge).
Hahn L.E. and Schilpp P.A. (1986) (eds.). *The Philosophy of W.V. Quine*. (La Salle: Open Court).
Hodgkinson C. (1991). *Educational Leadership: A Moral Art*. (Albany: State University of New York Press).
Kuhn T.S. (1970).*The Structure of Scientific Revolutions*. (Chicago: University of Chicago Press, second edition).
Lakomski G. (1991). *International Journal of Educational Research*. **15**(6), pp. 499–597.
Lakomski G. and Evers C.W. (1994). Greenfield's humane science, *Educational Management and Administration*, **22**(4), pp. 260–270.
Maxcy, S.J. (ed.) (1994) *Postmodern School Leadership*. (Westport, Connecticut, London: Praeger).
Walker J.C. (1988). *Louts and Legends*. (Sydney: Allen and Unwin).
Walker J.C. and Evers C.W. (1988).The epistemological unity of educational research, in J.P. Keeves (ed.) *Educational Research Methodology, Evaluation, and Measurement: An International Handbook*. (Oxford: Pergamon Press).

20

Knowledge, Certainty, and Openness in Educational Administration—*Martin Barlosky*

Educational administration is perceived by many to be in a state of crisis. Speaking within the Canadian context, Dolmage (1992, p. 90) traces this sense of crisis to two primary issues that will be familiar to educators and administrators in other national settings: '[e]ducational administration has been unable to define itself as a unique discipline and has not demonstrated an ability to improve the practice of administration in Canadian school districts'. Despite a series of ambitious efforts, educational administration has been unable to lift itself above its status as a 'timorous and subaltern discipline' (Hodgkinson 1982, p. 64) that consists of a highly contingent amalgam of theory and practice existing 'at the applied end of an applied field' (Holmes 1986, p. 86). Further, it appears unable to develop a coherent body of expertise that could be made manifest in a uniform professional practice.

Unlike the cognate management sciences, which have come to play a central role in the contemporary university and in the definition of an applied profession, educational administration has not acquired the accoutrements necessary for a distinctive disciplinary identity. It has yet to produce a comprehensive theoretical framework capable of directing a profession toward the demonstrable improvement of educational institutions. The ironic coincidence of educational administration's appearance as a distinctive area of inquiry with the apparent disorientation and decline of the North American public school has militated against the viability of its knowledge claims. Within the field, administrators and scholars appear to have difficulty making sense of one another's experience, and the relationship between practice and research is often uneven and unclear.

To make matters worse, those who identify themselves as scholars of educational administration can be a rancorous lot. The diversity of their viewpoints is often matched by the irascibility of their debate. Within the field, the Weberian dichotomies of understanding and explanation, subjectivity and objectivity, and value and fact have crystallized into oppositional perspectives that resist both interchange and reconciliation. The seemingly irreducible differences among partisans of these perspectives have led to the widely accepted conclusion that the field is in reality an assemblage of mutually exclusive paradigms whose individual epistemological territories are impervious to criticism and inimical to dialogue (see Dolmage 1992).

In *Knowing Educational Administration* Colin Evers and Gabriele Lakomski (1991) would redress the conceptual fragmentation that has been left in the wake of positivism's demise as a unifying framework for research and practice. As a remedy to the fractiousness that now characterizes the field, they offer the post-positivist epistemological perspective of coherentism. Developed from the thinking of W. V. Quine, coherentism combines an interrelated set of superempirical criteria to be used in the development and selection of theories claiming empirical adequacy. Through the employment of these criteria, theory choice is to supersede positivism's problematic dependence on the facticity of empirical verification.

> Theory choice needs to be guided by a consideration of the extra-empirical virtues possessed by theories. These virtues of system include simplicity, consistency, coherence, comprehensiveness, conservativeness, and fecundity, though they are often referred to collectively as coherence considerations or as the elements in a coherentist account of epistemic justification (Evers and Lakomski 1991, p. 4).

Lest it be misunderstood as merely a philosophical exercise removed from the press of practice, the authors' coherentist project has very concrete ends in view. Coherence considerations are applied to provide a unitary, encompassing framework for the development of a knowledge pragmatically useful in the organization of human experience and action. Nothing less than 'a new science of administration' (Evers and Lakomski 1991, p. 5) is promised through the adjudication of theory in accordance with coherence criteria.

To assess the value and viability of the authors' project, we must examine the adequacy of their epistemology and its ability to do the work they propose. We must also explore the ramifications that the doing of this work might have for the development of research and practice in educational administration. As part of this exploration, we shall give extended consideration to coherentism's capability to develop a conceptual framework appropriate to a field in which inquiry and practice are inextricably tied to the profound complexities of human experience and interaction. The potential contribution of alternative epistemological and methodological positions will also be examined, as will their potential role in creating new forms of knowledge in the field.

To set the stage for their coherentist project, Evers and Lakomski present the modest history of educational administration as a conflicted subdiscipline now emerging from the positivist assumptions that once promised certainty in thought and action. After a postmortem examination of positivism's weaknesses, the authors undertake a comprehensive critical review of the major contemporary perspectives that seek to define the nature of educational administration and the kind of knowledge that it may make available. They begin with the Theory Movement, that now widely suspect attempt to derive law-like prescriptions for administrative action through the application of the hypothetico-deductive method. Successive chapters are devoted to Greenfield's subjectivist critique of administrative science, Hodgkinson's value-centred philosophy of administration, cultural theories of organization and leadership, and critical theory perspectives for educational emancipation. After each of these perspectives is found lacking (i.e. theoretically inadequate), the inquiry is concluded with an uncompromising critique of the Kuhnian notion that knowledge can only exist within sovereign, epistemologically hermetic paradigms.

In preparation for this critical survey, the reader is reminded of the centrality of epistemic considerations in the development of educational administration as a field of inquiry. The importance of epistemology is evidenced by citing the influence of logical empiricism on the generation of the Theory Movement and the subsequent role played by Kuhn (1970/1962) and Feyerabend (1975) in developing the widely held postpositivist view that discrete epistemological paradigms define divergent and incommensurable forms of knowledge. Although the authors accept the nullifying critique of positivist empiricism made by Kuhn, Feyerabend, Greenfield, and others, they stress the inadequacy of the 'subjectivist' alternatives that are offered as its successors. The 'anything goes' relativism that is associated with subjectivism is regarded by the authors as the cure that is worse than the disease of positivism's naive truth claims. Alternatively, the authors would show that the jettisoning of mistaken positivist assumptions need not lead to an abandonment of objectivity and realism.

With this end in mind, coherentism is presented as the superior successor to positivism. Coherentism's superiority is located in its avoiding positivism's uncritical empiricism and in its surmounting the ungovernable pluralism that is said to attend the various manifestations of subjectivity. This double ambition is accomplished through the introduction of extra-empirical coherence considerations for the selection of theory in place of positivism's exclusive reliance on empirical observation for theory verification. The retention of objectivity and realism is to be secured by maintaining the requirement of empirical adequacy — a requirement that will be examined in greater detail below.

Although the current volume restricts itself to matters concerning educational administration, it is made clear that the coherentist agenda has broader objectives. The authors note that coherentist principles have an immediate application in the social science disciplines that share educational administration's philosophical pluralism, methodological conflict, and intractable uncertainties. Those acquainted with the thinking of W. V. Quine (1969), however, will not be surprised to find suggestions of a project far more vast: the reintegration of the social and natural sciences through the rephysicalization of the former and a concomitant naturalising of epistemology (i.e. the grounding of epistemology in a physicalized psychology). Coherentism, then, is to present a post-positivist opportunity for an ultimately physicalist reconstruction of the encompassing and unifying scientific certainty that had foundered with the collapse of positivism.

But in presenting this opportunity, coherentism would overcome the presumptive empiricism of positivism and bypass the foundationalism that would make theory dependent on immediate and underived knowledge. Thus coherentism claims to be a *post-positivist* and *non-foundationalist* epistemology that can support a renewed scientific realism through its requirement of *empirical adequacy*. It is post-positivist because it accepts the terminal criticisms of positivism including the theory-ladenness of all observation and the underdetermination of theory by evidence. In admitting that empirical observation is 'constructed' rather than discovered and that the same evidence may support a diversity of theories, the authors concede the inability of empirical observation to either prove or disprove theory. But they contend that the absence of determinate facts and the compatibility of evidence with a multiplicity of explanations need not undermine the discernment of better theory. This 'betterness', however, is now to be determined through the application of the superempirical criteria associated with coherentism rather than exclusively through the matching of hypothesis with a now attenuated empirical 'reality'.

Coherentism is able to claim the status of being non-foundationalist because it does not seek to ground epistemological considerations in an irrefutable and inherently certain premise from which all other knowledge claims may be derived. Foundational antecedents are replaced by the same criteria that allow coherentism to escape dependence on empirical observation.

> We believe the whole quest for some antecedently certain or reliable foundation for knowledge, such as sensation, observation, intuition, or introspection, upon which all knowledge may be based, inferred, or deduced, is a mistake. Instead, we claim that the proper justification of knowledge is structured by coherentist considerations such as theoretical simplicity, consistency, comprehensiveness, conservativeness, and fecundity (Evers and Lakomski 1991, p. viii).

Coherentism thus avoids a dependence on ontological foundations, which are perpetually vulnerable to a Cartesian regress to uncertainty, while offering what seem to be self-evident superempirical criteria for theory determination.

The authors' mitigation of the ontological primacy of evidence (i.e. reality) is consistent with the theory dependence of empirical observation and with coherentism's antifoundationalism. This mitigation, however, has consequences for the renewed realism and objectivity that would be realized through the coherentist project. In distinguishing the theory of evidence from the theory of truth, the authors elaborate Quine's pragmatic equation of evidence-reality with what must be assumed in order to organize experience effectively:

Once we have our best theory according to these coherence criteria, it is the resulting theory itself that we use to tell us what exists and how the theory's sentences match up with that posited reality. What corresponds to true sentences is therefore something that is determined after the theory of evidence has done its work. It is not something that figures a priori, or in some privileged foundational way in the determination of the best theory (Evers and Lakomski 1991, p. 229).

But it is difficult to see how this relationship between theory, evidence, and truth avoids a circularity that would foreclose the promise of objectivity and realism. The only apparent check to ensure the realism that the authors espouse would be to allow the concept of 'empirical adequacy' to admit a determinative empirical confirmation of theory — a position that contradicts coherentist tenets and would stymie the coherentist agenda with the problems made explicit in the failure of positivism.

The authors indirectly address the role that empirical adequacy would play in theory selection through the introduction of the metaphor 'benchmark'. In each of two apparent references to Greenfield's (1980) signal paper, 'The man who comes back through the door in the wall: discovering truth, discovering self, discovering organisations', we are reminded that doors are a better means than walls for exiting an office. In the second iteration of this reminder, the authors note that 'success in getting in and out of rooms is *something of a benchmark for objectivity*' (Evers and Lakomski 1991, p. 229, emphasis mine).

The apparent reliance of theory objectivity on benchmarks, which evoke a sense of determinate facticity and an echo of foundationalism, would seem to cut against the post-positivist assertion that all observation is theory dependent. If 'there is no reality that can be seen independent of competing theoretical perspectives' (Evers and Lakomski 1991, p. 228), benchmark observations would seem to offer little more than a tautological compression of the theory that determines their character. It is also unclear how such benchmark observations would prove resistant to the interpretative appropriations that destabilize the certainty that positivism tried to locate in determinative and impregnable empirical 'facts'.

To accomplish the renewed objectivity and the unifying realism promised by coherentism, benchmark observations would seem to require an exemption from the previously admitted theory dependence of observation. But even if we were to entertain this possibility, how would we know when and on what grounds the asserted theory dependence of observation might be occluded? What special characteristics would separate benchmarks for objectivity from observations of the theory-laden sort? And although few would contest the pragmatic superiority of the doorway to the wall as a means of room entry, the discernment of which empirical observations are to have benchmark status will be far less clear in the complex, unstable, and contested realities that concern educational administration.

As the above example suggests, when the authors do allude to the sorts of 'posited reality' that might constitute 'benchmarks for objectivity', their examples are so incontrovertible as to be unhelpful. 'What is given by our most coherent account of our (interpreted) experience' is indeed self-evident when it concerns the existential efficacy of 'sticks and stones, bricks and bones' (Evers and Lakomski 1991, pp. 86–87). Surely it is not surprising that such observations coincide with an easily elicited consensual pragmatism and with 'folk theory' (Evers and Lakomski 1991, p. 86).

But this quality of self-evidence pales quickly when we must search for benchmarks among the values and meanings that motivate choice and action in the complex social world where educational administration is practised. Pragmatic clarity becomes elusive amid the contending values and viewpoints at play in an inherently unstable social environment that is coursed with ambiguity and the unpredictabilities of circumstance. As Cibulka, Mawhinney, and Paquette (1991, p. 7), citing the work of March and Olsen, remind us, there is very little simplicity, stability, or clarity in the world of administrative decision-making:

> It has, however, become clear that decision making in education rarely addresses predictable and unambiguous issues with clear outcomes. Instead, it is often characterized by multiple goals and inconsistent preferences, unclear technologies for producing the outcomes, and the fluid participation of actors in the decision-making process.

The invocation of benchmarks for objectivity and the definition of empirical adequacy, then, are far more contentious, complex, and pivotal issues than are presented to the reader. Their importance is underscored by coherentism's claimed sufficiency to generate 'a new science of administration' characterized by sound theory and a unifying realism adequate to the educational and administrative realities it would address. As it stands, the problematic and unresolved relationship between theory, evidence, and truth imposes on coherentism's claims to a post-positivistic objectivity. Short of allowing benchmarks to assume the status of irrefutable empirical warrants, a position that would contravene coherentist strictures, it is unclear how these observations are to escape the taint of interpretation. Like the weakness the authors tellingly discern in Kuhnian paradigm theory, the accomplishment of an efficacious realism appears to bring coherentism uncomfortably close to a reliance on a variant of privileged empirical observation — a reliance that sits uneasily with their uncompromising critique of positivism.

Before considering the self-evident soundness of the coherence criteria themselves and the possible outcomes of their application to theory development in educational administration, it will be useful to examine a peculiar use of language in the framing of educational theory. There is a marked rhetorical tendency in educational research to capture conceptual high ground and to pre-empt substantive debate through the strategic appropriation of self-affirming vocabulary. For example, the use of 'effective schools' to define a body of research would seem to place any and all critics in the untenable position of advocating ineffective schools. The resultant pre-emption of discourse has militated against the open debate that could clarify thinking and unburden research of faulty presuppositions. It has also invited a premature identification with seemingly attractive positions that are often revealed as inadequate and insubstantial. This, in turn, has led to scepticism about education as an academic discipline capable of improving an enterprise that is so often unresponsive to occupational argot, however euphemistic it may be.

A similar, if inadvertent, tendency is present in the selection of coherentism to define the epistemological principles advocated in the present volume. To differ with the authors would seem to place one in the unenviable position of advocating incoherence in place of knowledge. And this is precisely the charge made against coherentism's subjectivist antagonists! In contrast to subjectivism's alleged nihilistic invitation to incoherence, the criteria of coherence, comprehensiveness, conservativeness, consistency, and simplicity are presented as self-evident goods capable of supporting a unitary body of theory and knowledge. They are the common touchstone 'virtues' that can only enhance the development of a body of theory that would improve our understanding of educational organizations and that would lead axiomatically to their improvement. But this remains so only if we withhold the crucial observation that has been phrased by Perrow (1986, p. 117):

> It is quite possible that our social theories in general, and organizational theory in particular, have been altogether too rational . . . what if much of our world exhibits low coherence, accidental interactions and consequences, highly situational (rather than enduring or basic) determinants of behaviour, and very specific rather than broad cultural reinforcements and demotivators?

Perrow's observation implies that a discontinuous social reality may demand theory virtues of complexity, situatedness, fluidity, and divergence rather than those associated with coherentism. A reality imbued with the discontinuities and pluralistic tensions endemic to contemporary experience (Harvey 1989; Jameson 1991) may require the fractionation of the traditional, unifying theories that have sought to define the horizons of human thought and action. Hargreaves' (1995) critique of 'dead certainties', through his presentation of the value of *situated certainty* over *scientific certainty*, suggests the superiority of provisional and context-dependent knowledge for assisting educational practitioners in their actual work situations. And Greenfield's (1978, p. 7) note that 'our conceptions of organizations must be as complex as the reality we try to understand' is a terse reminder that the presumptive virtue of theory simplicity may jeopardize rather than advance a compelling sense of realism.

But even if we were to concede the possibility and the appropriateness of realising a grand unifying theory like that proposed by the Quinean coherentist project, our expectations about what such a theory could yield might well be disappointed. In his *Brief History of Time*, Stephen Hawking (1988, p. 12) presents the possibility that the fruition of the search for a complete unified theory of the human and natural sciences could lead to one of three normally unanticipated outcomes: tautology, error, or vacuity:

> Now if you believe that the universe is not arbitrary, but is governed by definite laws, you ultimately have to combine the partial theories into a complete unified theory that will describe everything in the universe. But there is a fundamental paradox in the search for such a complete unified theory. The ideas about scientific theories outlined above assume we are rational beings who are free to observe the universe as we want and to draw logical deductions from what we see. In such a scheme it is reasonable to suppose that we might progress ever closer toward the laws that govern our universe. Yet if there really is a complete unified theory, it would also presumably determine our actions. And so the theory itself would determine the outcome of our search for it! And why should it determine that we come to the right conclusions from the evidence? Might it not equally well determine that we draw the wrong conclusion? Or no conclusion at all?

Although it is often ignored, the construction of such theory also involves implicit moral considerations. A sub-theme in Eco's (1990) novel *Foucault's Pendulum* concerns the moral impact of the often unanticipated outcomes of theory construction. In a recent paper, Ryan (1993, p. 8) quotes the enlightening remark made by Casaubon, a central character in the novel, who has come to regret the dire consequences of his contributions to an idiosyncratic variant of a grand unifying theory:

> I have come to believe that the whole world is an enigma, a harmless enigma that is made terrible by our own mad attempt to interpret it as though it had an underlying truth.

Casaubon's reflections and the novelistic circumstances that give rise to them introduce the appropriateness of a decidedly moral constraint on theory making. Similarly, Greenfield's outline of an alternative epistemological perspective to that of grand explanatory systems suggests that the intertwining of moral and epistemic considerations in the selection and the building of theory is unavoidable. As Greenfield's (1980, p. 55) remarks make clear, this intertwining has special poignancy in the case of theories that would address the phenomenon of human organization:

> The paradigm I argue for in the exploration of social realities is certainly against overarching systems of unilateral explanation; it is one that admits

the many voices of truth and recognises them as attached to self, to individuals. My chief argument is against those who would fit all truth about organizations into a single, objective, non-political, self-less truth called science.

The speculations of Perrow, Hargreaves, Greenfield, Hawking, and Eco emphasize the possibility that coherent and encompassing explanations may produce results of questionable empirical and moral value. Their criticisms caution that the project of theory generation may by its very nature loose a host of presumptive virtues that can lead to faulty thinking and to problematic research. The Borges (1967) short story 'Averroes' Search' is a reminder that much may be gained through reconsidering the assumptions that lead to the premature valorisation of coherence and to the ready debunking of its alternatives. We learn this in the story through the title of the book being redacted by the Islamic scholar protagonist. That volume is titled simply *Tahafut-al-Tahafut* (The Incoherence of Incoherence).

Although the diminution of the encompassing and instrumental certainties normally associated with science may initially seem perplexing and vertiginous, it need result in neither intellectual resignation nor despair. Such a reaction would be justified only if truth could be equated with surety and if knowledge were coextensive with mastery. Given that neither equation obtains, the sense of crisis in educational administration appears to be an over-reaction to the epistemological failure of positivism and to the inappropriateness of its attendant vision of scientific inquiry. Rather than anxiously seeking a replacement framework capable of achieving a theory-based human technology, researchers and practitioners might do well to pause and assess the opportunity with which they have been presented.

In the paper cited previously, Ryan notes the questionable value of constructing universal theory in educational administration and proposes an extensive agenda for the generation of a more particularistic form of research that can support the interests of educators and their clientele. In closing his synopsis of alternative research methodologies he observes:

> We need not, then view the receding prospects for generating a unified theory of school administration and organization with disappointment or despair. On the contrary, social scientists and practitioners should welcome the opportunities that accompany this view. At the very least, it frees us from the shackles imposed by the imperative to generate fictitious universals. (Ryan 1993, p. 27)

The opportunities of which Ryan speaks may become more apparent by exchanging the word 'theory,' with its implications of universal truths, for the more modest word 'narrative', with its connotations of a more local, contingent, and authored knowledge. To borrow from Bruner (1986), the establishment of the verisimilitude or lifelikeness associated with the narrative may be more appropriate to a human, and ideally humane, praxis like educational administration than are the putative, generalizable truths that might emerge from theory. In educational administration, a practised sensitivity to the particular may exceed the value of normative generalization.

Having addressed the epistemological limitations of coherentism, the empirical and moral constraints on the construction of grand unifying theory, and the plausibility of alternative frameworks for the generation of knowledge, we may return to an examination of what many consider a crisis situation in educational administration. A selective borrowing from Lyotard's (1984) usage of the term 'metanarrative' can help to clarify both the current status of educational administration as a discipline and the role that coherentism would play within it.

As the conceptual superstructures that would give sense and order to inquiry, metanarratives seek to determine the limits within which knowledge may be defined and experience understood. They shape and pervade the structures of inquiry, discourse, and practice. Within modern Western civilization, scientific rationalism has functioned as the formulaic supercode that would render the complexity of experience and event amenable to explanation, predictability, and control. Grant's (1969, p. 116) observation that 'the modern sciences can best be understood as a unity around the idea of mastery' underscores the centrality of mastery to the modern scientific enterprise. Through the accomplishment of mastery in the physical world, science has become identified with knowledge, and the metanarrative of scientific mastery has come to define the concept of 'truth'.

Despite a marked lack of accomplishment to date, it remains the hope of many that the human sciences can reflect the mastery realized in the physical sciences through the development of concordant epistemological frameworks and methodological procedures. In the context of a somewhat attenuated sense of philosophy, Evers and Lakomski (1991, p. 219) announce the role that coherentism would play in this enterprise.

> The task of philosophy, which is not sharply distinguished from but continuous with empirical science, is to find general principles common to all sciences. This means extending the use of such principles even to the regulation of human conduct and the organization of society. The move to a postpositivist philosophy of science is quite compatible with such a view of the nature of science and its role in human affairs.

Coherentism, then, would supply a renewed epistemic subtext to the metanarrative of mastery — a subtext that would claim efficacy in the regulation of human affairs through a procedural scientific rationalism. The attendant practice of administration would presumably consist of the expert application of the general principles that would effectively guide 'the regulation of human conduct and the organization of society'.

Working within a very different sense of philosophy's task, MacIntyre (1984,

p. 107) casts a definitive moral doubt upon the present or future viability of such social regulation and the concomitant notion of managerial effectiveness:

> The concept of managerial effectiveness is after all one more contemporary moral fiction and perhaps the most important of them all. The dominance of the manipulative mode in our culture is not and cannot be accompanied by very much actual success in manipulation. I do not of course mean that the activities of purported experts do not have effects and that we do not suffer from those effects and suffer gravely. But the notion of social control embodied in the notion of expertise is indeed a masquerade. Our social order is in a very literal sense out of our, and indeed anyone's control. No one is or could be in charge.

March makes a similar point in his comments on the symbolic role of executive behaviour in reinforcing what has come to be perhaps the primary shibboleth of modernism. In addition, he suggests how the mythic construction of an efficacious control of human affairs may drive an errant but functional belief system:

> Executive behaviour and management procedures contribute to myths about management that become the reality of managerial life and reinforce a belief in a human destiny subject to intentional human control. (March 1986, pp. 286–287)

The very awareness of the mythic character of control in human affairs begins to affect the undoing of the metanarrative of mastery. The destabilizing of metanarratives that is associated with the advent of postmodernism reflects their failing power to condition experience in accordance with the dictates of an often unconsciously accepted horizon. Indeed, the arrival of postmodernism may be understood as contemporaneous with the realization that rather than representing transcendental truths, the metanarratives that have conditioned modernism are themselves a set of metaphorical tropes that selectively amplify experience in accordance with their implicit presumptions. The hallmark of postmodernism is our understanding within the context of our lived experience that these metanarratives are social constructions rather than discovered truths, contrivances whose determinative character is subject to revision.

The increasing loosening of the rationalist metanarrative of scientific mastery has an especially forceful impact on those disciplines that are concerned with understanding organizations and their administration. Clegg (1990, p. 4) has suggested why this should be so:

> Representational modernism consisted in the sketching of a singular set of empirical tendencies which were imagined to be irresistible and inevitable. These were the famous 'rationalisation' of the world, the success of which

would be attributed to bureaucracy as the primary mechanism of its achievement. . . Organisation theory is in many respects a modernist discipline *par excellence*, springing as it does from Weber's modernist vision of the modernist world.

Given Clegg's observation, it is not surprising that the attenuation of the metanarrative of mastery may be experienced as a crisis in educational administration. As Evers and Lakomski make unavoidably clear, the central concentration of educational administration as a field of inquiry and practice has been on the primary artefact of this increasingly insubstantial modern metanarrative: the mastery of human conduct requisite to the regulation of social organization.

The failure of a rational instrumentalism driven by the metanarrative of scientific mastery to generate the principles sufficient for the instrumental 'regulation of human conduct and the organization of society' will, then, have an especially pronounced effect in the applied subdisciplines of the human sciences. But as the equation of knowledge with mastery is superseded, new prospects for the development of research and practice become more readily available. In the case of educational administration, the opportunity is presented to redescribe educational organizations and to reconceive our ways of understanding and acting within them. As the quest for certainty and a generalizable objectivity give way to a less prescriptive definition of knowledge, a knowledge sensitive to the particularity inherent in human affairs, multiplex forms of understanding may be expected to emerge from the conceptual fissures that seem so immediately perplexing.

Although the exercise is highly speculative, Lyotard has begun to fill in some of the possible content of emergent concepts of science and knowledge that reflect the postmodern condition:

> Postmodern science by concerning itself with such things as undecidables, the limits of precise control, conflicts characterized by incomplete information, *'fracta'*, catastrophes, and pragmatic paradoxes — is theorising its own evolution as discontinuous, catastrophic, nonrectifiable, and paradoxical. It is changing the meaning of the word *knowledge*, while expressing how such a change can take place. It is producing not the known, but the unknown. And it suggests a model of legitimation that has nothing to do with maximised performance, but has as its basis difference understood as paralogy (Lyotard 1984, p. 60; emphasis in the original).

As Lyotard indicates and as has been previously suggested, emergent forms of knowledge may have more to do with the discontinuous rather than with the coherent, with difference rather than with unity, with the unexpected rather than with the predictable, with the particular rather than with the general, and with complexity rather than with simplicity. And in place of the faltering reductive certainties of an explanatory scientific rationalism, the discontinuity that has become synonymous with postmodernism offers manifold opportunities for fresh thinking and practical innovation.

Although the delegitimation of the metanarrative of mastery may understandably trigger attempts to resuscitate exhausted forms of knowledge capable of supporting its increasingly tenuous continuity, it also invites a more interesting and creative use of intellectual energies. Relieved of the tired agenda of manipulation and control, researchers and practitioners of administration may begin to redefine both the outlines and the content of a discipline that has become open to new possibilities. A brief reference to Eco's writings on the 'open work' may furnish a metaphorical point of orientation. Although Eco (1989, p. 23; emphasis in the original) is addressing the practice of art, his remarks have equal relevance to those engaged in the study and the practice of educational administration.

> The situation of art has now become a situation in the process of development. Far from being fully accounted for and catalogued, it deploys and poses problems in several dimensions. In short, it is an 'open' situation, *in movement*. A work in progress.

The future of educational administration as 'a work in progress' may well depend on whether the problems posed by openness are regarded as a crisis necessitating rear-guard actions or as an opportunity that may be engaged through the development of new forms of knowledge.

In closing, a few general remarks are addressed to prospective readers of *Knowing Educational Administration*. The journey through this tightly written volume of extensive grasp will prove an arduous one. The intricacies of the philosophical arguments presented demand much from the reader. But the journey is well worth making. In spite of occasionally superficial renderings of perspectives differing from their own, the authors have unflinchingly presented the central dilemmas that have come to characterize educational administration as a field of inquiry. Obviously one need not agree with either the authors' proposed epistemology or the role that it would play in defining the purposes of educational administration to benefit from their work. Their text is a provocative stimulus to rethink the importance of epistemology and its relationship to theory and practice at a pivotal moment in the field's history.

Two distinctive authorial voices can be heard in the text. Evers' finely tuned grasp of the epistemological problems that have perplexed educational administration presents a welcome deepening of arguments that are often unconscious of the philosophical antecedents they invoke. Lakomski's unrelenting press to avoid the reduction of administrative choice making to simple preference is both pointed and appropriate, given the nature of the educational choices administrators are empowered to make on behalf of others. Through their combined work the authors have reinforced the importance, the richness, and the possible civility of the ongoing philosophical debate concerning knowledge and purpose in educational administration.

This reader, however, left the volume with some regret that the authors' considerable talents were used toward rather circumscribed ends. Neither appeared to have grasped the significance of the remarks of another student of Quine who offers a countervailing statement of how we might comprehend our collective cultural task:

> We need a redescription of liberalism as the hope that culture as a whole can be 'poetised' rather than as the Enlightenment hope that it can be 'rationalised' or 'scientised'. . . . A poetised culture would be one which would not insist we find the real wall behind the painted ones, the real touchstones of truth as opposed to touchstones which are merely cultural artefacts. It would be a culture which, precisely by appreciating that *all* touchstones are such artefacts, would take as its goal the creation of ever more various and multicoloured artefacts (Rorty 1989, pp. 53–54; emphasis in the original).

References

Borges J. L. 1967. *A Personal Anthology*. A. Kerrigan (ed.). (New York: Grove Press).
Bruner J. (1986). *Actual Minds, Possible Worlds*. (Cambridge, Mass: Harvard University Press).
Clegg S. R. (1990). *Modern Organisations: Organisation Studies in the Postmodern World*. (London: Sage).
Cibulka J. G., Mawhinney H. B. and Paquette J. (1991). Rationality and progress in North American education 1967–1991: Problems of administration and governance, paper prepared for presentation at the annual meeting of the American Educational Research Association, Chicago, April 1991.
Dolmage W. R. (1992). The quest for understanding in educational administration: A Habermasian perspective on the 'Griffiths-Greenfield debate', *Journal of Educational Thought*, 26(2), pp. 89–113.
Eco U. (1989). *The Open Work*. (Cambridge, Mass: Harvard University Press, translated by A. Cancogni).
Eco U. (1990). *Foucault's Pendulum*. (New York: Ballantine Books).
Evers C. W. and Lakomski G. (1991). *Knowing Educational Administration: Contemporary Methodological Controversies in Educational Administration Research*. (Oxford: Pergamon Press).
Feyerabend P. K (1975). *Against Method*. (London: Verso).
Grant G. (1969). *Technology and Empire*. (Toronto: Anansi).
Greenfield T. B. (1978). Organisations as talk, chance, action, and experience, paper commissioned for publication in *Die Psychologie des 20. Jahrhunderts, Band VIII: Lewin und die Folgen*, Annelise Heigl-Evers and Ulrich Streeck (eds.) (Zürich: Kindler Verlag).
Greenfield T. B. (1980). The man who comes back through the door in the wall: Discovering truth, discovering self, discovering organizations, *Educational Administration Quarterly*, 16(3), pp. 27–59.
Hargreaves A. (1995). *Changing Teachers, Changing Times: Teachers' Work and Culture in the Postmodern Age*. (Toronto: OISE Press).
Harvey D. (1989). *The Condition of Postmodernity*. (Oxford: Blackwell).
Hawking S. W. (1988). *A Brief History of Time*. (New York: Bantam Books).
Hodgkinson C. (1982). Educational administration in Canada: a conspectus, *School Organisation and Management Abstracts*, 1(2), pp. 61–67.
Holmes M. (1986). Comment [on T. B. Greenfield, The decline and fall of science in educational administration]. *Interchange*, 17(2), pp. 80–90.
Jameson F. (1991). *Postmodernity or the Cultural Logic of Late Capitalism*. (Durham, N.C.: Duke University Press).
Kuhn T. (1970/1962). *The Structure of Scientific Revolutions*. (Chicago: University of Chicago Press, second edition).

Lyotard J.-F. (1984). *The Postmodern Condition: A Report on Knowledge*. (Minneapolis: University of Minnesota Press, translated by G. Bennington and B. Massumi).
MacIntyre A. (1984). *After Virtue: A Study in Moral Theory*. (Notre Dame: University of Notre Dame Press).
March J. G. (1986). How we talk and how we act: administrative theory and administrative life, in M. D. Cohen and J. G. March *Leadership and Ambiguity: The American College President*. (Boston: Harvard Business School Press, second edition).
Perrow C. (1986). *Complex Organizations: A Critical Essay*. (New York: Random House).
Quine W. V. (1969). Epistemology naturalized, in W.V. Quine *Ontological Relativity and Other Essays*. (New York: Columbia University Press).
Rorty R. (1989). *Contingency, Irony, and Solidarity*. (Cambridge: Cambridge University Press).
Ryan J. (1993). Order, anarchy and inquiry in educational administration, paper presented at the Canadian Society for the Study of Education, June 1993, Ottawa, Canada.

21

Response to Barlosky: Methodological Reflections on Postmodernism

In his review of our book *Knowing Educational Administration* (Evers and Lakomski 1991), Martin Barlosky (Chapter 20) has given a sympathetic, informed and, for the most part, accurate account of our position. He has also made some important criticisms. Our response aims first to provide a little more context that may be of help in understanding what we are up to in the book, and second to offer a reply to some of the criticisms. Not all can be discussed here, but we single out for closer examination Barlosky's notion of 'benchmark observation' incorrectly attributed to us; the idea that our coherentism may oblige opponents to advocate incoherence as the appropriate alternative position; his assimilation of our fallibilist approach into a traditional, empiricist, model of prediction and control; and the status of coherence criteria mistakenly presumed to be self-evident. Finally, we raise some problems regarding the claims of postmodernism as a suitable alternative framework for educational administration. We conclude little evidence has been provided that would lead us to abandon the powerful resources of our naturalistic coherentism in our endeavour to provide a better account of administrative theory and practice.

Context

Theories about knowledge and its justification exert a powerful influence on both the content and structure of administrative theories. However, while there is wide agreement about the role played by logical empiricist epistemology in the development of traditional science of administration associated with the 1950s Theory Movement and its descendants, the place of epistemology across the field's current theoretical diversity is less transparent. Our book attempts to make matters explicit by focussing on the epistemological character of both the framework sustaining diversity and the nature of major differences among administrative theories.

There are at least three advantages that result from attention to theory of knowledge in educational administration. First, because of the centrality of epistemological issues, we can improve our understanding of much of the debate that contributed to the development of the field's principal theoretical perspectives. Thus arguments from Thomas Kuhn's (1970) attack on empiricist theories of science, which gave rise to the current vogue for paradigms, turn up in subjectivist, critical theory, and cultural theory approaches to administration. And Kuhn's arguments for the paradigm relativity of justification do service in postmodernism's classification of epistemic methodology as metanarrative, a position beginning to have consequences for the nature of theorizing in educational administration. (Note that postmodernist views about the epistemic status of metanarratives are themselves not arbitrary but rather a consequence of a preferred, and often argued, epistemological position.)

Second, what we might call the problem of knowledge loss can be addressed more fruitfully. Human learning occurs across an enormous range of conditions and circumstances — from learning a first language, to mastering cultural mores, to acquiring leadership skills, to learning the piano, to grasping a scientific theory. There would thus be a premium on models of knowledge that can account for the flexible and holistic nature of human learning. Consequently, models that result in a *partitioning* of the class of knowledge claims can result in loss of information. This occurs on a massive scale with logical empiricism. Driven by an implicit theory of human learning and cognition, its criteria for acquiring valid knowledge, modeled on physics, are so exclusive they create difficulties even for physics, let alone social sciences like educational administration. Our criticism is that the knowledge claims being excluded are often more justified than the theories of learning and cognition used to make the exclusions.

Knowledge can also be lost by an epistemology that assigns the same epistemic status to all sets of knowledge claims partitioned by orthogonal canons of justification. Extreme forms of relativism associated with the work of Kuhn and Feyerabend, and more recently Rorty, have this effect. Yet anyone who tries to cross a busy road, cook a meal, or drive a car will be better served by attention to some theories rather than others. Inasmuch as there are regularities in the natural (including the social) world, the practice of equating theories that capture these patterns with those that do not amounts to giving knowledge away. However, we would say that the strategy of surrendering on the problem of theory adjudication (as advocated by e.g. Rorty 1989, p. 54) is based, once again, on theories of human knowledge acquisition that are less plausible than the powerful scientific theories that end up being discounted (see Evers and Lakomski 1991, pp. 228–231).

To avoid this methodological difficulty, we have imposed the condition that our own epistemology be compatible with the best available scientific accounts of cognition: in our view, theories of brain processes. Of course, this naturalizing of epistemology imposes a weak provisional ranking on knowledge claims, by the corollary that such claims should cohere with natural science (hence the characterization of our research programme as the development of a new science of educational administration). It also commits us to a coherence view of justification, since in denying any *a priori* privileged mechanism for partitioning knowledge, all candidates for knowledge, including scientific claims, are adjudicated on their contribution to global excellence of theory. But our hypothesis, subject to the usual conditions of open debate, is that our coherentist principles of adjudication make better use of knowledge than positivist, subjectivist, or postmodern alternatives.

Knowing Educational Administration, in critically reviewing theory development in the field using coherentist criteria, and in defending the possibility of a broadened science of educational administration — one able to include ethics and human subjectivity — has focussed on the above two advantages of adopting an epistemological perspective. A third advantage is that such a perspective implies systematic consequences for what it means for administrative theory to guide practice. We will touch briefly on this matter at the end of our replies.

Replies

After a determined and painstaking attempt to describe, defend, and apply our *coherentism*, we were surprised to see Barlosky interpret the term 'benchmark' in a foundational way. The term 'benchmark observation' is his, not ours. Of course, we mean touchstone. He also seems to think that our coherentist criteria need to be 'self-evident', which is another foundational expression. They don't. They need to be touchstone. However, touchstone theory is no epistemically privileged endpoint in a regress chain of justification. It is merely that (shifting and variable) body of theory that is common, or shared. The book contains no general argument that coherence criteria *must* be touchstone for all epistemic positions. People can argue alternatives. Presumably, to be persuasive, it would be desirable for these arguments to be consistent, coherent, non-arbitrary, comprehensive... and so on. Such is our strategy in Chapter 10 of Evers and Lakomski (1991), 'Against Paradigms'.

We chose the example of leaving an office via the door because we were looking for a touchstone example of objectivity: that is, one that might be agreed upon across otherwise extravagantly differing theories. We have been successful to the extent that Barlosky finds the example 'incontrovertible'. Greenfield (1993, p. 242) was also persuaded, but thought it 'meager knowledge'. However, the point we want to make with such an example is entirely different from Barlosky's. Our point is that whatever objectivity is agreed to in this, and other humdrum examples, it cannot be the result of *foundational* evidence, for such evidence of doors and walls is as much subject to the charges of theory-ladenness and underdetermination, as evidence of molecules and atoms (see again Evers and Lakomski 1991, pp. 228–229). Rather, objectivity resides in the resulting massive coherence conferred on our implicit and explicit theorizing about these humdrum contexts if we are willing to posit the existence of walls, doors, and the like. The problem with 'complex, unstable, and contested realities that concern educational administration' is not that these are somehow more theoretical, or subject to different standards of evidence. The problem is that there is less agreed, or common, theory. But then here is the usual place for debate and discussion of alternatives to begin. The claim we defend in our book is that despite the complexity, unpredictability, and instability that attends social contexts, there is enough methodological touchstone for epistemically progressive debate to be possible.

Barlosky worries that our use of the term 'coherentism' may inadvertently capture the conceptual high ground by forcing opponents into 'advocating incoherence in place of knowledge'. It had not occurred to us that this might be a result since the contrast in *Knowing Educational Administration* is with *foundational* epistemologies. Undeterred, Barlosky captures the low ground and offers a defense of incoherence. Citing Perrow (1986, p. 117) the argument seems to be that if the social world is confused, inconsistent, and incoherent then such a reality may demand that our theories of it be confused, inconsistent, and incoherent. We are happy to surrender this ground to anyone who would care to occupy it. However, for those rushing in, we note a key assumption behind the argument. To sustain the required isomorphism between world and theory, an implausibly strong representational thesis is required that maps properties of the world onto properties of adequate theory. The trouble is, there is no principled way of demarcating properties. Thus if the world is ninety percent green then our theory of it should be ninety percent green; if the universe is mostly hydrogen then our theory of it should be mostly hydrogen. Industrial strength representationalism may have its supporters, but it is odd to find one among the ranks of those who, with Rorty, endorse postmodernism's anti-representationalist stance.

Our modest representationalism suggests a more benign line: if the world is green then it is a virtue for our theory to *say* it is green. And if the social world is irrational or incoherent then it is a virtue for our theory to say so, but to say it with clarity, coherence, and enough detail to foster the construction of explanations, or to permit fruitful engagement and debate with rival equally well structured views. For example, the work of Tversky and Kahneman (1981) on human irrationality enjoys a number of theoretical virtues, but irrationality is not prominent among them.

Barlosky observes that even if our project of structuring educational administration for a place within a more encompassing coherent global theory should be realized, it may lead to disappointment and results of questionable empirical and moral value. Stephen Hawking (1988, p. 12) is quoted at length in defense of disappointment. Readers aware that Hawking's whole professional life has been devoted to the task of constructing his own grand unifying theory will be reassured to know that, two pages later, he manages to rise above disappointment:

> But ever since the dawn of civilization, people have not been content to

see events as unconnected and inexplicable... Humanity's deepest desire for knowledge is justification enough for our continuing quest. And our goal is nothing less than a complete description of the universe we live in (Hawking 1988, p. 14).

The full weight of fiction is brought to bear in evidence against the ethics of scientific truthfulness — such are the ways of postmodernism — and Greenfield (1980, p. 55) is quoted against science. While we waive comment on the moral authority of characters from fiction, we would like to emphasize that Greenfield is criticizing a view of science quite different from ours. As a further caution, Barlosky warns that our view, with its constant harping on coherence and consistency, can lead to faulty thinking. Here his worry over occupying the conceptual low ground has some real purchase, for sound thinking is not transparently advanced by promoting inconsistency and incoherence. We should add, however, that his difficulty is the result of no mere terminological ploy. At issue is the deeper question of whether it is methodologically incoherent to expect the case for incoherence to be coherent.

The last part of Barlosky's review is given over to sketching an alternative vision for administrative theory. Thus 'theory' has unhappy implications of universal truths and should be replaced with the more modest term 'narrative'. Epistemology, which involves theorizing about theories, becomes 'metanarrative', and ours, evidently because of its links with science, becomes 'an epistemic subtext to the metanarrative of mastery'.

Although these differences in terminology run deep, reflecting major differences of viewpoint, some similarities, or at least our sympathies, should be noted. We applaud a growing awareness of 'the mythic character of control' and are happy to see an end to the metaphor of mastery applied to science. We draw our own distinction between scientific practice and its resulting theories on the one hand, and theories *about* science, on the other. Until the 1960s the most influential theories about science have been varieties of foundationalist empiricism (usually lumped together under the label of positivism). Many of the most familiar criticisms of science are plausible only on the assumption that positivism is, in fact, an accurate account of science. Nowadays, the inadequacies of positivism as an account not just of science, but of knowledge in general, are widely known. For example, it cannot explain the enormous growth in scientific knowledge over the last four hundred years. And our book traces many of its inadequacies as a model of how theory in educational administration is to be construed. Indeed, it is against this background that we urge the adoption of a *postpositivist* view of science. Ironically, unless Barlosky thinks that science is *essentially* positivist, or essentially concerned with manipulation and control, it is hard to see, after all, why he regards the epistemology we use to attack positivism as an epistemic subtext to the metanarrative of mastery.

We find further room for agreement when Barlosky sides against certainty and predictability. We follow Popper in calling ourselves fallibilists, seeing all knowledge as provisional. Our main dissent is to use coherence criteria as a way of adjusting theory in the face of experience contrary to expectation (Evers and Lakomski 1991, pp. 36–38). We also follow Popper in accepting that a measure of unpredictability invests all social systems, even where they are assumed to be completely deterministic (a very brief account of Popper's 1950 argument, which uses Gödel's undecidability theorem, is given on pp. 207–208 of our book). Two important consequences of Popper's key argument for an Open Society are incorporated into our position. First, at the level of theory: although we think our theory is true, fallibilism implies that we might be mistaken. To discover the errors and improve the theory, coherentism counsels (i) fostering the social conditions for maintaining a *diversity* of viewpoints within a community (which, in the absence of any algorithm for generating knowledge is a vital aspect in the creation of new ideas) and (ii) promoting the conditions for criticism and the critical engagement of these viewpoints. Barlosky gets the first part right, but his approach seems to rule out the possibility of learning, in any epistemically progressive way, from the diversity thus promoted. However, failure of the positivist 'metanarrative' to account for the growth of scientific knowledge should not function as a general argument against the possibility of any epistemic progress. We think that our coherentism, informed by naturalistic accounts of how the human brain learns, is a more promising approach (see Chapter 9).

A second consequence of fallibilism affects policy and administrative practice. On policy, we argue against the synoptic tradition and for incrementalism and piecemeal reform. The major payoff for 'a practised sensitivity to the particular' is that it is easier to trace the consequences of applying a global theory if it is applied locally and piecemeal. And where error lurks, there is a premium on providing theoretically interpreted consequences as a source for feedback-driven coherent theory adjustment. On administrative practice, we respond to fallibility and uncertainty by championing organizational learning models (Evers and Lakomski 1991, pp. 187–89; see also Chapters 5 and 11 of this volume) and a view of leadership that is primarily educative, namely, one concerned to maintain the social relations of individual and collective learning. Again, we puzzle over Barlosky's assimilation of our new science perspective with the old science view of educational administration.

Among areas of common ground between our naturalistic coherentism and the postmodernism Barlosky endorses, we would include anti-essentialism and anti-foundationalism. We conclude our reply with some reflections on an important difference: our respective stands over representation.

In recent times the question of whether our theories (or perhaps the contents of our minds) mirror, or represent the way the world is, has been the centre of renewed interest, both in epistemology and philosophy of mind. Unfortunately, there is an influential strand of the debate that runs together evidence of representation with foundational evidence of representation (see Rorty 1979, pp.

165–212, on 'privileged representations'). When foundational epistemologies are rightly found wanting, it is wrongly assumed that the notion of representation also becomes problematical. However, we think some weak form of representationalism is required in order to explain how creatures learn to negotiate their environments, how they are able successfully to engage in certain practices. Consider the simple case of a bat, which is able to fly in darkness *between* the bars of a cage instead of crashing into them. From a naturalistic perspective we have the resources to hypothesize that this is achieved because the statistical structure of the bat's accumulated acoustical and tactile experience is somehow represented in a causally efficacious way; for example, by some physical arrangement in the bat's brain that causally maps the perception of sounds onto the guided flapping of its wings. In other words, the bat has a representational structure that contains *information*. The structure is able to select from among an arbitrary number of flight trajectories just those that make for a successful flight. Generalizing a bit, it is helpful to see learning as a process that increases the amount of information in a representational structure, and neural network models of cognition are beginning to reveal significant details concerning how real brains are able to extract patterns from among data (see Churchland and Sejnowski 1992; also Chapter 8 of this volume).

Much shaping of the representations physically embedded in the neural wetware of creatures occurs via feedback from experiences associated with behaviour. A more efficient way of learning some things is to develop methods of compressing information and coding it in a publicly accessible form (usually symbols) for processing by neural nets. Given a suitably long training period for nets to be able to recognize and causally process code, symbols are able to go proxy for some experiences. Ironically, the capacity for human brains to detach learning from non-symbolic experience also permits representational failure. For example, providing there is a statistical anchor of regular symbolic usage and its circumstances, successful lying can be sustained (Arendt 1973, pp. 11–13); and so can fiction; also, the postmodernist case against representation and, of course, our reply. But now it would appear that the causal operation of the human cognitive machinery that permits the case against representation to be formulated and communicated depends on the existence of a weakly representational neural structure.

If this argument is accepted, we can make some progress on the matter of epistemic progress. The epistemic advantage of a representation (or theory encoded in wetware) that guides the bat between the bar, is of a piece with the superiority of a theory that successfully guides a person from an office. Although this sense of advantage is modest, and might be reckoned as meagre knowledge, as with the case of objectivity we can work our way up towards more controversial and complex versions. Thus a person's representation of their environment will be used to plot a trajectory of practice. Where mismatches occur between the feedback of experience and the template of expectation, holistic adjustments are made — a change of tack here, a revised goal there, sometimes a major shift in direction. Such are the circumstances by which we learn to cope, solve problems, and improve our representations. Further experience, perhaps augmented by knowledge deriving from symbolic representations of other successful practices, accelerates the learning cycle. Within the last ten years, new developments in cognitive science have begun to yield naturalistic accounts of how representational structures, modeled as neural networks, can be improved. We are gratified to see that these improvements, which seem to result from the application of global soft constraints on theory change driven by the gap between network outputs and required outputs, may provide much needed detail on the physical operation of the coherence criteria we recommend for theory choice (see Churchland 1989, pp. 153-196).

A deep account of these matters is some way off, but we hope we have provided enough background to make plausible an epistemological framework that systematically sustains the enterprise of trying to improve administrative theory and practice, of seeking more adequate policy advice for the improvement of schools, and of trying to do better with administrator training programmes.

Barlosky, and other critics of positivist accounts of science, are right to be concerned about the epistemic closure exhibited by traditional conceptions of science. For they exclude much that is of value to educational administration. However, the postmodernism Barlosky advocates, with its treatment of knowledge as narrative, can lead equally to closure. It threatens to detach educational administration from the most powerful and fecund source of knowledge available today. If our research programme has merit, it will be to permit the possibility of advancing administrative theory and practice in education by drawing on all facets of human experience.

References

Arendt, H. (1973). *Crises of the Republic*. (Harmondsworth: Penguin Books).
Churchland P.M. (1989). *A Neurocomputational Perspective: The Nature of Mind and the Structure of Science*. (Cambridge, Mass.: MIT Press).
Churchland P.S. and Sejnowski T.J. (1992). *The Computational Brain*. (Cambridge, Mass.: MIT Press).
Evers C.W. and Lakomski G. (1991). *Knowing Educational Administration: Contemporary Methodological Controversies in Educational Administration Research*. (Oxford: Pergamon Press).
Greenfield, T.B. (1980). The man who comes back through the door in the wall: discovering truth, discovering self, discovering organisations, *Educational Administration Quarterly*, **16**(3), pp. 26-59.
Greenfield T.B. (1993). Conversations, in T.B. Greenfield and P. Ribbins *Greenfield on Educational Administration: Towards a Humane Science*. (London: Routledge).
Hawking S. (1988). *A Brief History of Time*. (New York: Bantam).
Kuhn T. (1970). *The Structure of Scientific Revolutions*. (Chicago: University of Chicago Press, second edition).

Perrow C. (1986). *Complex Organizations: A Critical Essay*. (New York: Random House).
Rorty R. (1979). *Philosophy and the Mirror of Nature*. (Princeton: Princeton University Press).
Rorty R. (1989). *Contingency, Irony, and Solidarity*. (Cambridge: Cambridge University Press).
Tversky A. and Kahneman D. (1981). The framing of decisions and the psychology of choice, *Science*, **211**, pp. 453–458.

22
A Rejoinder to Evers and Lakomski—*Martin Barlosky*

The demise of foundational epistemology, however, is often felt to leave a vacuum which needs to be filled.
— Richard Rorty, *Philosophy and the Mirror of Nature* (1979)

In their 1991 book and in their reply (Chapter 21) to my review of it (Chapter 20), Colin Evers and Gabriele Lakomski have provocatively addressed the epistemological issues that underlie the more pronounced divisions within the field of educational administration. In doing so, they have not only spoken to issues internal to educational administration, but have also offered a comprehensive vision of human understanding and meaning. In raising questions concerning the adequacy of their proposals in each area, I will make the general argument that coherentism supersedes the means, but not the ambitions, of the positivistic epistemologies it would replace.

Before offering my rejoinder, I would like to share an anecdote once told by T. B. Greenfield (1980, p. 28). The story succinctly illuminates the nature and the conditions of difference.

> Observing two women standing at windows atop their ancient houses and seeing them shout at each other across the narrow street that separated them, Sydney Smith, the eighteenth-century English essayist, is said to have commented. 'These women will never agree; they are arguing from different premises.'

On the one hand, the anecdote underlines the fact that the division between the two antagonists was a very real one. They were indeed arguing from entirely different and long-standing premises. On the other hand, it notes that the street dividing the disputants is narrow. Further, it implies that the importance of the issue being disputed is shared, and that the dwellings exist within a community.

Like the women in the anecdote, those who engage in debates pertinent to educational administration are also co-inhabitants of a community that retains its vivacity because of, rather than in spite of, the possibility of spirited difference.

And although my rejoinder is intended to emphasize points of difference, what value it may have derives solely from the claim I share with the authors (Evers 1994) that 'Theories about knowledge and its justification exert a powerful influence on both the content and structure of administrative theories.' Such theories become significant, in turn, because they condition what administrators regard as practicable, possible, and worthy of commitment. Administrative theory and its epistemic antecedents matter to a still broader audience because they condition how people who possess power will act and how they will seek to justify their actions in the social world they share with us.

Rejoinder

In what follows, I will pass over the more frivolous and rhetorical issues raised in the Evers and Lakomski reply (e.g. can fictional characters say true things? can 'greenness' and 'complexity' be treated as similar properties? what constitutes 'conceptual low ground' in educational debates?) in order to make two substantive arguments:

(1) that coherentism is committed to an essentially reductionist enterprise that brings it, perhaps inadvertently, to share the aspirations of positivist and foundationalist perspectives, and
(2) that coherentism's emphasis on epistemic commensurability misunderstands the significance of difference by undervaluing the distinction between what I shall call, borrowing from Rorty, 'normative' and 'non-normative' discourses.

Although the two arguments are intertwined, I will begin by examining the reductive character of the authors' coherentist epistemology. This reductive quality, which is characteristic of coherentism in general, can be seen most clearly in the quest to rephysicalize the social sciences and to naturalize epistemology. The Quinean promise of a grand physical synthesis of the social and the natural sciences is driven by the wholesale reduction of complex human phenomena to what Rorty (1979, p. 338) has called 'atoms-and-the-void' terms.

In their response the authors use the engaging language of neural 'wetware' convincingly to explain how a bat manoeuvres between the bars of a cage. But can we really expect to 'work our way up' from this example to unravel the complex web of intellectual and moral decisions that result in administrative action? In fact, the dynamic is not really one of 'working up' at all; it is simply a *working down* of human phenomena to the level of physiology and biochemistry. This reduction can only be held worthwhile by preserving a fallacy explicated by R. D. Laing (1970/1959, pp. 22–23):

> There is a common illusion that one somehow increases one's understanding of a person if one can translate a personal understanding of him into the impersonal terms of a sequence or system of *it-* processes ... It seems extraordinary that whereas the physical and biological sciences of it-processes have generally won the day against tendencies to personalise the world of things or to read human intentions into the animal world, an authentic science of persons has hardly got started by reason of the inveterate tendency to depersonalise or reify persons.

The attempt to read distinctively human qualities as coextensive with neuro-chemical wetware can have decidedly peculiar consequences. As Christopher Hodgkinson (1993, p. 181) has pointed out, this physicalist reduction leads us to the odd conclusion that 'Truth, Beauty, and Goodness are not teleological ideals but mere epiphenomena of sub-molecular activity.' That such a conclusion can be reached through the exercise of seemingly sensible and, by definition, coherent theory adjudicators does not make it any less aberrant. And although a physicalist reduction may provide a more ordered presentation of the otherwise idiosyncratic operations of the human personality, it leads to an impoverishment of meaning. In concluding his book on the origins of the universe, the Nobel Prize-winning physicist Steven Weinberg (1993, p. 154) has given incisive expression to the purposelessness that so often attends physicalism: 'The more the universe seems comprehensible, the more it also seems pointless.'

The wish to attenuate systematically the ambiguities associated with human qualities also has a perhaps unintended side effect: it brings the coherentist project to share the epistemic desires and the tactical ambitions of the positivistic theory movement. These epistemic desires are grounded in the hope for a normative framework that can render divergent discourses, in this case discourses about administration, commensurable. The tactical ambitions are formed by a belief that a suitably neutral basis can be found for the derivation of principles and methods sufficient for the regulation of individual and social conduct.

When these hopes are combined in an unqualified fashion, we have the beginnings of a revised administrative manifesto for the scientific management of human affairs. Despite its derivation from non-foundational superempirical epistemic criteria, such management remains problematic on at least two counts. First, its putative quality of objectivity would obviate the responsibility attending individual and collective choice. Second, it impinges on the capacity of the human imagination to re-configure peaceably both what is possible and what may be considered of value.

In my review I quoted a statement by the authors that bears repeating here, especially given their applause for the undoing of 'the mythic character of control' and 'an end to the metaphor of mastery applied to science.'

> [T]he task of philosophy, which is not sharply distinguished from but continuous with empirical science, is to find general principles common to

all sciences. This means extending the use of such principles even to the regulation of human conduct and the organization of society. The move to a postpositivist philosophy of science is quite compatible with such a view of the nature of science and its role in human affairs (Evers and Lakomski 1991, p. 219).

In asking how this compatibility comes to be, we can turn to Nicholas Maxwell's (1984, p. 61) analysis of the generic conceptual legerdemain at work here:

> [I]n essence what is being advocated is this. First, the social sciences need to develop improved theoretical knowledge of laws governing human behaviour and social systems. This knowledge then enables us to predict that if such and such human, social circumstances are realised, such and such will reliably be the outcome. As a result we are in a position to develop useful social technology. But this amounts quite simply to developing techniques of human social manipulation. Built into the very enterprise of the social sciences, conceived of in this way, is the ideal of developing more effective techniques for manipulating people.

The authors might argue that they differ from Maxwell on two points. In addition to replacing his implied determinant foundations with 'touchstones', they would likely wish to substitute 'organizational and behavioural regularities' for Maxwell's empirically verifiable 'laws'. But even if both points are conceded, coherentism retains the promise of mastery (i.e., the reduction of human behaviour to physical variables whose interactions may be predicted and regulated) that provided the often tacit rationale for the development — not to mention the funding — of the positivistic social sciences.

Coherentism, then, would provide a revised legitimating framework for the scientific regulation of human affairs — an enterprise that had philosophically foundered with the demise of empirical positivism. Although forsaking the foundationalist methodology of its positivist predecessors, coherentism shares their intent to develop an objective knowledge base through which the exercise of power can be authenticated. By acting in accordance with a knowledge base that would naturalize axiology as well as epistemology, those who exercise power would absolve themselves of individual choice and responsibility. Fortunately, as Maxwell (1984, p. 61) goes on to remark in a manner paralleling the quotations from MacIntyre and March in my review, 'The saving grace of this procedure is perhaps its ineffectiveness.'

The second substantive argument I will make entails a reconsideration of 'commensurability' as a philosophical value. For present purposes, commensurability may be understood as the hope that a framework can be found to resolve the differences that separate disparate epistemological perspectives. The workings of such a framework would effect a double reconciliation. The first part of this reconciliation would bring conflicting epistemological constructs within a common conceptual orientation. The second would operationally define 'representativeness' as the relationship between the resulting theory and the referent reality. That is, through subscribing to roughly the same 'meta-rules' for discourse, harmony among epistemological perspectives and a consensual mapping of the 'real' would simultaneously be accomplished.

The possibility and the value of commensurability are challenged by nascent postmodernist philosophical perspectives that locate their origin in the delegitimation of the master commensurating discourses referred to as 'metanarratives' (Lyotard 1984). In contradistinction to the coherentist reticence to partition knowledge claims, postmodernism would abandon the wish for epistemic constancy as an uninteresting ambition of dubious worth. This is because postmodernism views the knowledge that is lost through epistemic partitioning as not real knowledge at all, but rather as an artifact of a misleading conceptual pretense. Such knowledge is a dissimulation that would paper over the aspects of difference that sit uneasily within commensurating epistemic categories.

Rorty has developed this postmodernist viewpoint by describing what he calls normal ('normative') and abnormal ('non-normative') discourse. Borrowing from Kuhn's concept of normal science, Rorty defines normal or normative discourse as inquiry that occurs within the limits of conventional knowledge categories. Abnormal or non-normative discourse sets aside those conventions and begins to talk in new ways about issues that do not fit within the existing network of presuppositions and methods. Normative discourse allows us to reduce the unknown to the familiar; it sustains the conceptual incorporation of diverse experiences through the exercise of our existing vocabulary. Non-normative discourse occurs when this incorporation is frustrated by that which is discordant and elusive. The instances that give rise to non-normative discourse ask 'whether we do not need to change our vocabulary and not just our assertions' (Rorty 1979, p. 353).

The construction of commensurating epistemologies is the signal activity of normative discourse; the contravention of commensurating categories is indicative of non-normative discourse. The former is preoccupied with developing epistemic constraints sufficient for the task of conceptual resolution, which in its ideal type is represented by grand unifying theory. It is concerned with closure. The latter is concerned with creating an openness amenable to new ways of describing experience and expanding what is conceivable. It matters little if the means to a synthetic normative discourse are located in a search for foundations, benchmarks, or touchstones. All are about the same work; all would find the set of constraints capable of effecting commensurability.

This being said, it was the authors and not myself who introduced the term 'benchmark'. As I pointed out in my review, they did so on page 229 of their text in the course of discussing coherentism's claimed retention of objectivity. Although this use of an epistemologically loaded term may have been an oversight that is subsequently mollified by a reiteration of the adjudicatory effectiveness of coherence criteria, its appearance is not inconsequential. Rather, it seems something of a Cartesian slip that reveals the desire to re-establish legitimacy for a reductive commensurability thought lost with foundationalist positivism. To borrow from Lakomski (1987, pp. 115 ff., Evers and Lakomski 1991, pp. 112 ff.), when the Geertzian explanation of 'turtles all the way down' is regarded as epistemologically unsatisfactory, we can expect the arrival of a surrogate that can do the legitimating work once performed by an obsolesced foundationalism. That is, if we continue to seek commensuration as the epistemic *sine qua non*, we will continue to invent successive machines to produce what has so far proven to be an unachievable product.

As I have suggested in my review, an abiding sense of philosophical insecurity combined with a lasting attachment to mastery may prohibit us from reconsidering the value of this attempt. Both factors can make it difficult to entertain the possibility that the equation of understanding with commensurability — and not their dissociation — might condemn us to a philosophical quagmire. As long as we understand philosophical progress as coextensive with the movement toward the double sense of commensurability discussed above, we may be diverted from more interesting and more productive work. At the risk of raising the authors' eyebrows once again, I will make reference to a parabolic synopsis of the Borges fiction 'Del Rigor en la ciencia' presented by Lyotard (1984, p. 55) as an illustration of this point:

> An emperor wishes to have a perfectly accurate map of the empire made. The project leads the country to ruin — the entire population devotes all its energy to cartography.

There is a final point to be made concerning commensurability. The possibility that normative and non-normative discourses are inherently incommensurable does not mean that we are in an either/or situation. That is, we are not required to dismiss one in order to salvage the other. Nor are we required to diminish the value of or to abandon 'conversation' between the two. Although they are different in substance, normative and non-normative discourses about educational administration are complementary rather than mutually annulling. This complementarity implies that, when held in creative tension, divergent forms of discourse can bring interest, knowledge, and perhaps wisdom to what would otherwise be an arid and wrong-headed technical enterprise of mechanistic administration.

In this regard, I would like to share Colin Evers' (1994) words from a letter he sent me together with the advance copy of the response manuscript. In it, he perceptively cautioned:

> Just as postmodernism sees science as narrative, so science returns the compliment by seeing narrative (reflexively) as information.

And indeed this is so. As non-normative discourse seeks to refract and to open the vocabulary of normative discourse, so normative discourse seeks to regard these refractions and openings as grist for new commensurating assertions. Like Evers, Rorty (1979, p. 318) has given similar expression to this interrelationship:

> For epistemology, conversation is implicit inquiry. For hermeneutics, inquiry is routine conversation.

Where Evers would advance the inquiries of science, Rorty would enrich the narrative of hermeneutics. And so the conversation that is human discourse continues by finding new ways to address an existential situation which providentially frustrates reduction to the predictable.

I will end with a final quotation from Rorty in which he expresses how an expansive understanding of human purpose and possibility is tied to the ability to continue a conversation that benefits from difference and from the paradoxical value of incommensurability:

> To see keeping a conversation going as a sufficient aim of philosophy, to see wisdom as consisting in the ability to sustain a conversation, is to see human beings as generators of new descriptions rather than beings one hopes to be able to describe accurately (Rorty 1979, p. 378).

May it be that those concerned with educational administration have the patience, the wit, and the wisdom to continue a sometimes troubling conversation and to elude the pre-emptive epistemic resolutions that are ultimately of less significance than the many questions waiting to be asked. As we learn to converse amiably rather than to shout across the divisions that separate our conceptual premises, perhaps we can come to appreciate the implications of William Blake's (1967/1793, pp. 254, 262) coincident aphorisms: 'Every thing possible to be believed is an image of the truth', and 'Opposition is true friendship'.

References

Blake W. (1967/1793). *The Marriage of Heaven and Hell*, in *The Portable Blake*. (New York: Viking).
Evers C. W. 1994. Private correspondence.
Evers C. W. and Lakomski G. (1991). *Knowing Educational Administration: Contemporary Methodological Controversies in Educational Administration Research*. (Oxford: Pergamon Press).
Greenfield T. B. (1980). The man who comes back through the door in the wall: discovering truth, discovering self, discovering organizations, *Educational Administration Quarterly*, 16(3), pp. 27–59.
Hodgkinson C. (1993) The epistemological axiology of Evers and Lakomski: some un-Quineian quibblings. *Educational Management and Administration*, 21(3), pp. 177–184.
Lakomski G. (1987). The cultural perspective in educational administration, in R. J. S. Macpherson (ed.) *Ways and Meanings of Research in Educational Administration*. (Armidale, Australia: University of New England Press).
Laing R. D. (1970/1959). *The Divided Self: An Existential Study in Sanity and Madness*. (Baltimore: Penguin).

Lyotard J.-F. (1984). *The Postmodern Condition: A Report on Knowledge*. Minneapolis: University of Minnesota Press, translated by G. Bennington and B. Massumi).
Maxwell N. 1984. *From Knowledge to Wisdom: A Resolution in the Aims and Methods of Science*. (Oxford: Blackwell).
Midgley M. (1989). *Wisdom, Information, and Wonder: What is Knowledge For?* (London: Routledge).
Rorty R. (1979). *Philosophy and the Mirror of Nature*. (Princeton: Princeton University Press).
Weinberg S. (1993). *The First Three Minutes: A Modern View of the Origin of the Universe*. (New York: Basic Books, second edition).

Conclusion: Doing Educational Administration — Preliminary Reflections on a Theory of Practice

In a book primarily concerned with the *exploration* of a set of ideas, any conclusion may seem out of place, or at least premature. What has gone before is very much work in progress. On some matters we will be obliged to change our minds simply as a result of further developments in our research program; for example, where it articulates with the constantly shifting frontier of cognitive science. On other matters, we have benefited from the very generous and detailed comments of our critics who have challenged features of our naturalistic coherentist program, ranging from its theoretical integrity through to its applicability to the real world of educational administration. With so much remaining to be done it therefore seemed appropriate for us not to tie up loose ends, or attempt a final word on any number of controversial issues, but to give some indication of how we see the task ahead.

Despite scope for ongoing debate over our theoretical position, we see the main line of development for our work as providing a systematic framework for approaching problems and issues in administrative practice; matters such as leadership, administrator training, decision-making, agency and structure, organizational design, planning and policy, ethics, power, culture, and organizational learning. The foregoing explorations give an indication of where we stand on some of these aspects of administrative studies. In this concluding chapter we would like to sketch a number of key elements that govern our approach to practice.

Epistemology, Social Theory and Practice

If the quest for context-invariant features of reality is a legitimate goal for a natural science such as physics then the same can hardly be said for most work in the social sciences. For when it comes to specifying generalizations in the social realm, particularities of context and circumstance crowd in. This point is freely acknowledged by all theorists, some even elevating it to the status of a purported partition between natural and social science. While we disagree with attempts to partition knowledge, we have been insistent on acknowledging certain limits to what can be known. When all knowledge is reckoned to be fallible, the quest for certainty needs to be replaced with the task of improving our theories, of correcting errors when they are discovered.

On the problem of improving our social theories, we accept two points made by Karl Popper many years ago. The first is a limitation result and claims that even for a totally deterministic physical system, where its future states are affected by the growth of knowledge, some of these states will be in principle unpredictable. (See Evers and Lakomski, 1991, pp. 207–208). Now, since the development of social systems over time is affected by new knowledge — for example, new discoveries in science or medicine being applied in technology, or new developments in economic theory being applied in social policy — our knowledge of the future behaviour of such systems is limited. Yet for all sorts of purposes, we need to make assumptions about what will happen; it is just that the assumptions are not known to be reliable. As a consequence, it is important to have an error discovery and elimination procedure that can be applied within the constraints of real social practices. Popper's second point addresses this problem. Essentially, any knowledge to be gained from a study of the consequences of theory implementation is likely to be lost if the full complexity of social causation has to be taken into account. That is, trying to improve theorizing of system-wide, or global change, *by examining the consequences of system-wide, or global implementations of theory* is methodologically too ambitious. Learning from social change is more likely to occur when that change is conducted on a small-scale, or piecemeal basis; where there are fewer variables to keep track of and less reliance is being placed on the accuracy of hypothesized law-like statements about the behaviour of an entire social system. The equivalent thesis in administrative and policy studies was advanced, as everyone knows, by Charles Lindblom.

Notice that accepting that practice should make possible the improvement of theory is tantamount to accepting a particular set of normative constraints on possible relations between theory and practice. Of course, if one is convinced that a social theory, or comprehensive set of policies, is absolutely correct then the learning constraint will have little purchase. Abundant evidence against the plausibility of this conviction, however, can usually be found in the historical ubiquity of error.

A more serious theoretical worry with the learning constraint can be found in the nature of theory itself. Because we use a finite vocabulary to refer, describe, and classify an infinite number of things around us, even quite ordinary, humdrum terms like 'chair', 'table', or 'freezer', are *general* terms satisfied by arbitrarily large classes of objects. Similarly, the terminology applicable to local and particular phenomena, the natural subject matter of piecemeal change, is mostly quite general, extending well beyond the boundaries of localized practice. Now while it is always possible to tag the particular, generality in theorizing is much prized, especially in the logical empiricist tradition. And so it should be, but not uncritically. Our point is that the press for generality in theory needs to be traded off against its relevance for practice. Abstract characterizations of practice, easily indulged in through the use of linguistic representations of knowledge, can be misleading in important ways when it comes to theorizing about the social. As a step towards correcting this bias, we emphasize a familiar insight that social practices, including learning, are accomplished by cognitively limited agents. Herbert Simon's work traces one important strand of this insight into its ramifications for decision-making. The strand we explore is an implication of a brilliant paper by Viviane Robinson (1994) that raises the possibility that many concepts that refer to the results of human cognition may be sufficiently localized in practice to bring about a fragmentation of these concepts.

Cognitive Constraints and Theoretical Fragmentation

Concerns about error and the conditions under which it can be corrected, imply that we should adopt caution in our theorizing of the social. Despite this limitation, the goal of achieving great generality still remains. Using our naturalistic epistemology perspective, however, we wish to urge some limitations on this goal. In what follows, we shall suggest that the cognitive conditions of practice, and in particular the Robinson constraint of limited cognition, make it plausible to posit the theoretical fragmentation of certain central concepts in administrative theory, to the extent that they are applicable only in local, or context-specific circumstances of practice.

To begin with an example from natural science, consider the concept of temperature, which flourishes in ordinary usage as, perhaps, 'intensity of heat and cold'. With a little ingenuity, and taking advantage of the expansion of some liquids when heated, modest apparatus will permit the construction of a simple thermometer that can be used to augment qualitative and subjective perceptions of temperature with reproducibly quantitative values. Although a theoretical instrument, we may even use the thermometer to revise the evidence of our inner feelings, especially if we know we are ill with a fever, or have been recently exercising. In this way, the concept enlarges, becoming richer in its inferential relations in our theory of the world. However, as Paul Churchland (1989, p. 285) points out, further very sophisticated theorizing results in temperature coming apart into three distinct concepts: mean molecular kinetic energy for gases; mean maximum molecular kinetic energy for a classical solid; 'and in a vacuum it is a specific wavelength distribution among the electromagnetic waves coursing through the vacuum'. Each concept is applicable only to its particular domain. This means that the enterprise of developing the most coherent systematic global theory, such as that to which physics aspires, needs to be construed as being compatible with the possibility that such a theory, or web of belief, could turn out to have a fairly modular structure rather than a top down generality-to-particularity pyramidal design as envisaged by logical empiricists.

Consider now the case of social competence. At the functional level of description, generality is easy to come by. It can mean skill at performing adeptly as the social occasion demands. But if social occasions do not have any particular essence, or universal common feature, if at best they disaggregate into different and distinct patterns of human interaction, then at the causal level, where learning can be found, the concept will come apart in ways that reflect the cognitive skills of the agent. Thus one may be socially competent when it comes to eating in restaurants or attending the theatre, and inept at hosting a small dinner party. As with games, which also have no essence or common feature, it may be more meaningful to speak of skill in particular games, or kinds of games, rather than some generic skill.

The basic strategy here is to resist the pull of top-down theory-driven generalization by utilising bottom-up considerations of particularity and circumstance. Where concepts enjoy close inferential relations with assumptions about human cognitive performance, in our view a bottom-up approach will be of more value in understanding administrative practices than the usual top-down functionalist taxonomies that abstract from causal detail.

To see how this might be so, consider two concepts that have figured prominently in theorizing about administrative practices: power and leadership. Whatever functional homogeneity attaches to each concept, there are good reasons for thinking that each is instantiated in practice in causally heterogeneous ways. In the case of power, there are any number of candidates for explaining why a social system has certain posited characteristics: the influence of ideology, ownership of the means of production, systematically distorted speech situations, the distribution of ability in the case of a meritocracy, and so on. However, regardless of the big picture, at some point in the detail, power will be mediated through the cognitive activity of agents; events, actions and happenings need to elicit responses of a relatively stable, predictable and coordinated kind for influence to be exercised in a non-random, purposeful fashion. The problem is that there do not appear to be common causal conditions for the exercise of power in such disparate contexts as classrooms, school staff rooms, boardrooms, bureaucracies, retail outlets, academic departments, research labs, or social clubs. Role theory can achieve some generality, but again at the expense of abstracting from relevant causal differences, and in any case, we know that particularity breaks out over the question of how roles are interpreted by agents whose interpretations make the role possible.

The belief that a theory of power can do useful explanatory work in advance of research on the microstructure of effective action seems to us to reflect an unwarranted confidence in essentialist assumptions. (We have raised a similar concern about the empirical content of systems theory in Chapters 1 and 4.) Waiving these assumptions means a shift in emphasis away from general theories of power and towards specific analyses of situated examples of effective action. We think this is cognitively more realistic since, with the possible exception of academics, an agent's theory in use is more likely to be attuned to the specificities of the context of action than to society-wide ramifications. If this advice is heeded, the concept of power will give way to the more fruitful but theoretically fragmented concept of effective action, with conditions for effectiveness varying from context to context, and with no expectation that anything common can be found in all such contexts that is responsible, or partly so, for that effectiveness.

While it might be premature to suggest that the quest for an essence to leadership has run into the sands, or that there is no causally homogeneous concept, or even that the structure of leadership events is not isomorphic across different material contexts, this research program's results have not been promising. (Gronn 1995, 1996). Management theory which is shaped by logical empiricist canons of theory construction can be tempted to discount the bad news here because of the theoretical premium on generic notions of leadership. Nevertheless, a practice-oriented approach will be sensitive to the relevant constraints on cognition. Thus, consider the case of a school principal responding to parent demands for better learning outcomes in the context of other senior staff pressing for the aggressive recruitment of a certain number of children with behaviour problems, in the name of social justice. Scope for the principal's action is significantly constrained by how other staff see matters and respond, which is in turn a function of their theories. If we see a person's theory as a kind of map by which they steer, that selects salient patterns from the passing flux of experience and causes characteristic responses to be initiated, and if we see coordinated group responses as a resultant net effect of people responding to each other in the interpreted circumstances, then the exercise of leadership is very much an outcome of a new resultant that includes the principal. That is, scope for leadership is causally dependent on patterns of issue-specific followership. A singularly strong attachment to the social justice condition can force the principal into reconceiving the entire problem if gains on behalf of aggrieved parents are to be made. Since some issues are more easily dealt with than others, success in leadership will be sensitive to variation of issue, or problem.

Once the relevance of cognitive specifics is factored in, a practice perspective favours relational rather than intrinsic (essentialist) analyses of social phenomena. This is not to say that intrinsic features are not at work, or are not relevant, but only that a theoretically unified account of disparate phenomena is something to emerge rather than something that is given as a starting point.

Knowledge Representation, Concepts and Practice

The foregoing reflections indicate some consequences of adopting a naturalistic, bottom-up framework to the understanding of administrative practice in contrast to a top-down context invariant perspective. In this concluding section we extend a little the naturalism given in earlier chapters to sketch the cognitive basics of what lies behind our concern adequately to theorize the particularity of practice.

Our central point has been that linguistic representations of knowledge mislead when it comes to understanding the causal dynamics of knowledge acquisition and theory-driven behaviour. The first lapse, which we dwelt on in Chapter 3, is that language is a very poor medium for representing practical knowledge, and we indicated our preference for neural network models since they can deal with both symbolic and non-symbolic representations. The second lapse, which we have been drawing attention to here is that the dominant theory of concepts in educational administration is language-oriented, and that this produces a bias towards context-free generalities. An alternative account of concepts that we would like to explore is the theory of prototypes, especially as located in neural network representations of knowledge. Because prototypes are more context-specific, and mesh well with the activity of learning from practice, we think that theorizing concepts as prototypes will lead to a more fruitful understanding of many of the most important terms in administration.

Although it is out of place to outline our theory of prototypes here, two important aspects can be noted. First, in common with other accounts, a 'prototype may be thought of as the central tendency of feature values across all valid members of the category' (Sternberg and Horvath, 1995, p. 10). Second, our account, because it draws on neural network models of representation, cashes out talk of 'feature values', 'central tendency', and the like in terms of model equivalents, such as characteristic patterns of activation across hidden layers of neurons. But as can be seen from the examples in Chapter 9, the capacity to recognize similarities among clusters of features is built up from the bottom through feedback-driven experience in localized contexts. Human knowledge of concepts of practice construed as prototypes has consequences for how professional training might be conducted, and also how practical competence might be understood — but these are developments that lie in store.

Naturalism in epistemology claims that theories of knowledge should be informed by our best natural science, particularly that branch of science that deals with human cognition. This perspective leads to a shift to holism in epistemology, and to coherentist procedures for justifying knowledge. As we have seen, the consequences of coherentism in administrative theory are extensive. Many of the explorations in this book seek to extend our naturalistic epistemology into areas to do with the representation of knowledge, noting limits to traditional ways of theorizing matters of practice. Our developing work on the nature of practice is aimed at extending our program to include the naturalizing of practical concepts in all their contextual particularity. The result promises to be no less than a unified account of theory and practice.

References

Churchland, P.M. (1989) *A Neurocomputational Perspective*. (Cambridge, Mass.: MIT Press).
Evers, C.W. and Lakomski, G (1991). *Knowing Educational Administration: Contemporary Methodological Controversies in Educational Administration Research*. (Oxford: Pergamon Press).
Gronn, P.C. (1995). Greatness re-visited: the current obsession with transformational leadership. *Leading and Managing*, 1(1), pp. 14–27.

Gronn, P.C. (1996) From transactions to transformations: a new world order in the study of leadership? *Educational Management and Administration*, **24**(1), pp.7–30.

Robinson, V.M.J. (1994) The practical promise of critical research in educational administration. *Educational Administration Quarterly*, **30**(1), pp. 56–76.

Sternberg, R.J. and Horvath, J.A. (1995) A prototype view of expert teaching. *Educational Researcher*, **24**(6), pp. 9–17.

Author Index

Abrahamsen, A. 35, 104, 111, 125–6, 138
Achinstein, P. 17
Adorno, T.W. 216, 217, 223
Alexander, J.C. 170
Arendt, H. 268
Argyris, C. 26, 39, 57, 63–5, 82–3, 105, 152
Armstrong, D.M. 93
Arrow, K.J. 154–63
Austin-Broos, D.J. 193–4
Avolio, B.J. 73–5, 76, 77, 79, 80

Bachelard, G. 224
Barker, F. 17
Barlosky, M. 247–60, 271–7
Barnard, C. 41
Barrett, R.B. 239
Bass, B.M. 73–5, 76, 77, 79, 80
Bates, R.J. 6, 23–4, 189–97, 213–14
Baudrillard, J. 223
Baxt, W.G. 110–11
Beare, H. 71
Bechtel, W. 35, 104, 111, 125–6, 138
Beer, S. 57
Begley, P.T. 73
Best, R. 182, 185
Bidwell, C.E. 48–9, 52
Bigum, C. 9
Black, D. 159
Blair, B.G. 9
Blair, D.H. 156, 157, 158, 160
Blake, W. 277
BonJour, L. 20, 94, 243
Boorse, C. 230
Borges, J.L. 255
Boyan, N.J. 171
Boyd, W.L. 87
Bozeman, B. 87
Branson, J. 194
Braybrooke, D. 59, 87
Bruner, J. 255–6
Bucuvalas, J. 90
Burnheim, J. 69
Burns, J.M. 72–3, 77
Burrell, G. 42, 46

Caldwell, B.J. 71
Callahan, D. 87
Campbell, D.T. 89, 95
Campbell, R.F. 42, 52, 165
Carnap, R. 218–19
Carr, E.H. 184
Chaitin, G. 134
Chapman, J.D. 59
Chisolm, G.B. 105
Chomsky, N. 235
Churchland, P.M. xvi, 18, 36, 82, 88, 94, 95–6, 98, 99, 107, 111, 121–3, 125, 137, 239, 269, 281
Churchland, P.S. 19, 80, 81, 82, 88, 96, 97, 98, 99, 120, 239, 268
Cibulka, J.G. 252
Clark, A. 137
Clegg, S.R. 257–8
Cohen, M.D. 43
Conger, J.A. 73
Conway, J.A. 58, 59
Cornman, J.W. 230–1
Correa, H. 106
Cousins, J.B. 73, 75
Culbertson, J.A. 16
Cusick, P.A. 168

Dancy, J. 230, 231
Davidson, D. 239
Davson-Galle, P. 223
Deleuze, G. 223
Dennett, D. 134
Devander, C.L. 77
Dewey, J. 31, 39, 165, 173
Dews, P. 223
Dickson, W.J. 41
Dilman, I. 226, 227
Dolmage, W.R. 247, 248
Duignan, P.A. 39, 216
Dunn, W.N. 87, 88–95

Ecker, G. 42
Eco, E. 254, 259
Eisner, E.W. 216
Estler, S. 104

AUTHOR INDEX

Feigl, H. 1, 17, 19, 22, 30
Feyerabend, P.K. 16, 17, 249
Fischer, F. 87
Fitzgerald, E. 199
Fodor, J.A. 97
Forester, J. 87
Foster, W.P. 6, 23, 105, 148
Foti, F.J. 77

Gaddes, W.H. 118
Garson, G.D. 16, 87
Gathorne-Hardy, J. 181
Geertz, C. 193
Gerth, H.H. 227
Getzels, J.W. 52
Gibson, R.F. 239
Giddens, A. 16
Gill, J. 92
Glaser, B.G. 168
Goodin, R.E. 66
Graff, O.B. 165
Grant, G. 256
Green, B. 9
Greenfield, T.B. 7, 8, 10, 16, 17–18, 31, 42, 129–40, 176–87, 211–12, 216, 233, 234, 240, 241, 253, 254–5, 264, 266, 271
Gregg, R.T. 165
Grice, H.P. 226, 227
Griffiths, D.E. xiii, 1, 2, 3, 9, 15–16, 22, 29, 48, 131, 176
Gronn, P.C. 38, 42, 71, 73, 77, 129, 176–87, 181, 211–12, 283
Grossberg, S. 124
Guba, E.G. 52, 88

Habermas, J. 6, 23–5, 148, 190–3, 196–7, 213–14
Hacking, I. 96
Hahn, L.E. 239
Hallinger, P. 71, 75
Halpin, A.W. 21, 34
Hargreaves, A. 9, 253
Harmon, M.M. 42, 47–8
Harris, W.T. 165
Harsanyi, J.C. 106
Hart, A.W. 9
Harvey, D. 253
Hawking, S. 254, 265–6
Heap, S.H. 106
Hebb, D. 119–20
Held, D. 166
Hempel, C.G. 19, 22, 131
Hesse, M. 20, 24
Hintikka, J. 239
Hinton, G.E. 110, 113
Hirst, P.H. 117, 127

Hodgkinson, C. 5, 29, 33, 38, 105, 142–52, 177–9, 199–210, 212–13, 216, 225, 241, 247, 267, 273
Holmes, M. 247
Honey, J.R.de S. 181
Hooker, C.A. 18, 22, 100
Horvath, J.A. 284
House, R.J. 72
Howe, W. 77
Hoy, W.K. 1, 2, 42, 50–4, 79, 132, 170
Hume, D. 228–30
Hunt, J.G. 72, 73, 77

Jameson, F. 253
Jantzi, D. 73
Jennings, B. 87

Kahn, R.L. 47, 49, 50
Kahneman, D. 265
Kandel, E.R. 119, 120
Kant, I. 7
Kanungo, R.N. 73
Katz, D. 47, 49, 50
Koch, Ch. 82, 98
Kuhn, T.S. 16, 17, 31, 240, 249, 263

Laing, R.D. 272–3
Lakatos, I. 221–2
Lancaster, K. 67
Landsbergen, D. 87
Lasswell, H.D. 87
Lawrence, P.R. 48
Leithwood, K. 73, 75–7, 79, 80
Lepsius, M.R. 73
Lerner, D. 87
Lincoln, Y.S. 88
Lindblom, C.E. 59, 60, 87
Lipham, J.M. 52
Lipsey, R.G. 67
Litterer, J.A. 46–7, 52
Locke, E. 58
Lord, R.G. 77
Lorsch, J.W. 48
Lycan, W.G. 20, 183–4
Lyotard, J.-F. 223, 256, 258, 275, 276

McClelland, J.L. xvi, 35, 98, 104, 109, 111, 123–4, 137
McDermott, J. 224
MacIntyre, A. 256–7
McKee, M. 68
McNamara, J.F. 105
Macpherson, R.J.S. 39, 216
MacRae, D. 87
Maddock, T.H. 6, 10, 215–36, 238–9, 242–6
March, J.G. 43, 257
Margolis, J. 217, 220, 230–1, 232, 235
Mason, R.O. 91

Mawhinney, H.B. 252
Maxcy, S.J. 241
Maxwell, N. 274
Mayer, R.T. 42, 47–8
Meyer, M.W. 50
Miller, D. 194
Millikan, R.H. 71
Mills, C.W. 227
Minsky, M. 110
Mishan, E.J. 68
Miskel, C.G. 1, 2, 42, 50–4, 79, 132, 170
Mitchell, D.E. 87
Mitroff, I.I. 91
Monahan, W.G. 165
Moore, H.A. 1, 21
Morgan, G. 42, 46
Mueller, D.C. 156–9, 161, 163
Murphy, J. 71
Musgrave, A. 221
Musil, R. 224

Newell, A. 106, 107
Newton-Smith, W.H. 229
Ng, Y.K. 68
Nietzsche, F. 199
Nisbett, R.E. 80, 97

Ogawa, R. 42
Ohm, R.E. 165

Papert, S. 110
Paquette, J. 252
Perrow, C. 42, 253, 265
Peters, R.S. 117–18, 127
Peterson, M.F. 77
Phillips, D.C. 16
Plato 37–8, 104, 142–4
Plott, Ch.R. 155, 156–7, 159, 160–1
Pollak, R.A. 156, 157, 158, 160
Popper, K.R. 19, 20, 39, 57, 60–1, 95, 221, 230, 280
Putnam, H. 213

Quine, W.V. 18, 20, 21, 23, 38, 60, 88, 202, 213, 217–21, 223–4, 226–8, 230–3, 235, 249

Rawls, J. 151
Ribbins, P. 7, 8, 176–87, 211–12, 240, 241
Rizvi, F. 33
Robinson, V. 83, 281
Roethlisberger, F.J. 41
Rorty, R. 260, 263, 267–8, 272, 275, 277
Rosenberg, C.R. 35, 99, 111
Rosenblatt, F. 109
Roth, P.A. 218, 230, 231–2
Rowan, B. 75
Rumelhart, D.E. xvi, 35, 98, 104, 109, 110, 111, 123–4, 137
Ryan, J. 254, 255

Scheffler, I. 116, 230, 234
Schilpp, P.A. 239
Schoderbek, P.P. 62
Schön, D. 26, 39, 57, 63–4, 82–3, 105, 152
Schwartz, J.H. 119
Schweiger, O. 58
Scott, W.R. 41, 43, 46, 47
Sejnowski, T.J. 35, 82, 98, 99, 111, 268
Sen, A.K. 156, 157, 158
Simon, H.A. 25–6, 37, 48, 49–50, 57, 59–60, 62, 65, 66, 105, 106, 107, 142, 144–8, 216
Singer, P. 151
Slovic, P. 106
Smelser, N.J. 169, 172
Smith, E.E. 108
Smith, P.B. 77
Steinbach, R. 73
Sternberg, R.J. 284
Stich, S. 88, 96, 97, 107
Strank, R.H.D. 57, 61
Strauss, A.L. 168
Strawson, P.F. 226, 227
Street, C.M. 165
Sugden, R. 67, 68

Taylor, F.W. 41, 48
Thelen, H.A. 52
Toulmin, S. 88, 90, 91
Tversky, A. 265

Ullian, J.S. 18, 38, 88, 232

von Bertalanffy, L. 43–5, 54

Walker, J.C. 30, 31, 33, 66, 69, 88, 94, 211, 215, 217, 222, 230, 238, 240, 241, 242
Wallerstein, I. 166
Watanabe, S. 61
Watkins, P.E. 191
Weber, M. 48, 73, 130, 133
Weick, K.E. 43
Weinberg, S. 273
Weiss, C.H. 90
West, E.G. 68
Wildavsky, A. 87
Wilenski, P. 66
Williams, M. 18, 93, 94
Williams, R.J. 110
Willower, D.J. 16, 165–74, 211, 215–16, 233
Wilson, T.D. 80, 97
Wittgenstein, L. 226–8
Wonham, W.M. 61

Yukl, G.A. 73, 77

Zuckerman, H. 172

Subject Index

Administration theory
   developments 1–12, 15–16, 129–31
   educational theory congruence 32–4
   moral perspective 5–7, 37–9
Administrative Behaviour 25–6
AI *see* artificial intelligence
Analytic-synthetic distinction 219–20
Anthropology, cultural perspective 193–5
Artificial intelligence (AI) 107, 109–13
Autonomy, personal 33
Axiology, epistemological 199–210
Axiomatic social choice theory 155

Behaviour
   organizations 52–3
   rational 117–18
Behavioural studies 41
Belief-desire theory 107
Benchmark observation 251–2, 264, 275
Bounded rationality 106
Brain processes 118–20

Certainty 247–61
Chaos theory 8–9
Cognition 104–6, 142–6, 281–3
Coherentism 10–12, 15, 32–9
   authors' defence 243–4, 264–9
   Barlosky's view 248–60, 271–7
   epistemology 11, 18–21
   Maddock's view 222–5, 243–4
   scientific approach 11
   Willower's view 166–7
Commensurability 274–6
Compression algorithms 134–6
Connectionism *see* parallel distributed processing
Control theory 61–3
Critical theory 6–7, 214
   Bates' defence 189–93, 196–7
   critique 23–5
Cultural perspective 167–9, 193–6
   Sergiovanni 195

Decision-making
   decentralized 57–8
   errors 59–61
   feedback loops 39, 60, 61–3
   impossibility theorem 154–63
   means-end relationship 147–8
   moral 5
   neural network application 112–13
   organizational learning 63–5
   participative 58–9
   rational 145
   relevant information 60
   Simon's theory 25–7, 37, 144–8
   theory 105–7, 144–8
   top-down approach 59
Disjointed incrementalism 60
Durable agency 230

Education, epistemology relationship 116–18
Educational theory, administrative theory congruence 32–4
Efficiency
   educational organizations 65–8
   ethics 5
Empirical adequacy 250–1
   leadership 79
Empiricism
   classical 20
   logical 1–4, 17, 19–20, 22–3, 30, 132, 240–2
   weaknesses 31
   Willower's view 166–7
Entropy, systems 47
Epistemology 16–18, 30, 172–3
   axiology 199–210
   coherentist 18–21
   conclusions 279–81
   education relationship 116–18
   foundationalist 15
   moral 150–2
   naturalism 242, 249–50
   Theory Movement 130–1
Equifinality, systems 47–8, 49

SUBJECT INDEX

Errors
  decision-making 26, 57, 59–61
  judgement 61
Ethical theory 37–9, 173–4
  authors' defence 244–5
  Hodgkinson's views 200–5
  justification 230
  see also values
Extensionalism 219

Fallibilism 10, 39, 266–7
Falsificationism 221–2
Feedback loops
  decision-making 39, 60, 61–3
  error correction 26
  organizations 26, 63–5, 105
  systems theory 50
Field research 167–9
Folk psychology 88–9, 95–6, 100
Foundationalism 15, 21, 217, 230–3

General systems theory *see* systems theory

Hierarchy, organizations 57, 105
Holism 209
  Lakatosian 221
  Quinean 220
  systems 46
Human Relations Movement 41
Hypothetico-deductive structure 1–2, 17, 20, 31

Impossibility theorem 154–63
Instrumental reason 146–50
Is/ought dichotomy 201–2, 225–6
Isomorphism
  knowledge theory 143, 152
  systems theory 54

Job satisfaction, participative decision-making 58
Journal of Speculative Philosophy 165
Judgement, errors 61
Justified true belief 88, 93–5, 115

Knowing Educational Administration
  authors' defence 211–14, 238–46, 262–9
  Barlosky's review 247–60
  Bates' commentary 189–97, 213–14
  Gronn and Ribbins' commentary 176–87, 211–12
  Hodgkinson's commentary 199–210, 212–13
  Willower's commentary 165–74, 211
Knowledge
  accessing 97
  acquisition 35–7, 81–2

conclusions 283–4
decision-making 59–61
justification 18–19
justified true belief 88, 93–5
loss 263
procedural 116
transactional model 89–91

Language 226–7
  acquisition 97
  learning 123–4
Leadership 71–84
  charismatic 73, 75–6, 79
  critique of theories 77–81
  educative model 39
  empirical adequacy 79
  four Is 73–4
  models 77–8
  plausibility judgements 80–1
  theories 77–81, 143–4
  top-down 71–2
  transactional (TA) 72–7
  transformational (TF) 71–7
Learning
  human 82
  language 123–4
  neural networks 120–7, 137–8
  organizational 26, 63–5, 81–4
Linguistic behaviour 96–7
Logic 131–3, 201–2
  extensional 132–3
Logical derivation 230
Logical empiricism 1–4, 17, 19–20, 22–3, 30, 132, 240–2
Logical positivism 218–19, 233–4

Majority rule 155–6
Marxism 169
Mastery concept 256–7
Materialist pragmatism
  defence 238–46
  Maddock's critique 215–36
Mentalese 97
Metanarratives 256–9
Moral judgements 173–4
Moral knowledge 5, 37–9

Natural science
  policy analysis 99–101
  social sciences dichotomy 14–15, 32, 134–6, 169–70
Naturalism 151, 218, 242
Naturalistic coherentism 239
Naturalistic fallacy 150, 202–3, 225–6
Neo-pragmatism, Quine's theory 217–21
Neural networks 11–12, 35–6, 98–9
  learning 120–7, 137–8
  medical applications 110–11

SUBJECT INDEX

models 243
multilayer 110–11
NETtalk 35, 98, 99
reasoning 125–7
rock/mine distinguishing 121–2
symbolic processing synthesis 112–13
two-layer 109–10
Non-symbolic representation 109–13

Observation statements 20
Openness 239–42
Organizations
  behaviour 52–3
  closed systems 48–9
  cognitive capacity 142–3
  decentralized 57–8
  efficiency 65–8
  feedback loops 26, 105
  hierarchy 57, 105
  human relations movement 41
  learning loop 63–5, 82–4
  neo-classical view 41–2
  open systems 50–5
  organizational learning 26, 63–5, 81–4
  social systems approach 42
  structure 41

Paradigms approach 17, 31, 130
Parallel distributed processing (connectionism) 98, 99–100, 104, 120–7, 137
Pareto efficiency 67
Participation, decision-making 58–9
Philosophy 173, 239–42
  educational administration links 30–2, 165, 170
Physical-symbol system hypothesis 107
Physicalism 219
Plato, just society 38, 104, 142–4
Plausibililty judgements, leadership 80–1
Policy analysis 87–101
  Dunn, W.N. 88–95
  Hodgkinson's views 205–9
  scientific status 87, 99–101
  transactional model 89–91
Positivism 15, 233–4, 242–3, 249–50
  inadequacies 266
Post-positivism 15, 32
Postmodernism 9–10, 257–9, 262–70
Practical derivation 230
Practical reason 88, 89
Practising Administrator 216, 238
Pragmatism, materialist 215–36
Productivity, participative decision-making 58–9
Prototype activation model 98

Questionnaires, leadership studies 77, 81

Randomness 134
Rationality 37, 106–7, 117–18
Reasoning
  instrumental 146–50
  neural networks 125–7
Regress of reasons 93–5
Relativism 9

Satisficing 49–50, 59–60
Schools
  behaviour prediction 53
  closed system theory 49
  decision-making 58–9, 60
  educational aims 33
  leadership 75–7
  open systems theory 50–1
  voting procedure 161–3
Scientific approach 1–4, 30–2
  criticisms 6, 14, 29–30, 48
Second best theory 67–8
Sentential paradigm 96–8, 115, 231–2
Social reality, theories 139–40, 279–81
Social sciences, natural science dichotomy 14–15, 32, 134–6, 169–70
Subjectivity approach 7–8, 211–12
  critique 129–40
Symbolic representation
  limits 108–9
  neural network synthesis 112–13
  processing 107
Synoptic tradition 87
Systems theory 3–4
  closed 48–50
  critique 53–5, 171–2
  definition 46–7, 54
  entropy 47
  equifinality 47–8, 49
  general systems theory 43–6
  holism 46
  isomorphism 54
  open 3, 47–8, 50–5
  open/closed demarcation 47
  von Bertalanffy's theory 43–5, 54

Theories
  formulation 131–4
  fragmentation 281–3
  learnable 21
  logical 131–3
  social reality 139–40
  testing 7–8, 19–20, 30–1
Theory Movement 1–2, 16, 31
  critique 21–3, 32–4, 170–3
  educational organization 32–3
  Greenfield, T.M. 130–1

leadership studies 77
logical empiricism 17, 21–3
values 4
Willower's views 170–3
Theory-in-use 82–3
Theory/practice relationship 12, 22, 34–7, 136–40
Touchstone theory 221–2, 264
Transactional model 89–91
Transcendental argument 6–7
Truth tests 90, 93

Universal theory 255–6

Values 4–7, 203–4
  decision-making 173–4
  efficiency 5
  hierarchy 146
  Maddock's views 244–5
  see also ethical theory
Victoria, Schools of the Future project 5, 49
Vienna Circle 17, 131
Voting, impossibility theorem 155–63